System-Level Synthesis

NATO Science Series

A Series presenting the results of activities sponsored by the NATO Science Committee. The Series is published by IOS Press and Kluwer Academic Publishers, in conjunction with the NATO Scientific Affairs Division.

A. Life Sciences	IOS Press
B. Physics	Kluwer Academic Publishers
C. Mathematical and Physical Sciences	Kluwer Academic Publishers
D. Behavioural and Social Sciences	Kluwer Academic Publishers
E. Applied Sciences	Kluwer Academic Publishers
F. Computer and Systems Sciences	IOS Press

1. Disarmament Technologies	Kluwer Academic Publishers
2. Environmental Security	Kluwer Academic Publishers
3. High Technology	Kluwer Academic Publishers
4. Science and Technology Policy	IOS Press
5. Computer Networking	IOS Press

NATO-PCO-DATA BASE

The NATO Science Series continues the series of books published formerly in the NATO ASI Series. An electronic index to the NATO ASI Series provides full bibliographical references (with keywords and/or abstracts) to more than 50000 contributions from international scientists published in all sections of the NATO ASI Series.
Access to the NATO-PCO-DATA BASE is possible via CD-ROM "NATO-PCO-DATA BASE" with user-friendly retrieval software in English, French and German (© WTV GmbH and DATAWARE Technologies Inc. 1989).

The CD-ROM of the NATO ASI Series can be ordered from: PCO, Overijse, Belgium.

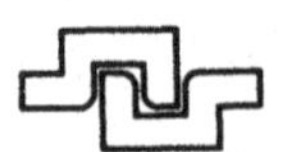

Series E: Applied Sciences - Vol. 357

System-Level Synthesis

edited by

Ahmed A. Jerraya
TIMA Laboratory,
Grenoble, France

and

Jean Mermet
TIMA Laboratory,
Grenoble, France &
ECSI, Gières, France

Kluwer Academic Publishers

Dordrecht / Boston / London

Published in cooperation with NATO Scientific Affairs Division

Proceedings of the NATO Advanced Study Institute on
System Level Synthesis for Electronic Design
Il Ciocco, Lucca, Italy
11–21 August 1998

A C.I.P. Catalogue record for this book is available from the Library of Congress

ISBN 0-7923-5748-5 (HB)
ISBN 0-7923-5749-3 (PB)

Published by Kluwer Academic Publishers,
P.O. Box 17, 3300 AA Dordrecht, The Netherlands.

Sold and distributed in North, Central and South America
by Kluwer Academic Publishers,
101 Philip Drive, Norwell, MA 02061, U.S.A.

In all other countries, sold and distributed
by Kluwer Academic Publishers,
P.O. Box 322, 3300 AH Dordrecht, The Netherlands.

Printed on acid-free paper

Printed in the Netherlands

TABLE OF CONTENTS

Preface vii

MODELS FOR SYSTEM-LEVEL SYNTHESIS

1. *Rolf Ernst*
Embedded System Architectures 1

2. *Luciano Lavagno, Alberto Sangiovanni-Vincentelli,* and *Ellen Sentovich*
Models of Computation for Embedded System Design 45

3. *A.A. Jerraya, M. Romdhani, Ph. Le Marrec, F. Hessel, P. Coste, C. Valderrama, G.F. Marchioro, J.M.Daveau and N.-E. Zergainoh*
Multilanguage Specification for System Design 103

4. *Carlos Delgado Kloos, Simon Pickin, Luis Sánchez* and *Angel Groba*
High-Level Specification Languages for Embedded System Design 137

5. *Steven E. Schulz*
Towards A New System Level Design Language -- SLDL 175

TECHNIQUES FOR SYSTEM-LEVEL SYNTHESIS

6. *Wayne Wolf*
Hardware/Software Co-Synthesis Algorithms 189

7. *Wolfgang Rosenstiel*
Rapid Prototyping, Emulation and Hardware-Software Co-Debugging 219

8. *Giovanni De Micheli, Luca Benini* and *Alessandro Bogliolo*
Dynamic Power Management of Electronic Systems 263

9. *Masaharu Imai, Yoshinori Takeuchi, Norimasa Ohtsuki, and Nobuyuki Hikichi*
Compiler Generation Techniques for Embedded Processors and their Application to HW/SW Codesign 293

METHODOLOGIES AND TOOLS FOR SYSTEM-LEVEL SYNTHESIS

10. *Daniel D. Gajski, Rainer Dömer* and *Jianwen Zhu*
IP-Centric Methodology and Design with the SpecC Language
System Level Design of Embedded Systems 321

11. *James Shin Young, Josh MacDonald, Michael Shilman, Abdallah Tabbara, Paul Hilfinger,* and *A. Richard Newton*
The JavaTime Approach to Mixed Hardware-Software System Design 359

12. *Ross B. Ortega, Luciano Lavagno, Gaetano Borriello*
Models and Methods for HW/SW Intellectual Property Interfacing 397

Index 433

PREFACE

System level synthesis has become the bottleneck for the design of electronic systems including both hardware and software in several major industrial fields including telecommunication, automotive and aerospace. The major difficulty is the fact that system level synthesis fields require contributions form several research fields covering system specification, system architectures, hardware design and software design. This book collects contributions to system-level synthesis that were presented at an Advanced Study Institute (ASI) sponsored by the Scientific and Environmental Affairs Division of the North Atlantic Treaty Organization (NATO). The Institute was held in Il Ciocco, Italy in August 1998. The Advanced Study Institute provided an opportunity for internationally recognized researchers from these different disciplines to discuss all the aspects of system-level synthesis. The main outcome of the meeting is a common understanding and a clear definition of a global design methodology for system level synthesis starting from a high level specification down to an implementation that may include both hardware and software. The Institute covered the multi-language approaches for the specification of electronic systems as well as the main techniques of the system level synthesis flow including partitioning, hardware design, software design, low-power optimization and prototyping. This book also details several codesign approaches.

The first five chapters address issues of modeling electronic systems. Three classes of models are discussed : architectures, models of computations and languages. In each class several models exist and are specialized for specific applications. Complex electronic systems may require to combine several models in each class.

The next four chapters detail the steps of a system-level synthesis flow. They detail hardware/software partitioning algorithms, software synthesis techniques, low-power optimization and prototyping. The discussions of these techniques show clearly that different techniques exist for each of the system level synthesis steps. It also appears clearly that there will not be a unique system-level synthesis flow covering all application domains and that different flows for different application domains may be required.

The final three chapters detail specific system-level synthesis flows. These are aimed to give the reader a global view of the overall system level synthesis domain. These chapters also show that the gap between the system specification and the implementation makes it difficult to build a completely automatic synthesis flow ; interactivity is required.

The editors are very grateful to the NATO/Scientific and Environmental Affairs Division for the financial support that made the organization of the ASI "system-level synthesis for electronic design" possible and that allowed many graduate students and junior researchers to attend. Up to one half of the attendees took advantage of a NATO grant.

The editors must also thank TIMA Laboratory and ECSI, the most important co-sponsors of the ASI. Similar thanks go to "Il Ciocco", the Electronic Research Laboratory of the University of Berkeley, and the companies SYNOPSYS and MENTOR GRAPHICS for their precious complementary contributions. The editors have highly appreciated the kind hospitality and arrangement of the "Il Ciocco" hotel management team that made smooth the organization of the event.

The editors would also like to thank Mrs Florence Pourchelle for her assistance throughout the organization of the meeting as well as Mrs Sonja Amadou for her effort and patience in assembling this book.

But this ASI has been a success mainly because of the outstanding courses delivered by Carlos Delgado Kloos, Rolf Ernst, Daniel Gajski, Masaharu Imai, Ahmed Jerraya, Luciano Lavagno, Giovanni De Micheli, Richard Newton, Wolfgang Rosenstiel, Alberto Sangiovanni-Vincentellli, Steven Schulz and Wayne Wolf. The scientific value of this book is their intellectual property.

The Directors of the 1998 NATO/ASI on System-Level Synthesis,
Ahmed A. Jerraya
Jean Mermet

EMBEDDED SYSTEM ARCHITECTURES

ROLF ERNST

Technische Universität Braunschweig

Institut für Datenverarbeitungsanlagen

Hans-Sommer-Str. 66

D-38106 Braunschweig

Germany

1. Introduction

This chapter surveys the target architectures which are typically found in embedded systems and explains their impact on the system design process. Embedded systems have a high impact on the overall industry defining the features of telecommunication equipment, consumer products, or cars, to name just a few industrial sectors where these systems are ubiquitous. There is a large variety of embedded systems architectures which significantly differ from general purpose computer architectures. Despite their importance, these architectures are typically neglected in university curricula.

Embedded system architectures are determined by **circuit technology**, **application requirements**, and **market constraints**. We will briefly review the trends in each of these areas to understand the resulting trends in target architectures and their design.

The trends in circuit technology are best explained using the 1997 SIA roadmap [1]. This roadmap forecasts the development of the semiconductor industry in the next 15 years based on the expertise of a large part of the semiconductor industry. It predicts the reduction in minimum feature size (transistor gate length) from 200nm in 97 to 35nm in 2012. Considering the increasing maximum economically useful chip size, standard microprocessors are expected to grow from 11M transistors today up to 1.4 billion transistors in 2012 while DRAMs will reach 256Gbit capacity.

Besides general purpose processor architectures, memory intensive embedded system applications, in particular multimedia and embedded image processing (3D), high throughput telecommunication, or powerful PDAs are most likely to be the technology drivers for this development. On the way to 2012, an increasing number of complex systems consisting of many processors, coprocessors and memory of different types and sizes will be integrated on a single chip (systems on chip SOC). The integration of dynamic memories and standard logic has been demonstrated in commercial designs (e.g.

A. A. Jerraya and J. Mermet (eds.), System-Level Synthesis, 1–43.

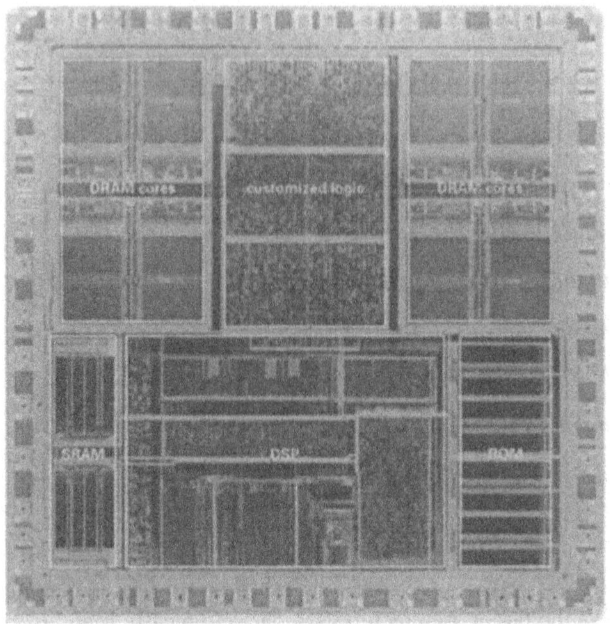

SOC with 16 bit DSP, 15k gate equiv. custom logic, SRAM, ROM, 1Mbit DRAM

Figure 1: Audio processing system (Siemens) [2]

[2], see fig. 1) and will enable further system integration, but requires new tradeoffs in VLSI system design, as will be explained later. It is not unlikely that memory issues will eventually dominate system design.

In parallel to the growth in system complexity, the design cycle time is expected to decrease to 55% when compared to 97 [1]. To reach this cycle time at increasing system sizes, the individual design productivity must grow 160 times over the next 15 years. This tremendous productivity goal which faces increasing physical design problems seems only possible when more than 90% of a design can be reused from previous designs or from a library [1]. These reused components will come from different sources with exchange across companies. The transfer of intellectual property between companies has coined the term **IP** for these reused parts.

The reuse requirements will have a large impact on the design process, in particular since reuse must be combined with a significant rise in the level of design automation to account for the remaining productivity gain requirements (90% reuse alone provides a maximum productivity gain of 10).

On the other hand, many embedded system applications will not require such a huge circuit capacity. Already today, the vast majority of embedded systems is integrated on single chip microcontrollers with a 4 to 16 bit processor as core. Some of these systems, however, implement rather complex functions, such as in automotive control, where distributed embedded systems are required to closely cooperate. Active suspension or drive-by-wire are just two examples of such functions. So, a discussion of embedded system architectures should not be restricted to single chip systems. Nevertheless, research should focus on the large circuit technology drivers to enable the required productivity gain.

System complexity will even grow faster than VLSI complexity since an increasing part of embedded system functionality will be implemented in software. Software can be used to adapt an SOC to different applications or for product diversification, i.e. for creating products with different features based on the same hardware platform. Examples are mobile phones with different features or automotive controllers for different motor types and different countries (emission control). The higher volume of the common hardware platform can reduce the overall system cost. Therefore, additional flexibility may account for slightly higher system complexity, but without compromising system constraints, such as minimum battery lifetime. On the other hand, higher system integration gives more opportunities for system specialization and cost reduction. Also, there are new forms of flexibility, such as FPGA components and reconfigurable systems which must be taken into account. Flexibility of system components or subsystems may also be used to enable the inclusion of IP. The controlled use of flexibility will, hence, become an important factor in system design.

Much more than general purpose computer design, embedded system design is affected by non functional market constraints, most importantly:

- **Strict cost margins** Often, the design objective is not only to minimize cost but to reach a certain market segment consisting of a price range and an expected functionality. So, if cost is above the intended price margin, functionality and performance may have to be redefined. This is different from typical general purpose microprocessor design for PC and workstation applications, where performance plays a dominant role.
- **Time-to-market and predictable design time** This constraint also applies to general purpose processor design except that in embedded system design, the specification includes all software functions. Furthermore, the development can lead to a family of similar products or include other embedded products, which might not yet be completely defined.
- **(Hard) time constraints** Many embedded systems are subject to hard time constraints. By definition, when hard time constraints [3] are violated, a major system malfunction will occur. This is often the case in safety critical systems. For such systems, timing predictability is of primary importance. Modern RISC processors with long pipelines and superscalar architectures, however, support dynamic out of order execution which is not fully predictable at compile time. So, the obtained speedup is to a large part useless in the context of hard real-time requirements. This is not an issue in a PC environment.

- **Power dissipation** is certainly also a problem in PC processor design. In portable systems, such as PDAs and mobile phones, however, battery life time is also an important quality factor.
- **Safety** Many embedded applications such as automotive electronics or avionics have safety requirements. There is a host of work in self checking and fault tolerant electronic systems design [4, 5] but, in embedded systems, mechanical parts must be regarded as well. Safety requirements can have very subtle effects on system design when we take EMI (electromagnetic integrity) into account [6]. As an example, the clock frequency may be constrained since (differential mode) radiation of a digital system increases roughly with the square of the clock frequency.
- **Physical constraints**, such as size, weight, etc...

We may assume that designers will only accept computer aided design space exploration and system optimization if the resulting target architectures show a quality comparable to manual designs. We need a thorough understanding of these architectures and their underlying design techniques to be able to match their quality in computer aided codesign. This not only holds for an automated design process, but even more for design space exploration where the designers would expect to have the same design space available as in manual designs. Moreover, target architecture knowledge is a prerequisite for efficient and accurate estimation of performance, cost, power dissipation and safety.

While current system synthesis tools are limited to a small set of architectures and optimization techniques, future CAD tools for 100 million transistor IC designs will have to cover a wide range of architectures for efficient exploration of a huge design space. To explore the inhomogeneous system design space, we will need design transformation, which allows to map a system to different types of target architectures such as a parallel DSP, a microcontroller or an ASIC or an application specific processor. However, we will show that the wide variety of different target architectures and components can be explained with a small set of specialization techniques and communication principles which might be exploited in system design space exploration.

The next section introduces the specialization techniques which will then be applied to explain the different classes of target architectures in use today. We will use small systems which will be sufficient to show the main target architecture characteristics. We will focus on the main principles and will exclude other aspects such as test, fault tolerance, or production. The examples are selected to demonstrate the concepts and shall not imply any judgment on design quality. We will assume a working knowledge in general purpose computer architecture, such as represented e.g. by [7].

The rest of this chapter will be organized as follows. We will first explain how embedded system design exploits specialization techniques to meet the design objective and reduce cost. In this step, we will assume that all system design parameters are free corresponding to an ideal system design from scratch. We will show practical examples of specialization using different techniques for different levels of flexibility using an important application, multimedia processors. In the last section, we will briefly summarize the influence of IP reuse and the relation between IP reuse and flexibility. We will conclude with proposals for research topics.

2. Architecture Specialization Techniques

2.1. COMPONENTS AND SYSTEMS

Hardware design develops components and integrates them into larger entities. The size of the components depends on the design task. For the purpose of this paper, system design components shall be defined as function blocks, such as data paths, buses, and memory blocks, or combinations of these elements with a single thread of control, such as core processors, coprocessors or memories. Systems have multiple threads of control requiring load distribution and optimization of control and data flow.

We can classify architecture specialization in 5 independent specialization techniques for components and additional techniques to support load distribution and to organize data and control flow in a system of specialized components. Components shall be defined to have a single control thread while systems may have multiple control threads. The reference system used for comparison is always a general purpose programmable architecture. We start bottom up with specialization techniques for individual components and will then derive specialization techniques for systems.

2.2. COMPONENT SPECIALIZATION TECHNIQUES

The first technique is **instruction set specialization.** It defines the operations which can be controlled in parallel. It can be adapted to operations which can frequently be scheduled concurrently in the application or which can frequently be chained. An example for the first case is the next data address computation in digital signal processors (DSPs) which can mostly be done in parallel to arithmetic operations on these data, and an example for the second case are multiply-accumulate or vector operations which can also be found in signal processing applications.

The potential impact of instruction set specialization is the reduction of instruction word length and code size (chaining) compared to a general purpose instruction set for the same architecture which in turn reduces the required memory size and bandwidth as well as the power consumption.

The second technique is **function unit and data path specialization**. This includes word length adaptation and the implementation of application specific hardware functions, such as for string manipulation or string matching, for pixel operations, or for multiplication-accumulation. Word length adaptation has a large impact on cost and power consumption while application specific data paths can increase performance and reduce cost. As a caveat, specialized complex data paths may lead to a larger clock cycle time which potentially reduces performance. This problem, however, is well known from general purpose computer architecture. In embedded systems, however, a slower clock at the same performance level might be highly desirable to reduce board and interconnect cost or electromagnetic radiation.

The third technique is **memory specialization**. It addresses the number and size of memory banks, the number and size of access ports (and their word length) and the supported access patterns such as page mode, interleaved access, FIFO, random access, etc. This technique is very important for high performance systems where it allows to increase the potential parallelism or reduce the cost compared to a general purpose system with the same performance. The use of several smaller memory blocks can also reduce power dissipation.

The fourth technique is **interconnect specialization**. Here, interconnect shall include the interconnect of function modules, as well as the protocol used for communication between interconnected components. Examples are a reduced bus system compared to general purpose processors to save cost or power consumption or to increase performance due to a minimized bus load. Interconnect minimization also helps to reduce control costs and instruction word length and, therefore, increases the concurrency which is "manageable" in a processor component. One can also add connections to specialized registers (memory specialization) which increases the potential parallelism. Examples include accumulator registers and data address registers or loop registers in digital signal processors.

The fifth technique is **control specialization**. Here, the control model and control flow are adapted to the application. As an example, microcontrollers often use a central controller without pipelining. Pipelining is employed in high performance processors while distributed control can be found in complex DSP processors as we will see later. The individual controller can be microprogrammed or an FSM, fully programmable or hard wired.

2.3. SYSTEM SPECIALIZATION TECHNIQUES

System specialization has two aspects, **load distribution** and **component interaction**.

Unlike load distribution in parallel systems, embedded systems are typically heterogeneously structured as a result of component specialization. Unlike load distribution in real-time systems (which may be heterogeneous as well), the target system is not fixed. In hardware/software co-design, the distribution of system functions to system components determines how well a system can be specialized and optimized. *So, hardware/software co-design adds an extra dimension to the cost function in load distribution, which is the potential specialization benefit.*

Two load distribution objectives seem to be particularly important to embedded system design, control decomposition and data decomposition.

Control decomposition (which could also be called **control clustering**) partitions system functions into locally controlled components *with the objective to specialize component control.* As a result, some of the components are fully programmable processors, some are hard coded or hard wired functions without a controller (e.g. filters and periph-

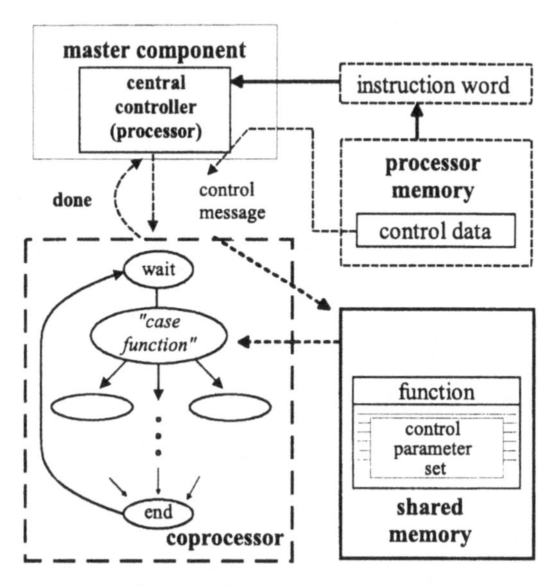

Figure 2: Dependent coprocessor

erals) and some components are implemented with a set of selectable, predefined control sequences, possibly with chaining. An example for the latter case are floating point coprocessors, DMA coprocessors or graphics coprocessors. The components control each other which requires a global control flow. Again, the global control flow is specialized to the system and is heterogeneous in general. We will explain the global control flow techniques in the next section.

Data decomposition (or **data clustering)** partitions the system functions and their data with the objective to specialize data storage and data flow between components. This leads to different, specialized memory structures within and between the components, as we have seen in the context of component memory specialization.

In the next sections, we will have a closer look at the techniques of global control flow and global system communication which are needed to support global specialization techniques.

2.4. SYSTEM SPECIALIZATION TECHNIQUES - A CLOSER LOOK

Control specialization which is restricted to individual components results in a system of communicating but otherwise *independently controlled components*. Control specialization in a system can go further and minimize component control such that it becomes dependent on other components. Such asymmetric control relationships require global control flow. Again, there is a large variety of global control mechanisms, but they adhere to few basic principles. We will distinguish independently controlled components, depend-

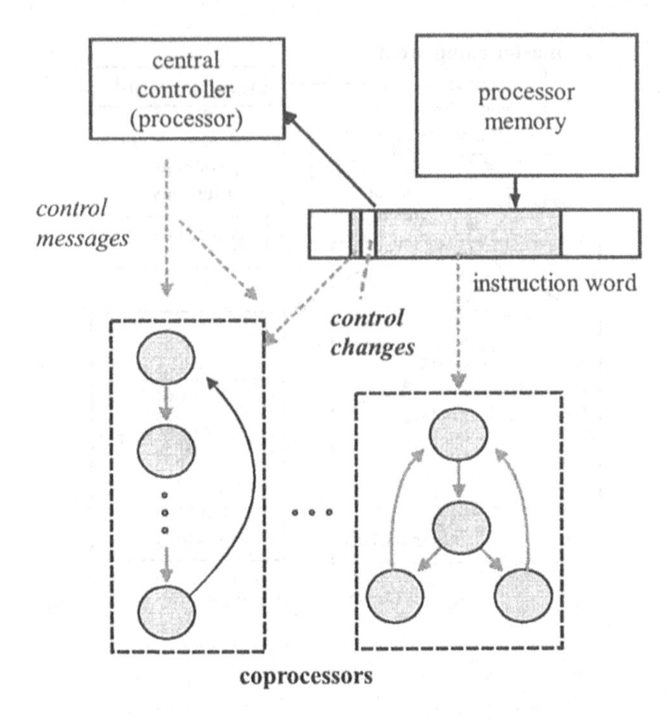

Figure 3: Incrementally controlled coprocessors

ent coprocessors, incrementally controlled coprocessors and partially dependent coprocessors.

The first class are **independently controlled components**. If the components are programmable processors, we would classify the system as MIMD [8]. There is no global control flow required between such components.

On the other extreme are components which completely depend on the controller of another component. They shall be called **dependent coprocessors**. The controlling component shall be the *master component*, usually a processor, with the *central controller*, usually a processor control unit. Fig. 2 shows the typical implementation of the global control flow required for dependent coprocessors. The master component sends a control message to the *coprocessor* which it derives from its own state (in case of a processor: instruction). As a reaction, the *coprocessor* executes a single thread of control and then stops to wait for the next *control message*. So, the *coprocessor* behavior can be modeled with a "wait" and a "case" statement as shown in fig. 2.

If needed, additional control parameters (control data) are passed in a *control parameter* set, typically via *shared memory* communication. This alleviates the *master component* from control data transmission. Examples are small peripheral units such as UARTs (serial I/O) or PIOs (parallel I/O). From an instruction set point of view, we can consider the coprocessor functions as complex, multi-cycle microcoded processor functions.

Master component execution can continue in parallel to the coprocessor (multiple system control threads) or mutually exclusive to the coprocessor (single system control thread). Dependent coprocessors can also be found in many general purpose architectures, such as floating-point coprocessors, or graphics coprocessors [9]. It is also the coprocessor model of many co-synthesis systems, such as COSYMA [10].

The second class shall be called **incrementally controlled coprocessors**. Incrementally controlled coprocessors can autonomously iterate a control sequence. A central controller (the master component frame is omitted for simplicity) only initiates changes in the control flow and in the control data of the *coprocessor(s)* (fig. 3). This technique reduces the amount of control information for locally regular control sequences. In case of such regularity, the same instruction word can be shared among the main component (e.g. a *processor*) and several coprocessors such that the control "effort" is shifted between the units in subsequent instructions using different instruction formats. A good illustration of this approach is found in some digital signal processors with incrementally controlled concurrent address units which exploit highly regular address sequences (see examples below).

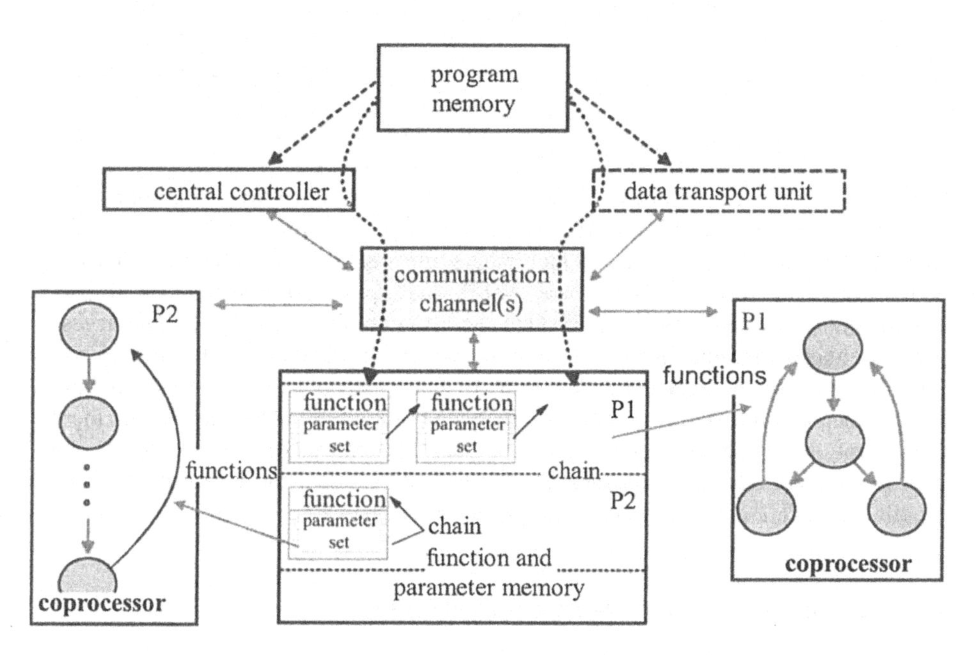

Figure 4: Partially dependent coprocessors

Both dependent coprocessors and incrementally controlled coprocessors are a part of a control hierarchy. Even if components are independent with their own control memory, there can still be a control hierarchy where a master component defines the program to be executed by a coprocessor. Such coprocessors shall be called *partially dependent coprocessors*. Partially dependent coprocessor implementation is strikingly similar in very different architectures, such as digital signal processors and complex peripheral controllers. The central controller typically downloads a linked set of function blocks each consisting of a function designator and a parameter block. The links can form a circle for iterative execution. The mechanism only allows very simple round robin linear sequences of functions but each function may correspond to a complex (micro)program with data dependent control flow. This is outlined in fig. 4. Obviously, these independent coprocessors are well suited for shared memory architectures.

The control hierarchy can be extended to several levels as we will see in some of the more complex examples.

2.5. MEMORY ARCHITECTURES

Memory architecture design is subject to system specialization techniques as well. The application data set can be decomposed and mapped to several specialized memories to optimize cost, access time or power consumption. A priori knowledge of application data structures and data flow can be used in several ways. Besides extending the component memory specialization techniques to systems, **system memory specialization** exploits the data flow within a system. A priori knowledge of data flow is used to insert and optimize buffers, assign memories for local variables and distribute global data such as to minimize access time and cost. This is known from DSP design, such as the TMS320C80 [11], which will be discussed as an example, or the VSP2 [12], but has recently also drawn some attention in general purpose processor design, where part of the memory architecture is specialized to multimedia application, e.g. using stream caches in addition to standard caches. There are even formal optimization techniques, in particular for static data flow systems [13]. One can also exploit knowledge of data access sequences either to reduce memory access conflicts with memory interleaving or e.g. stream caches reduce address generation costs by taking advantage of linear or modulo address sequences typical for many signal processing or control algorithms. Again, there are optimization approaches, e.g. [13, 14]. Reducing memory conflicts, however, can lead to less regular address sequences. So, regular address sequences and memory access optimization are potentially conflicting goals.

3. Target Architectures and Application System Classes

3.1. THE SPECIALIZATION PROBLEM

The main difference between general purpose highest volume microprocessors and embedded systems is specialization. We have seen a variety of specialization techniques on the component and system levels to optimize an embedded system for a given target application or a class of target applications. Nevertheless, efficient specialization requires suitable system properties. So, given the specialization techniques, a designer should identify the specialization potential of a target application with respect to the specialization techniques, before a decision on the architecture is made. Moreover, specialization must not compromise flexibility, i.e. it must possibly cover a whole class of applications (DSP) and include some flexibility for late changes. This also limits specialization. What is needed is a target application analysis which

- identifies application properties that can be used for specialization,
- quantifies the individual specialization effects, and
- identifies and takes flexibility requirements into account.

Unfortunately, there are no formal methods today which would support the three tasks, except for quantitative techniques from computer architecture [7], which may help in the second issue. Therefore, system specialization and design is currently for the most part based on user experience and intuition.

3.1.1. *Application System Classes and Examples*
To give an impression on how specialization is done in practice, we will discuss target architectures for several important but very different application system classes. We selected high volume products which have been used for a few years, where we may assume that a large effort went into careful manual system optimization, and where the design decisions have already been validated by a larger number of applications. We will distinguish four application system classes,

- **computation oriented systems** such as workstations, PCs or scientific parallel computers,
- **control dominated systems**, which basically react to external events with relatively simple processes,
- **data dominated systems** where the focus is on complex transformation or transportation of data, such as signal processing, or packet routing, and
- **mixed systems**, such as mobile phones or motor control which contain extensive operations on data as well as reaction to external events. Here, we will only look at the latter three classes.

4. Architectures for Control Dominated Systems

Control dominated systems are reactive systems showing a behavior which is driven by external events. The typical underlying semantics of the system description, the "input model of computation", are coupled FSMs or Petri-Nets. Examples are StateCharts or Esterel processes ([15]). Time constraints are given as maximum or minimum times. These shall be mapped to communicating hardware and software components possibly after a transformation which decomposes and/or merges part of the FSMs (of the Petri-Net).

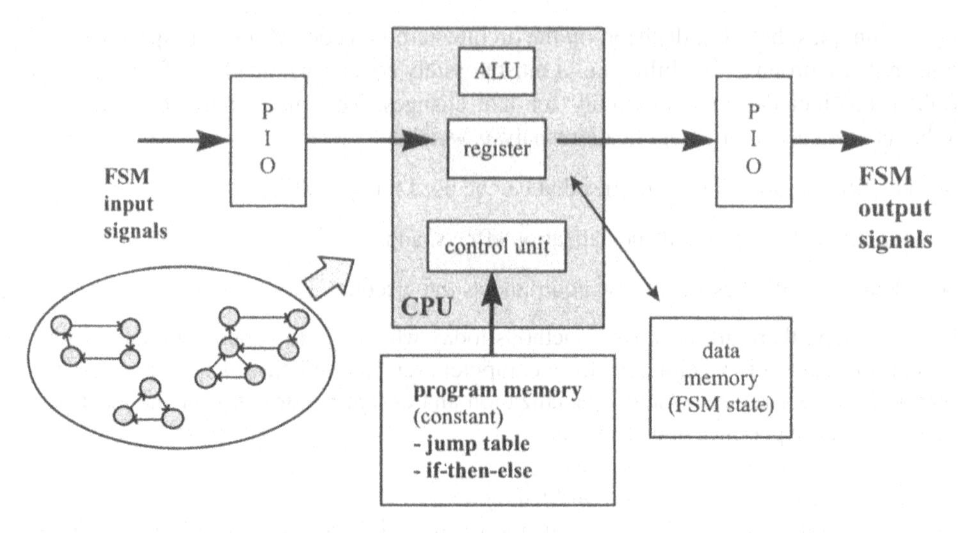

Figure 5: Microcontroller FSM implementation

The resulting co-design problems are the execution of concurrent FSMs (or Petri Net transitions) reacting to asynchronous input events, which have an input data dependent control flow and are subject to time constraints. Very often, control applications are safety critical (automotive, avionics). So, correctness (function and timing), FSM transition synchronization and event scheduling are the major problems while computation requirements are typically very low.

Fig. 5 shows the example of a minimal software implementation of a set of coupled FSMs on a **microcontroller**. Microcontrollers are small systems-on-a-chip used in many low-cost applications from simple functions in automotive electronics to dish washer control. Parallel input and output units (PIO) which can be found in any microcontroller latch the FSM input and output signals. FSM transitions must be serialized on the processor which is a resource sharing problem. After transformation, each FSM is mapped to a process. These processes consist of a small variable set containing the FSM state and the FSM transition functions encoded as short program segments in chains of branches or in jump tables. Program size grows with the number of states and transitions. In microcontrollers, program memories often require more space than the CPU. So, code size is an important issue. In microcontroller architectures, FSM communication uses shared memory. Since each FSM transition typically requires a **context switch**, context switching times are critical.

This short application overview reveals a number of specialization opportunities:

- most of the operations are boolean operations or bit manipulation,
- there are few operations per state transition,
- the data set containing the FSM states is typically small.

We will see in an example that this observation is used to apply all component specialization techniques plus control decomposition with different types of coprocessors.

4.1. 8051 - AN 8 BIT MICROCONTROLLER ARCHITECTURE

The 8051 is a quasi standard 8 bit architecture which is used as a processor core (CPU) in numerous microcontroller implementations by many companies. It is probably the highest volume processor architecture overall. Its main application are control dominated systems with low computing requirements. Typically, it is combined with a number of peripheral units most of them having developed into standard coprocessor architectures. It has 8 bit registers and a complex instruction set architecture (CISC).

We will characterize the 8051 architecture and its peripheral coprocessors using the specialization techniques derived above.

Instruction set specialization. The 8051 (e.g. [16, 17]) is a 1 address accumulator architecture [7]. It has 4 register banks with only 8 registers each and uses a highly overlaid data address space for 8 bit data addresses. Both features together enable very dense instruction coding of 1 to 3 bytes per instruction. This corresponds to the need for dense encoding and makes use of the observation that the variable set is typically small. To avoid a performance bottleneck and to allow short programs, the single address architecture and address overlay are complemented with a rich variety of data addressing modes (direct, indirect, immediate for memory-memory, memory-register, and register-register moves) which make it a CISC architecture [7]. Unlike other CISC architectures, however, there are no complex pipelined multi-cycle operations making interrupt control easy and fast. Also, instruction execution times are data independent making running time analysis easier. It is rather slow with 6 to 12 clock cycles per instruction cycle and clock frequencies up to currently 30Mhz. Among the more application specific operations are registered I/O with direct I/O bit addressing for compact transition functions and branching relative to a "data pointer" as a support of fast and compact branch table implementation.

```
MOV   P0,#8FH;    load P0 with #8FH                  (2 bytes)
CPL   P0_1;       complement bit no. 1 of port P0    (2 bytes, 1 cycle)
JB    P0_2,100;   jump to PC+100 if P0_2 is set      (3 bytes) (P0_2 is input)
```

Figure 6: 8051: Direct I/O bit addressing

Fig. 6 shows three examples with timing data taken from [18]. The first instruction is an immediate store to a registered 8 bit output: The output port P0 is loaded with the hex value 8F. Execution time is 1 instruction cycle. The second operation is a registered I/O operation to a single output bit, where a value of an output bit is complemented. Again, the execution time is a single instruction cycle only. And finally, in the third example, a

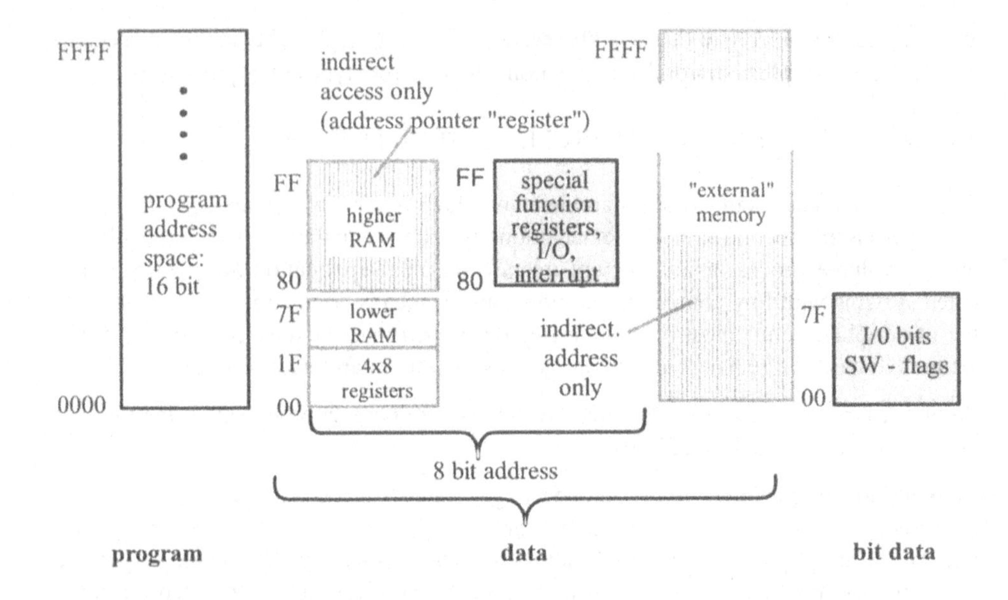

Figure 7: 8051 address space

conditional jump is executed on a single bit of an input port, executed in 2 instruction cycles. Especially this last instruction shows how the I/O signals are directly included in the control flow, very much like a carry bit of an accumulator. Such an integration of I/O events into the instruction set reflects the FSM view of the processor illustrated in fig. 5.

Function unit and data path specialization. The data path is only 8 bit wide. This is completely sufficient for boolean operations and bit manipulation required for FSM implementation and simple I/O protocols. Multiply and division functions are software implemented. Some of the typical peripheral coprocessors have specialized data paths.

Memory specialization. Fig. 7 shows that the lower 8 bit *address space* is overlaid with 5 different memory spaces. Which address space is selected depends on the operation and the address mode. The CPL operation e.g. always addresses one of the 128 *I/O bits* and *SW flags*, while MOV operations to the lower 128 addresses access the data RAM and register banks and the higher addresses either access the I/O ports, interrupt registers or extended data RAM, interestingly with higher priority to I/O. The register set consists of 4 banks of 8 bit registers. Bank selection is part of the program status word, so switching between banks is done with interrupt or subroutine calls in a single cycle. Complementing the short instruction execution times, this allows an extremely fast context switch for up to 4 processes. To give an example, even the slow instruction cycle times [18] still lead to context switching times of less than a microsecond. All processes use a single memory space and share the same local memory such that communication between software FSMs can easily be implemented in shared memory.

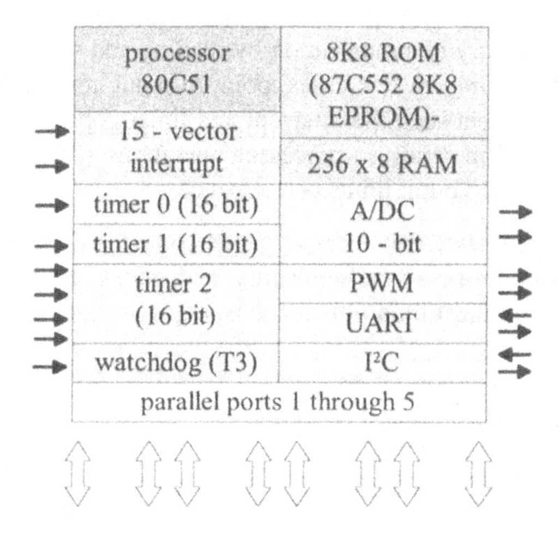

Figure 8: Philips 80C552

Interconnect specialization. Except for small 8 bit word length buses, interconnect is not particularly specialized but is based on a single bus due to the low data transportation requirements in such a system.

Control specialization. Component control is specialized to low power consumption and fast reaction to external events. Registered input signals can directly be linked to vectored interrupts which supports fast FSM process invocation on external events. Several power down modes which disable part of the microcontroller function can be applied to reduce power consumption.

Fig. 8 shows a complete 8051 microcontroller [17]. Two memories are on chip, an 8 kByte *EPROM* and a small 256 byte data *RAM*. The small size of both memories reflects the extreme cost sensitivity of high volume control dominated applications. The 5x8 bit *parallel I/O ports* and the universal asynchronous receive transmitter (*UART*) can be considered as dependent coprocessors. 3 *timers* can be used to capture the time between events or to count events, or they can generate interrupts for process scheduling. A pulse width modulation (*PWM*) unit is an FSM with a counter and a comparator which can generate pulses of a programmable length and period. Both can be classified as *incrementally controlled coprocessors*. PWM is a popular technique to control electrical motors. Combined with a simple external analog low pass filter, it can even be used to shape arbitrary waveforms where the processor repeatedly changes the comparator values while the PWM unit is running. There is also an I^2C unit to connect microcontrollers via the I^2C bus. A peripheral *watchdog* is a timer coprocessor which releases an interrupt signal when the processor does not write a given pattern to a register in a given time frame. It can be classified as an independently controlled component. Such watchdogs are used to detect whether the program is still executed as expected which is a helpful feature for safety critical systems. Finally, there is an analog digital converter (*ADC*) on the chip (a dependent coprocessor).

There are two types of system communication mechanisms in this microcontroller. The processor uses shared memory communication by reading and writing from or to memory locations or registers. Only in few cases, point-to-point communication is inserted between coprocessors, e.g. between the interrupt and timer coprocessors, i.e. almost all of the system communication relies on processor operations. This is flexible since the system can easily be extended but it limits performance.

Since several software processes can access a single coprocessor, there are potential resource conflicts requiring process synchronization and mutual exclusion. Peripheral resource utilization is an optimization problem and is, e.g. addressed in [19].

5. Architectures for Data Dominated Systems

5.1. DATA DOMINATED SYSTEMS

The term **data dominated systems** can be used for systems where data transport requirements are dominant for the design process, such as in telecommunication, or where a stream of data shall be processed, such as in digital signal processing. In this overview, we want to focus on the latter application. Often, such systems are described in flow graph languages, such as in signal flow graphs which are common in digital signal processing or control engineering (see e.g. [20]). Similar to data flow graphs, signal flow graphs represent data flow and functions to be executed on these data.

Such data dominated applications typically have a demand for high computation performance with an emphasis on high throughput rather than short latency deadline. Large data variables, such as video frames in image processing, require memory architecture optimization. Function and operation scheduling as well as buffer sizing are interdependent optimization problems [13].

The flow graph "input model of computation" offers many opportunities for specialization:

- Periodic execution often corresponds to a (mostly) input data independent system data flow.
- The input data is mostly generated with a fixed period, the sample rate.

Such systems can be mapped to a static schedule thereby applying transformations for high concurrency, such as loop unrolling. The resulting static control and data flow allows to specialize control, interconnect function units and data path while, at most, a limited support of interrupts and process context switch is required. Furthermore, static scheduling leads to regular and a priori known address sequences and operations which gives the opportunity to specialize the memory architecture and address units.

Digital signal processors (DSP) are the most prominent examples of target architectures for this class of applications. The various signal processing applications have led to a large variety of DSPs, with word lengths of 16 to 64 bit, fixed or floating point

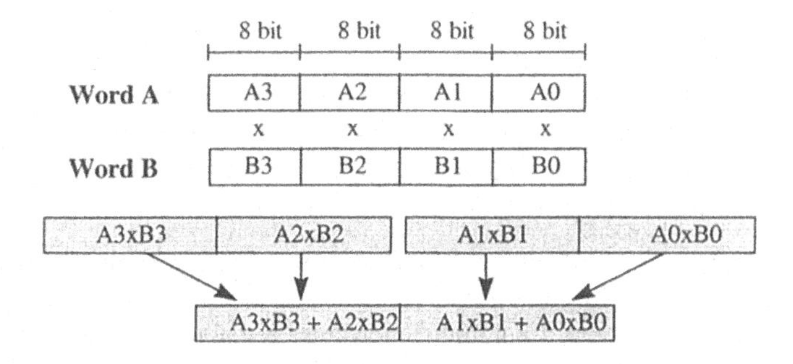

Figure 9: MMX subword operation: Multiply-add

arithmetic, with different memory and bus architectures targeting high performance, low cost or low power.

Specialization for data dominated systems has lead to a large variety of architectures. We will select an important application area, image processing for multimedia, and will discuss examples with a different degree of specialization. We will start with multimedia extensions of general processors and take MMX as an example. Next we will look at the Hitachi SH-DSP, following a completely different concept of a combination of RISC processor and digital signal processor specializing to a wider range of applications. Then, we will analyze a VLIW processor with specialization and coprocessors, namely the Philips TriMedia. Finally, we will look at multiprocessors with specialized data flow and memory architectures using the TI320C80 as an example.

5.2. MMX

MMX is an extension of the Intel (Pentium) architecture (IA). It defines parallel subword operations on the floating point registers. Each 64 bit floating point register can be split in 2, 4, or 8 subwords of 32, 16, or 8 bits. Additional instructions control the execution of parallel operations on the subword operands. Fig. 9 shows an example of a multiply-add operation. Controlled by a first instruction, the 8x8 bit subwords of two 64 bit registers are multiplied as 4 independent operations resulting in 4x32 bit results in two registers. With a second instruction the adjacent subwords of the two registers are added to 2x32 bit subwords which are then stored in a single floating point register. MMX introduces a total of 57 additional IA instructions to control arithmetic, boolean, and shift operation on floating point register subwords as well as move operations to support the conversion between subword and standard data representations. The combination of subword operations of a register is fixed for each instruction. This allows to control up to 8 parallel subword operations without changing the instruction format. As a tradeoff, branching on results of individual subword operations is not possible any more. Instead, boolean operations can be used to mask operations on subwords leading to conditional subword operations. This approach is known from SIMD array processing. MMX, furthermore, supports specific digital signal processing arithmetic, such as saturation. All operations are executed in an extended floating point unit.

These MMX operations are well suited to the matrix and vector operations on pixel arrays in image processing and image coding, e.g. for the JPEG or MPEG standards. The typical pixel size in these applications is 8 to 16 bits which fits the subword length. For these applications, performance improvements between 1.5 (MPEG 1 decoding) and 5.8 (matrix-vector product) have been reported [21]. As an additional effect, the code size is drastically reduced which may lead to fewer cache misses.

MMX is an example of function unit and data path specialization combined with instruction set specialization. In [21], the authors emphasize that the register model and the control for all other operations (including interrupts) remains unchanged which is required to stay compatible to the standard IA. MMX is specialized to a certain but time consuming part of image processing algorithms which is demonstrated by the difference in speedup between the complete MPEG 1 decoding and a matrix-vector operation which is an MPEG 1 subfunction (DCT). To summarize, MMX is a highly specialized but very area efficient extension of a general purpose processor.

This type of extension is not unique, but can be found in many architectures, such as the MAX and VIS extensions of the hp PA or the Sun Ultra SPARC architecture [22, 23].

5.3. DIGITAL SIGNAL PROCESSORS - HITACHI SH-DSP

While MMX extends the data path, it leaves most of the processor architecture unchanged. In digital signal processors (DSP), specialization goes further and includes the control unit, busses and memories. There are many different DSP architectures. For easy comparison with MMX, we pick the SH-DSP, a standard RISC microcontroller which is enhanced by a DSP unit.

The SH-DSP [24, 25] is based on the Hitachi SH instruction set architecture (ISA) which was extended in each generation, from the SH1 to the recent SH4 which is a superscalar architecture. The SH-DSP is a branch of this ISA supporting instructions which are not found in the SH4. Fig. 10 illustrates the processor architecture. The CPU consists of 2 main units, the CPU core with 32 bit integer arithmetic and a 16 bit DSP unit with extended internal word length. The SH-DSP runs at a clock cycle of 20 MHz.

The SH-DSP is targeted to single chip systems, but with higher computation requirements than the 8051. There are several dependent and partially dependent coprocessors for peripheral functions connected by a peripheral bus. For larger systems or for system development, up to 4 Mbytes of external memory can be accessed through an external bus interface (BSC). The DSP extension has lead to a quite complex bus architecture and memory access procedure. There is no on-chip cache function as some of the earlier SH2 architecture versions but 4 single-cycle on-chip memory banks (X and Y) and 3 processor busses. The CPU core is responsible for instruction fetch and program control. Program and data can reside anywhere in the 4 memory blocks. Instruction fetch and data transfers for the CPU core share the 32 bit I bus which would become a bottleneck when operating on array variables such as in digital signal processing. To avoid such conflicts, the DSP unit uses the 16 bit X and Y buses to transfer a maximum 2 data words in one cycle in parallel to the instruction fetch on the I bus, as long as the 3 transfers use different memory blocks (Harvard architecture). While the I bus supports 32 bit addresses in a unified address space, the X and Y busses use 15 bit addresses only to address a 16 bit word in the X and Y memory blocks. 15 bit addresses allow for a maximum of 32 K words in each memory block which closely fits the internal memory size. The DSP unit can also access the I bus but at the cost of conflicts with instruction fetch. The extra memory blocks and the X and Y buses were added to the SH architecture as specialized memory and interconnect closely focused on internal memory access. Concurrent memory access requires concurrent address generation. Since the CPU core would be idle during DSP operations, its ALU is used to generate the X memory address while an additional 16 bit ALU generates the Y memory address. This minimizes address generation overhead and stays close to the original CPU core architecture. It also allows to use the general register file for address generation which should simplify compilation. Address generation is incrementally controlled by the instruction word supporting increment of the type $addr = (addr + incr) \bmod m$, where *incr* is either 2 or 4 or an explicitly specified integer value. Such addressing patterns are typical digital signal processing, such as in the Fast Fourier Transform. The modulo operation is used to generate address patterns in 2 dimensional arrays.

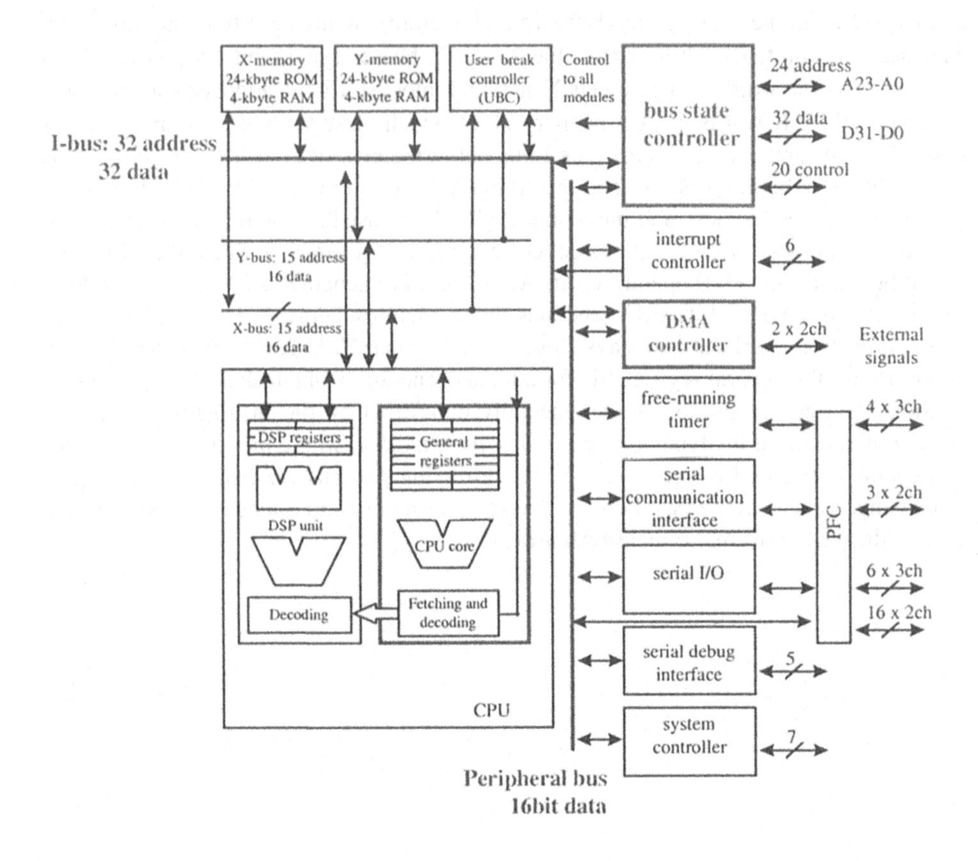

Figure 10: Hitachi SH-DSP

The internal memories are far too small to store larger array variables such as complete video images. Such applications require external memories which, in turn, requires fast memory I/O. The SH-DSP bus interface (BSC) contains an external memory controller supporting several memory types including DRAM. It can be configured to different memory word lengths and wait states with differing row and column access times. A direct memory access controller (DMA) can transfer a coherent block of data between the external memory and the internal memory blocks as a dependent coprocessor. It mainly consists of a set of address counters supporting 4 independent memory transactions (i.e. 4 DMA channels). Since it uses the I bus, only one memory transfer can be active at a time which is, furthermore, in conflict with CPU core memory access. This is a limitation for high throughput applications but fits slow external DRAMs in low cost systems. Both DMA and external memory interface are versatile enough to be able to optimize the external memory to a given low cost application but are less specialized to the DSP extensions. The overhead for higher throughput external memory access was obviously found to be inadequate for the SH-DSP class of applications.

The DSP unit has an internal register file of six 32 bit registers and two 40 bit registers. It can execute 32 or 16 bit operations on these registers using fixed point arithmetic. Like MMX, it supports saturation and conditional execution. Here, conditional execution rather than branching avoids an increase in the control delay of the CPU core which controls all branches. Also, for many DSP applications conditional operations are more appropriate, such as in conditional shift operations to normalize operands in fixed point arithmetic. Other specialized DSP unit operations are shift-by-n (barrel shift) and most significant bit detection, both needed for fast normalization of arrays to minimize rounding errors.

The 32 bit DSP instruction word allows combinations of up to 4 instructions, one X memory load or store operation, one Y memory load or store operation, a multiplier and an ALU operation. The load and store operations can be connected with a simple address increment/decrement of 1, 2, or 4, while other linear address sequences reduce the available parallelism, just as explained for incremental controlled units. This type of instruction level parallelism (ILP) is much more flexible than the matrix-vector type subword parallelism of the MMX allowing many different memory access patterns and operation combinations at different levels of parallelism, but this comes at the price of an extended specialization including memories, busses, data paths and instructions.

In general purpose computing, an ILP of 4 would already lead to a high impact of control hazards due to conditional branches [7]. In signal processing, most of the branches are found in loops with fixed iteration counts. Instead of using conditional branches, DSPs use a specialized control unit and loop instructions. It requires specialized control registers and a specialized control unit. At loop entry, a loop start address register is loaded with the start address of a loop, a loop end register is loaded with the last address of a loop and a loop counter is loaded with the number of loop operations. Then, the instructions between loop start and loop end are executed thereby decrementing the loop counter for each iteration. So, there is no loop overhead except for the loop instructions at the beginning. While the other operations of the DSP unit use one of the register files, the loop operation uses dedicated registers which must be saved for nested loops. Dedicated registers with special interconnect usually make compilation harder [26] and are therefore avoided in general purpose computing. The loop operation is an example of specialized control.

To summarize, the SH-DSP introduces a dedicated DSP unit which effectively operates as a dependent coprocessor which has no direct influence on the control sequence. The SH-DSP exploits all specialization techniques to optimize the DSP unit and the global architecture for maximum ILP in fixed loop operations on 16 bit fixed point arithmetic with normalization. The memory architecture assumes either small data arrays or high locality of the operations. We will see even higher specialization in a later example.

In the specialization process, the SH instruction set is maintained as well as the overall CPU core architecture and its memory architecture. So, the added functionality does not reduce the flexibility of the original core and keeps binary compatibility, like in case of MMX, but leads to circuit overhead which increases cost.

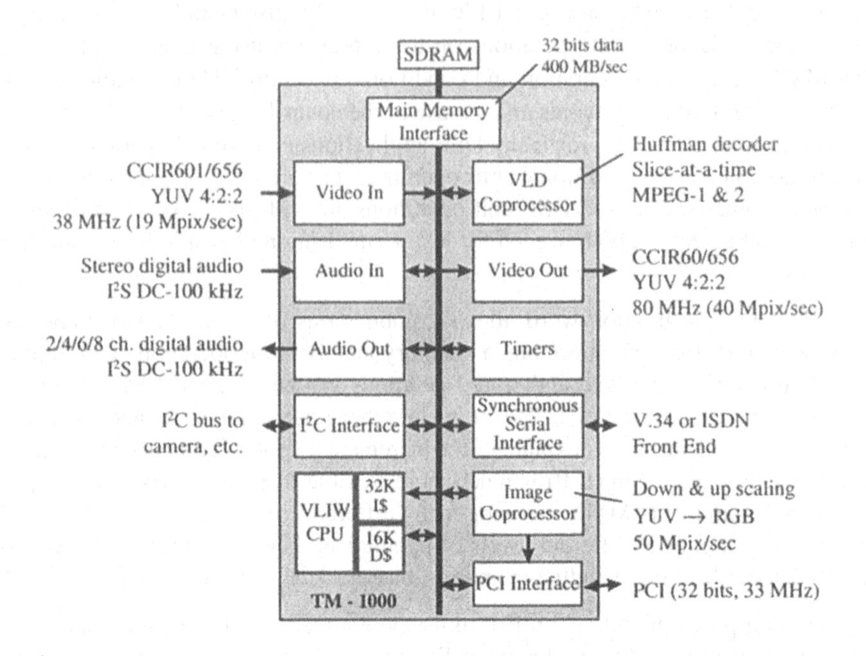

Figure 11: Philips TriMedia TM1000

5.4. VLIW SIGNAL PROCESSORS - PHILIPS TRIMEDIA TM1000

Adding function units, memory blocks and specialized functions does not arbitrarily scale for higher performance requirements and does not always lead to a cost effective solution. The Philips TriMedia TM 1000 processor [27, 28] is a specialized DSP architecture targeted to multi-media applications but still with emphasis on flexibility to support a wider range of applications in this domain. Fig. 11 gives an overview of the architecture. It consists of a VLIW CPU with cache, an SDRAM main memory interface and a set of peripheral interfaces and coprocessors connected over a 32bit internal bus with block transfer. A PCI bus interface simplifies the use of the TriMedia processor as a media coprocessor in workstations or PC systems. The serial I²C bus interface is standard in Philips processors and can be used for external device control. The processor clock frequency is 100 MHz.

To understand the overall concept, we will first discuss the coprocessors which are highly specialized to the application domain. Most notably in this context is the partially dependent Image Coprocessor, which can resize video images. Resizing requires undersampling or oversampling of video images. An important application is the display of one or more video images on a PC window interface. In the simplest way, this can be done by pixel repetition or omission, while for higher picture quality interpolation or filtering in vertical and horizontal dimensions is required. The microprogrammed Image Coprocessor (ICP) uses a 5 tap digital filter with programmable coefficients for higher quality requirements. From the TM 1000 manual, we can conclude that in the worst case of a 1024x768 picture size, this coprocessor executes more than 500 Mio. multiplications/accumulations plus the corresponding address generation and scaling operations thereby requiring 150Mbyte/s memory bandwidth for 16 bit operands. The coprocessor can be programmed to scaling factors larger or smaller than 1 and different filter operations in horizontal or vertical directions. For each of these operations, a different memory access sequence must be generated. With a second set of functions, the Image Coprocessor can convert a standard digital video signal stream (YUV 4:2:2 or 4:2:0) to the three RGB signals used in PC monitors which also requires filtering.

The VLIW CPU alone would hardly be able to reach this performance. The same holds for MMX. So, the Image Coprocessor enables the implementation of higher quality picture scaling thereby off-loading the CPU which can now be used for other parts of a video application. New functionality becomes feasible. When we consider flexibility as the set of functions which a system can implement, then the highly specialized Image Coprocessor has increased the flexibility of the TM 1000 over, e.g., an IA with MMX given the domain of consumer video applications. So, in embedded system design, flexibility should be defined in context of the application.

The architecture of the other TM 1000 coprocessors follows the same concept of implementing widely used and static video functions. Programmability is mainly restricted to select between a few function alternatives. The Video In and Video Out as well as the Audio In and Audio Out coprocessors generate or accept different audio and video standards using predefined internal representations in SDRAM. The VLD supports variable length decoding for the MPEG 1&2 video standard which mainly requires table based string operations.

The VLIW CPU (very long instruction word, [7]) is less specialized than a DSP. The architects avoided special registers to simplify compilation and context switch. The CPU contains 27 function units which are summarized in fig. 12. The selection of these function units is based on benchmarks of video and postscript processing applications [27].

The CPU can issue five 42bit instructions per clock cycle. These 5 instructions can be arbitrarily arranged in one VLIW instruction word. Each instruction is extended by a 2 bit instruction type tag leading to a 220 bit instruction word. To reduce memory cost, compile time instruction compression removes unused 42 bit instruction slots. The decompression step simply uses the unmodified instruction type tag to detect and reinstall the unused slots. This way, decompression is very fast and can be executed at instruction fetch time so that even the instruction cache holds compressed instructions.

The CPU is a load/store architecture using a large 128 entry register file. The register file and the CPU buses support 15 register read and 5 register write operations at a time. The processor uses register forwarding to reduce data hazards as known from general purpose processor design. It supports 64 bit floating point and fixed point arithmetic with specialized DSP operations. This architecture has much similarity with a general purpose VLIW architecture, with the main differences in quantitative details such as the selection and data path optimization of the function units, the size of the register file and the number of buses. Such detailed optimization of a general architecture can be ob-

functional unit	quantity
constant	5[1]
integer ALU	5
load/store	2
DSP ALU	2
DSP multiplier	2
shifter	2
branch	3
int/float multiplier	2
float ALU	2
float compare	1
float sqrt/div	1

[1] The "*constant*" units extract literals from the VLIW instructions.

Figure 12: TM 1000 CPU function units

tained with benchmarking [7]. A difference is the missing hardware branch prediction support which can be explained with the fixed loop structure of many signal processing applications while conditional (guarded) operations can be used to handle small if-then-else constructs like in the SH-DSP. The CPU also supports specialized subword instructions for media applications as discussed previously.

Memory specialization is rather different to general purpose processors and the previous examples accounting for the development of fast DRAMs. First, the large multiport register file replaces the multiple memory banks or multiport memories found in many DSPs. The data cache allows program controlled allocation of lines, e.g. for filter coefficients and program controlled prefetch of cache blocks. Both features allow program control of cache behavior which is effective for predictable address and operation sequences, as typically observed in time consuming core parts of DSP algorithms. This way most of the data cache misses which are frequent in large image processing slowing down hardware controlled caches can be hidden [27]. The designers decided to install a complex 8 way set associative cache both for program (32kByte) and data (16kByte) which helps to avoid bottlenecks due to concurrent program and hardware cache control. The memory interface supports synchronous DRAMs (SDRAMs) with 400Mbytes/s maximum throughput.

The designers might have decided to use a DMA unit as discussed before instead of a cache. We will see in the next section that an older processor, the TMS320C80, uses a powerful DMA unit which can generate complex memory access patterns even for several processors. Such a unit is appropriate for fast static SRAMs. Today, pipelined DRAMs, such as SDRAM, support block transfers at similar throughput and emerging standards such as SLDRAM [30] and DirectRambus [31] can remove the transfer setup time overhead. The block transfer of a cache utilizes this property, while the additional flexibility of the DMA is of little use. Furthermore, the cache line allocation can use part of the cache as a static RAM. On the other hand, the mixture of hardware and programmable cache control requires user controlled adaptation of the input program to fully exploit the system performance. Moreover, if there are coprocessors which simultaneously access the same memory, the memory throughput can drop due to conflicting row accesses (not allowed for the TriMedia). Such conflicts could be solved by a specialized DMA which is attached to the memory rather than to the processor, like in the case of the TMS320C80. DMA operation, however, is affected by access pattern predictability. In summary, the memory-processor cooperation will most likely be one of the dominant problems of multimedia embedded system design.

5.5. SINGLE-CHIP MULTIPROCESSOR-TEXAS INSTRUMENTS TMS320C80 MVP

The fourth example is a single chip multiprocessor, the TMS320C80, shown in fig. 13. The target domain is further narrowed down to pure video processing. This multiprocessor IC consists of four 32 bit integer DSPs (Advanced DSP) with 4 single-cycle RAM blocks of 2 Kbytes each and a 2 Kbytes instruction cache. The processors are connected by a programmable crossbar switch. A single master processor (MP) for general purpose computing and floating-point operations is coupled to this processor array. The transfer controller (TC) is a DMA processor which is programmed through the MP and is responsible for the data and program I/O over a 64 bit external bus. For highly integrated solutions, a video controller (VC) is implemented on chip. The clock frequency is 50MHz.

Data path and function units of the 4 *Advanced DSPs* support subword parallelism and special image processing operations. The ADSP ALU has 3 inputs to support sum operations on arrays. Several multiplexers in the data path allow chaining of operations. Each *Advanced DSP* has 2 address ALUs for index or incremental addressing corresponding to the 2 data buses of each *Advanced DSP*. Again, they are incrementally controlled coprocessors. Like the SH-DSP, the *Advanced DSP* sequencer supports zero overhead loops but adds a loop stack for 3 levels of nested zero overhead loops and a single zero overhead subroutine call. The stack extends the processor state which slows down context switching. There is a second register set for fast context switching between two contexts, e.g. for debugging purposes. This specialization increases performance but introduces many programming constraints.

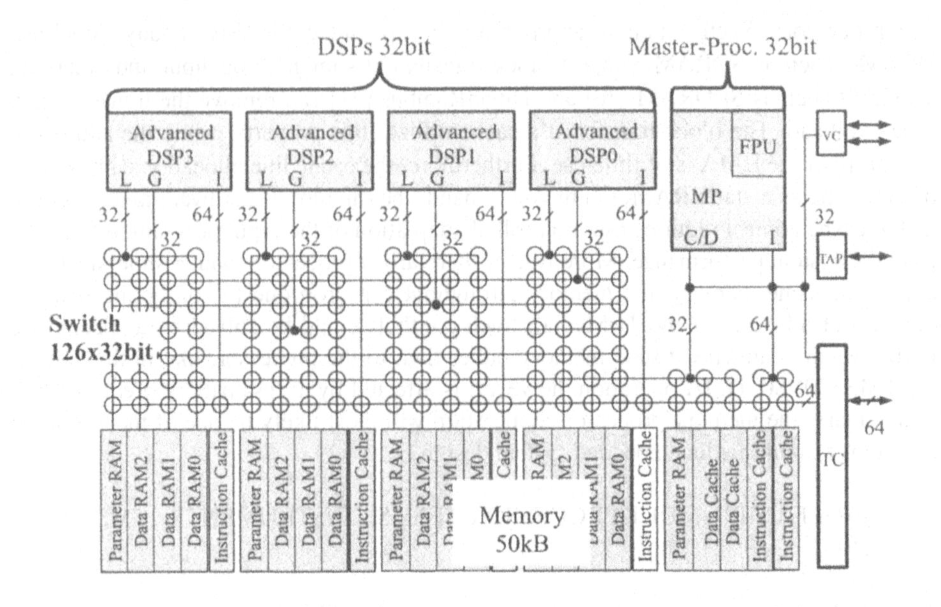

Figure 13: TMS 320C80 MVP

All instructions are given in a single 64 bit instruction word, but in a fixed format of 3 independent fields, one of them for the Advanced DSP operation, one for a local data transfer, and one for a global data transfer from the Advanced DSP to any of the memory blocks. Therefore, TI regards the Advanced DSP as a VLIW processor. Each Advanced DSP has a very small cache of 256 instruction words. Unlike the SH-DSP or the TM1000, on-chip program memory is restricted to very small caches. All instruction loads must be routed through the transfer controller.

The *memory architecture* with a crossbar interconnect seems very flexible. Obviously, there was high emphasis on minimum memory access time. On the other hand, the memory architecture is extremely specialized. Depending on the resolution, video frame sizes are on the order of several hundred Kbytes, such that not even a single frame could be stored in all RAM memories together. Since all RAM blocks can only be accessed through the TC bottleneck, this would be an unacceptable limitation in general purpose signal processing. So there must be an algorithmic justification for this memory size. In fact, video algorithms typically use some kind of spatial or temporal locality which is exploited here. As shown in fig. 14, the video signal scans the image, such that algorithms which operate on few neighboring lines can use a line buffer keeping a small number of recently scanned lines. Such a buffer could fit into one RAM block. If the algorithm operates on windows, such as in standard image compression applications [32], then only the current window could be loaded into internal RAM blocks. This also gives an impression of the prospective role of these RAM blocks. They can be used as buffers between the operations of a signal flow graph (SFG) [20] or a similar representation, such as a synchronous data flow graph (SDF) [13], while the operations or functions are mapped to Advanced DSPs. The crossbar interconnect allows to implement arbitrary flow graph topologies as long as the buffers are large enough. In this context, the TC will have to read and write the windows or lines to or from an external buffer (or the internal video interface) and, at the same time handle potential instruction cache misses.

This is certainly a potential bottleneck even if the internal data flow is larger than the external one. So, again, we observe flexibility in the one place (internally) and constraining effects of specialization on the other, this time in system interconnect. Here, only interconnect and memory specialization together make sense. So when specializing, the designers need a clear understanding of the application domain, which in this case leans itself to data decomposition of a single frame. The small cache size corresponds to view operations on each of the small data objects which is realistic in many video applications. We will see the same assumption in the PADDI and VSP architectures which are discussed later.

To reduce the constraining effects of the single I/O bus, the TC must be a powerful and versatile I/O processor. The TC is programmed in a very compact "Guide Table" where the windows to be transferred are defined as "patches" (fig. 15). Since the TC must manage cache misses and data transfers, it needs an internal context switching mechanism, which must support external memory delays. So, extra buffers are inserted in the TC.

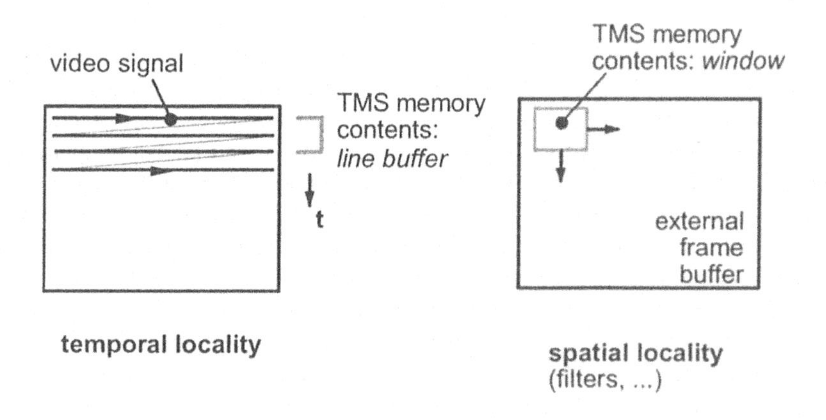

Figure 14: Locality in video algorithms

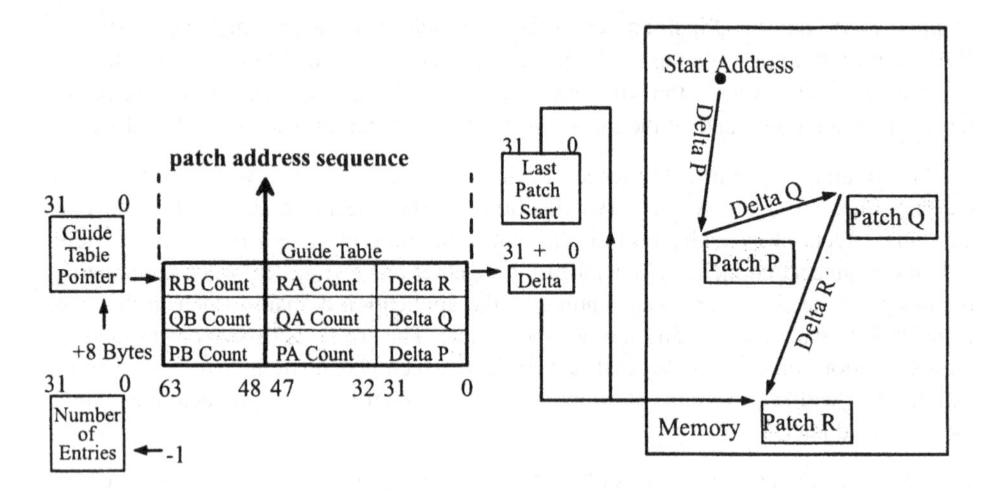

Figure 15: Patch addressing in the TMS320C80

The TMS320C80 is a good example for how to apply the classification scheme of system specialization techniques to understand adaptation and flexibility of an architecture.

6. **Systematic Approaches to HW/SW Flow Specialization**

6.1. DESIGN WITH ARCHITECTURE TEMPLATES

The more interacting components a system has, the larger the design space for optimization and specialization. To make the design manageable, design automation approaches for systems with many interacting components are usually based on architecture templates. These templates support global flow optimization. A well known template is the homogeneous multiprocessor [7] which allows to scale performance with size (in certain limits) and which is supported by parallel compiler and operating system technology. Even though there are different multiprocessor architectures, specialization is very limited and typically restricted to the communication network topology. Embedded system design includes component specialization but the design principle of separating component design and global flow optimization has successfully been applied to embedded system design as well. This is also reflected in the separation of component and system specialization techniques, as described above. The concept of separated component and global flow specialization can easily be extended to a system hierarchy, where a set of interacting components may be a component (i.e. subsystem) of a larger system. For simplicity, we will focus on a single level of hierarchy.

The PADDI [33] and VSP systems [12] address the implementation of static and cyclic signal flow graphs which describe the system data flow at the level of individual operations on data variables. Both architectures consist of small function units with a pro-

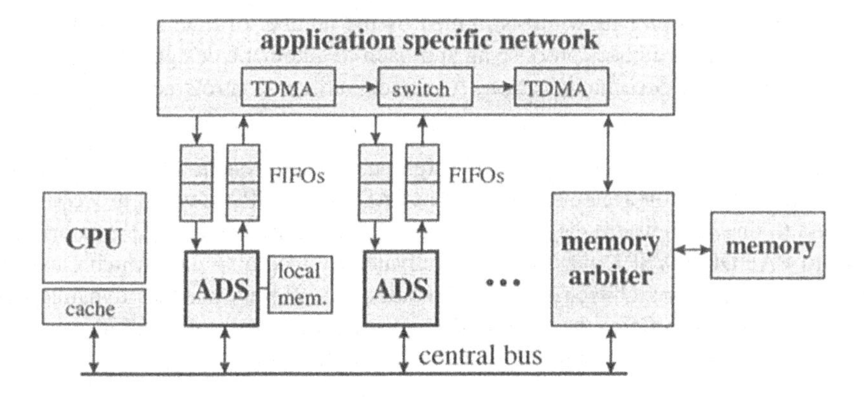

Figure 16: Prophid architecture template

grammable interconnect and buffer memories as local variable memories. The VSP2 local programs iterate over a small number of instructions without branches but with conditional execution while PADDI uses a small local nanostore to decode global 3 bit instructions. The second generation VSP2 has been used in commercial designs and integrates 12 ALUs and 4 memories on a single chip using 0.8 micron technology. Like PADDI, the VSP2 can be extended to multichip systems.

PADDI and VSP are specialized multiprocessors which can be programmed by a design automation system which includes component programming as well as global flow optimization. In contrast, the TMS320C80 programming environment contains a compiler for the individual signal processor but leaves global flow optimization and Transfer Controller programming to the user.

6.2. PROPHID

In contrast to all previously described systems, the more recent Prophid (fig. 16) is a configurable architecture template rather than a hardware concrete implementation [34, 35]. The Prophid template consists of a single core processor which controls several application specific independent (co)processors, ADSs, via a central standardized bus. This block structure and the communication protocol as well as the communication network topology are fixed, while the ADSs and the communication network parameters are application specific.

The ADS architecture can range from a simple data path to a programmable processor with program memory. Each ADS is equipped with at least one input and one output FIFO which are attached to the application specific communication network.

The application specific interconnect network is, again, generated from a template consisting of an application specific number of busses interconnected by programmable switches (space switch). The ADSs are assigned time slots (time division multiple access - TDMA) in which the FIFOs can read from and write to these busses. The switch arrangement enables point-to-point interconnection between any ADSs (a Clos network) while the communication bandwidth is limited by the number of time slots and the bus bandwidth. In [35], the authors propose an approach to automatic design of this network given the required ADS communication. A memory arbiter controls access to a larger memory.

Scheduling and allocation in the Prophid design automation system are executed at the process rather than on the statement level as in PADDI or VSP2. Several processes can be mapped to one ADS where they are activated by input data in the FIFO memories. In contrast to PADDI or VSP2, the ADSs are activated by incoming data which classifies Prophid as a data flow architecture. This activation rule closely matches dynamic data flow graphs [36] and reflects the data dependent execution times of Prophid, i.e. complex video algorithms, such as MPEG 2 video signal compression.

The Prophid template nicely demonstrates the separation of component specialization and global flow specialization and optimization. Almost any ADS can be inserted, as long as it adheres to the interface protocols of FIFO memories and central bus. This approach also enables component reuse, including reuse of programmable components.

Prophid is an example of a parametrized architecture template which has been targeted to support reuse and design automation. The hardware configurability has increased design space compared to the programmability of the VSP or PADDI architectures without loosing the design automation support. The efficiency of global flow optimization will obviously depend on the a priori knowledge of ADS communication pattern and timing.

6.3. EVENT FLOW ARCHITECTURE

The next example is a similar approach for a different domain, i.e. control dominated systems. In contrast to Prophid where a process is activated when all of its input data ("tokens") are available corresponding to an AND function over all input data, process activation in control dominated systems may dependent on arbitrary functions of external events. Events can consist of signal sequences, such as the occurrence of a certain protocol header in a bit stream or the activation condition for fuel injection in a combustion engine requiring the observation of sensor signals. In microcontroller architectures, determination of activating events and the activated process itself are both executed on the same processor. On the other hand, the architecture in [38] used in automotive engine control, employs peripheral coprocessors to detect complex sequential events such as needed for spark plug firing and fuel injection and, even more complex, for sensor failure detection.

The Event Flow Architecture [37] introduces a specialized processor for complex event detection as part of a general architecture template. Fig. 17 highlights the different communication network and (co)processor structure. Like in Prophid, buffers are used to hold input data, but input data evaluation and storage is performed by the specialized programmable event detection unit. The event detection operations are mostly boolean which is exploited to specialize the processor for a high word level parallelism not available in microcontroller cores. This parallelism cuts reaction times (latency) allowing to use programmable architectures in domains which have so far been reserved to hardware coprocessors. The event flow architecture has been applied to an industrial field bus design.

The event processing unit can range from a simple data path or finite state machine to a programmable microcontroller core which uses the local variable memory for storing and communicating data, activated by the event detection unit. Like in the Prophid architecture, the template allows to include reused components. In our terminology, the event processing unit is an independent coprocessor of the event detection unit which reflects the direction of control. The communication network (TDMA) is simpler than in Prophid due to the lower data throughput requirements of control dominated systems.

The Event Flow Architecture is supported by a design automation system which supports event flow graph mapping and component programming.

6.4. RECONFIGURABLE ARCHITECTURES

In all previous examples, the hardware architecture is fixed throughout the lifetime of a product. The hardware structure of a reconfigurable architecture can be changed in the field. Reconfiguration can be used to adapt the architecture specialization to the current application profile, to switch specialization to a different application or simply to update the architecture. It is a possibility to increase system flexibility which can be used in addition or as an alternative to programming. The price for this flexibility is, in general, a larger chip area, larger delay times and/or higher power consumption than a specialized

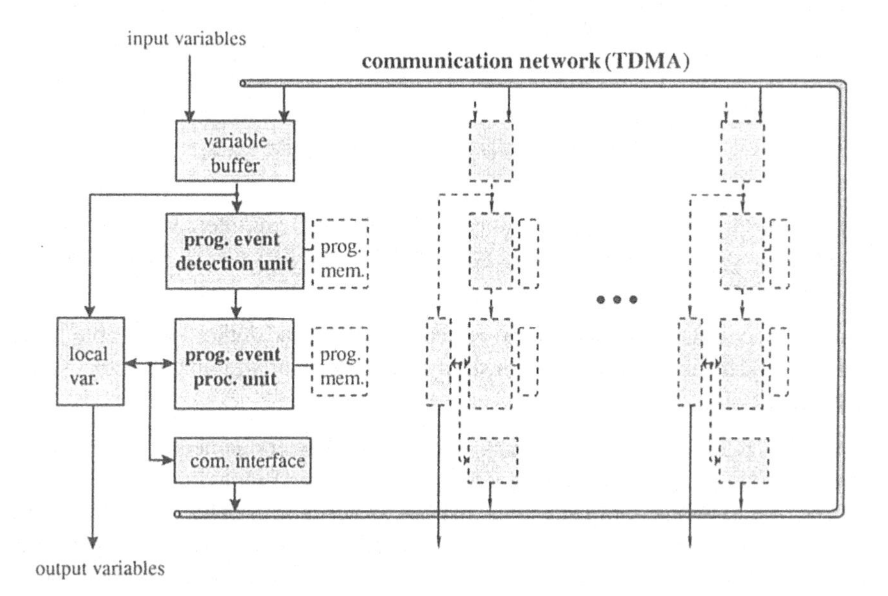

Figure 17: Event Flow Architecture template

custom architecture for a single application.

Reconfigurable architectures are dominated by field programmable gate arrays (FPGAs) [39], since these circuits allow to change the structure down to the logic level. FPGAs can be used to implement complete systems or system components, mainly constrained by the available logic level resources and memory. This logic level reconfiguration does not add new architectures but enables the implementation of alternative architectures on the same circuit.

Reconfiguration can also occur at the function block level. In the Pleiades architecture

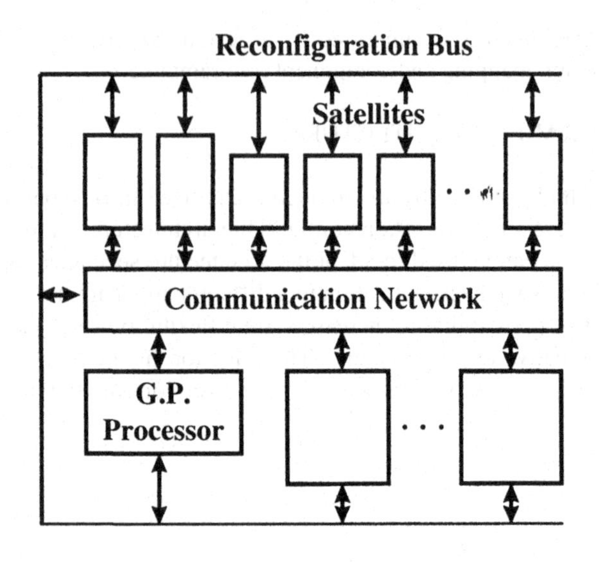

Figure 18: Pleiades architecture template

template [40], application specific coprocessors (Satellites) are interconnected through a reconfigurable communication network (fig. 18). The satellites can be fixed components or reconfigurable data paths.

Focusing on reconfiguration for low power consumption, the authors were able to demonstrate significant improvements compared to FPGA or programmable processor implementations [41].

The Pleiades satellites may be fixed or reconfigurable components which must be adapted to the set of applications similar to the ADSs of the Prophid architecture. On the other hand, the assignment of time slots and bus switches in Prophid can be considered as reconfiguration since it is not part of the instruction set architecture. Comparing Prophid and Pleiades, it becomes clear that programming and reconfiguration are closely related. In [40], the Pleiades authors introduce a classification which distinguishes logic, data path and data flow reconfiguration and where programming is just another type of coarse grain reconfiguration. Despite the similarities, we will keep the distinction of programming and reconfiguration, here, since the instruction set architecture, today, is the major dividing line between hardware and software defining binary code compatibility. We note, however, that reconfiguration is a very active research area which has lead to complex architectures, such as proposed by [42]. Detailed examples and discussions are beyond the scope of this chapter.

7. System Architectures with Complex IP and Flexibility Requirements

7.1. MOTIVATION

While microcontrollers show a clear distinction between a programmable core and relatively simple peripheral coprocessors, the larger multimedia architectures include flexible coprocessors, such as the TriMedia ICP or specialized multiprocessors such as the TMS320C80 or core processor extensions such as in the Hitachi SH-DSP. We have seen that the specialization techniques are typically more effective when used in combination. System level specialization, e.g. to optimize memory access, gives additional freedom which further complicates design decisions. A comparison of MMX and TriMedia also shows that an extension of the application domain favors a completely different architecture, while narrowing down the TriMedia application domain to reduce cost might lead to a Prophid type of architecture. So, while a microcontroller can be used for a wide domain of applications typically only limited by the selected coprocessors and the available memory size, domain specific design for multimedia applications requires a thorough analysis of the application domain.

Moreover, microcontroller components can easily be reused. Reuse of multimedia processor components is less obvious since they were specialized in context of the other system components.

Given the complex design problem, we might feel inclined to simplify the design task by restricting specialization and focusing on a single application rather than on a domain. The comparison of MMX, SH-DSP and TriMedia demonstrates that restricting specialization cuts the available performance and increases cost and will, therefore, not be competitive in practice. A set-top box using general purpose processors would simply be too expensive or too slow. Reducing the domain to a single application reduces the volume and limits the future reuse of parts of the architecture, both increasing cost. Finally, this approach does not account for products with different variants which exploit hardware flexibility to minimize the number of hardware platforms. Examples were given in the introduction.

As a consequence, we will need co-design and system synthesis approaches which create architectures for a set of applications rather than for a single application thereby exploiting hardware flexibility.

So far, we have assumed complete freedom in system specialization. Introducing larger IP components for reuse constrains the design space. Such components have been specialized in a preceding design process with possibly different design objectives and constraints.

Reused IP components can be fixed or adaptable. Components can be adapted by hardware modification, by programming, or by reconfiguration. Hardware modification changes the original specialization while programming keeps the original specialization and makes use of component flexibility. Reconfiguration makes use of component flexibility which is not represented in the original instruction set architecture.

Obviously, reusability and flexibility are closely related. In the remainder of this chapter, we will develop a taxonomy of flexibility and adaptability which we will then use to analyze the current design automation approaches and propose improvements.

7.2. FLEXIBILITY AND ADAPTABILITY - A TAXONOMY

For clarity, we will need some definitions. In the following, the system or component **behavior** shall include its **timing**.

Dynamic adaptability shall contain all possible run-time behaviors of a component or system with no change of the hardware structure or program. Examples for dynamic adaptability include instruction reordering, caching, or speculative computation [7]. This definition of dynamic adaptability also includes the case that a hardware realizes a set of different behaviors with dynamic selection at run time dependent on system state and input data. Examples are the simple dependent coprocessors of the 8051 which select from different behaviors depending on control register contents, or a filter circuit which selects between different filter structures depending on the input signal characteristics.

Component or system **programmability** shall contain all possible behaviors of a component or system with no change of the hardware structure and instruction set architecture. Please, note that the inclusion of timing is necessary since otherwise processor behavior is only delimited by determinism and memory, as known from computational theory. This definition of programmability includes dynamic adaptability at run time, which seems intuitive.

Component or system **configurability** shall contain all possible behaviors of a component or system as a result of hardware parameter modification **at design or production time**. Examples are memory size, word length, or bus width. This definition also includes (EP)ROM configuration and all board level configurations used to adapt a larger system at production time.

Component or system **reconfigurability** shall contain all possible behaviors of a component or system as a result of reconfiguration without change of the physical hardware. Reconfiguration is limited to changing the logic structure or interconnect of a component or system as seen from the instruction set model while the physical hardware remains unchanged. Examples can be found in the previous section on system specialization.

Dynamic adaptability, programmability and reconfigurability shall be summarized in the term **flexibility,** while **adaptability** shall summarize all four definitions.

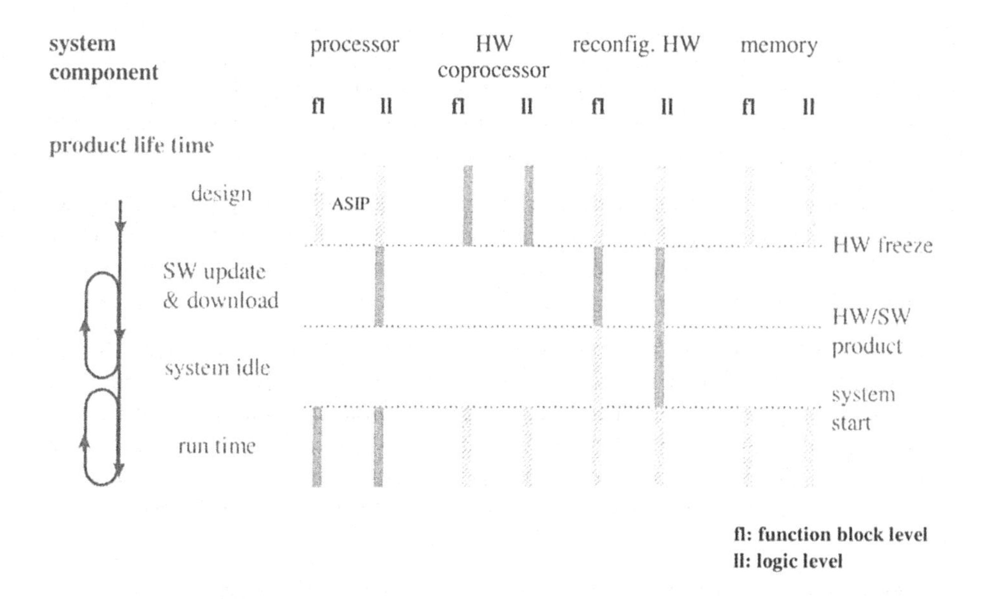

Figure 19: Adaptability during product life time

A closer look at the definitions shows that we have just introduced sets of behaviors. To be a little more formal, we define the terms **dynamic adaptability set**, **programmability set**, etc.. In practical cases, these sets are finite but hard to determine. As a next step, we should restrict the system behavior in the sets. The comparison of MMX and TM1000 suggested that we should define an application set of behaviors which is of interest to the intended embedded system plus the required constraints (max. power consumption, etc.) and then build cut sets with the adaptability set, programmability set, etc.. What we obtain shall be called **application programmability set**, **application flexibility set**, etc..

The effects of IP adaptability and adaptability requirements on system optimization are comprehensive. For clarity, we will take a two step approach. We will, first, consider the adaptability of different architecture in context of the product life cycle and then outline their impact on the design process.

We use a simplified model of the product life cycle and a simple classification of component types. The product life cycle starts with the design which may be a co-design process or a conventional sequential process. At some time in the design process, the hardware is fixed ("hardware freeze") and deployed for production. The software can still be modified until it is downloaded for product delivery. The software can also be used to create several variants of a product (see introduction), in which case there are several software variants for production. After product delivery, the software is fixed ("SW freeze"), but may be updated later, e.g. as part of a maintenance procedure.

After the product is delivered, the embedded system can be idle (turned off) or running. These two phases will alternate after delivery. When the system is idle, it can be reconfigured. It is, as well, possible to reconfigure at run time, such that the configuration itself becomes part of the run-time system function. A closer look reveals that even then the respective system part usually goes through a (short) idle period. Similarly, software can be downloaded, compiled and executed at run time ("just-in-time"), but this is not the standard case in embedded systems and shall be omitted, here.

Fig. 19 summarizes the relation between component adaptability and product life cycle. It distinguishes function block level (*fb*) and component level (*cl*). The bars in the figure denote adaptability in the respective phase of a product life cycle, a dotted bar shows that only part of the components are adaptable.

We will distinguish 4 classes of components, processors, hardware coprocessors, reconfigurable hardware, and memory. **Processors** are either given as fixed, non configurable cores or are configured or developed as part of the design process for the specific application (Application Specific Processors: **ASIPs**). Processors are programmable, but not reconfigurable. The individual function blocks of a processor (ALU, CU, ...) are not programmable but their function adapts dynamically to the current control messages (derived from the instruction).

The simplest case of a **hardware coprocessor** has a fixed, data independent control and data flow, such as a filter processor consisting of multipliers and adders. Its functionality is fixed at the function block and the component levels. Coefficients and data may change but the set of operations to be executed remains unchanged. A coprocessor can, however, change its behavior dependent on control messages and data, just as explained above when discussing the specialization techniques. Hardware coprocessors are developed during the design process possibly using IP function blocks.

Reconfigurable hardware summarizes FPGA based designs and reconfigurable processors. They can be based on fixed components with fixed interconnect, such that there is no adaptability effective in hardware design. The FPGA function is downloaded similar to a software program, both for the function block level and the component level, while in some reconfigurable computers, the function block level is fixed and only program and interconnect are reconfigured. Since an FPGA can be used to implement a simple filter, the reconfigurable hardware might have no dynamic adaptability.

Finally, **memories** must not be omitted from this consideration. Memories can be used as fixed entities or can be configurable at design time. They are usually not programmable, but complex dynamic memory circuits, such as SLDRAMs can even have reconfigurable bank assignment and protocols, as we have seen. Dynamic adaptability is, e.g. found in cache memories.

Obviously, the different forms of adaptability are not equivalent and lead to different target system properties with respect to market value, reuse, maintainability, etc.. Fig. 19 suggests that an embedded system specification should, therefore, contain adaptability requirements with respect to the different phases of a product life cycle. To be used in optimization and verification, these requirements should be formal. That this is not the case, today, can easily be explained with the observation that most specification approaches originate in software engineering.

7.3. IMPACT ON THE DESIGN PROCESS AND DESIGN OBJECTIVES

As a second step, we will consider the impact of adaptability on the design process. Configurability is a requirement for components to be reused, e.g. in libraries, but not for the target embedded system to be delivered, which requires flexibility. So, rather

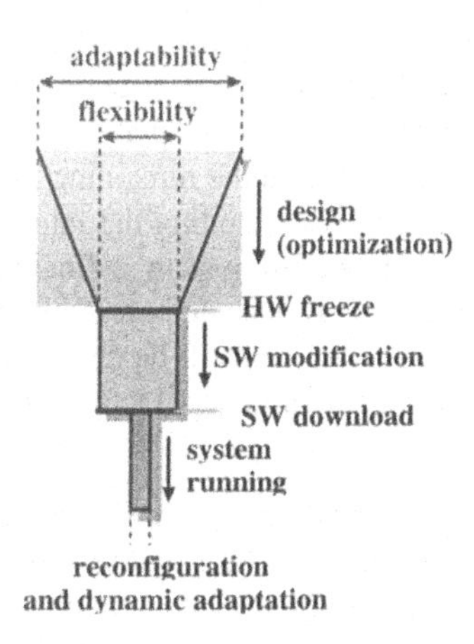

20a: Flexibility and adaptability-
a pictorial representation

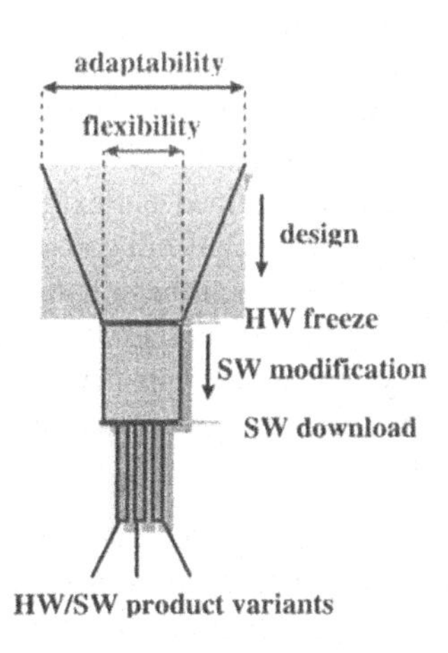

20b: Design with product variants

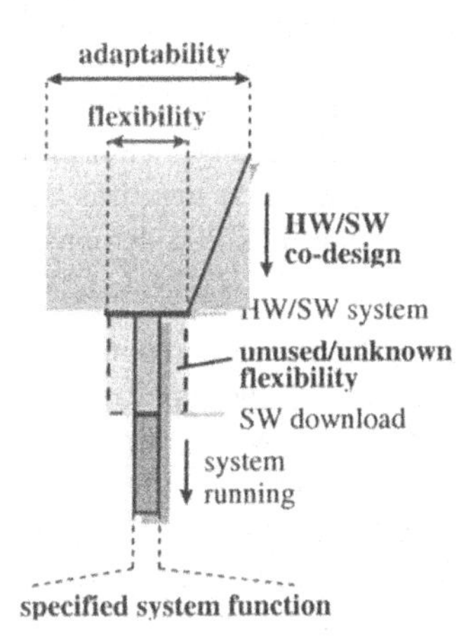

20c: Flexibility and co-design

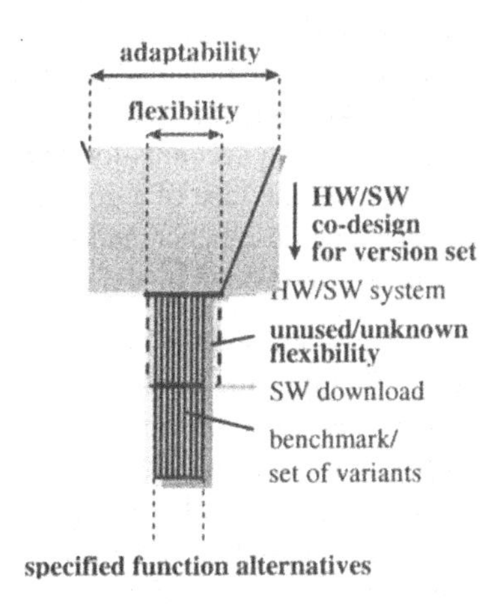

20d: Design with mutual exclusive specifications

Figure 20: Flexibility and adaptability - a pictorial representation

than discussing the challenges of IP based design (e.g. synthesis with IP), we will consider the impact of target architecture **flexibility requirements** as a **design objective** or even **design constraint** on the design process. We will, hence, consider the questions:

- How do the current approaches to co-design influence the target system flexibility?
- What techniques shall be used in the design process to control flexibility?

We will use the picture in fig. 20 to visualize the problem. Fig. 20a introduces the pictorial representation. The design process adapts the system components and architecture to the application. We have seen that this corresponds to a specialization i.e. a reduction of the adaptability. When the HW design is forwarded to production, the adaptability is reduced to what we defined as flexibility of the system. Depending on the remaining flexibility, the hardware platform can still be used for many purposes including the intended application. Only when the software is downloaded, the embedded system is finalized, and what is left are reconfigurability and dynamic adaptability. If designed correctly, the resulting system implements the required behavior. The software can still be changed or updated or it can be used to create different embedded system variants which is shown in fig. 20b.

The current co-design process starting from a single specification does not take such mutual exclusive software variants or late changes into account. The flexibility of the design, which comes from programmable processors or reconfigurable units, is not fully explored, not even documented. This holds for manual designs, but even more so for automated co-synthesis [43]. Fig. 20c visualizes this situation.

In contrast, general purpose computing has always aimed at system optimization under flexibility requirements. For this purpose, it uses benchmarks, which are sets of representative applications which are, one at a time, analyzed as running on the computer [7]. The systematic definition and evaluation of benchmarks ("quantitative" design approach) was found to be a major driver for progress in computer architecture. The benchmark approach accounts for software variants as well as for changes after product delivery which is the idea of a general purpose computer. So, as a first step, it is a good idea to define and use benchmarks for embedded system design. Fig. 20d demonstrates the effects on our abstract model.

There are however, significant differences. First, in general purpose computer design, the target system is the hardware platform with a rough idea on the type of software which will later run on the system. Embedded system design, however, aims at hardware/software systems which means that all constraints (timing, power, ..) must be analyzed and verified including the software. There is no such application verification in general purpose computers, except for specialized system parts, e.g. multi-media function accelerators. Interestingly, most papers on multimedia accelerators for general purpose computers (e.g [21]) give much space to the selection and detailed manual analysis of the intended applications rather than relying on benchmark simulations.

The second issue is the selection of benchmarks. It seems obvious that the higher the specialization, the larger the effect of benchmark selection on the target architecture and the more risky the exclusive use of benchmarks as a quantitative measure in processor design. Since benchmarks measure system properties, which are defined in the system specification, they become part of the system specification. So, we can *incompletely specify* a system with a benchmark set which misses important examples, we can unnecessarily *constrain* system design by an extended set of benchmarks, and we can *incorrectly weigh* the benchmark set if the members of the set overemphasize a certain system property. It would, e.g., be easy to define a benchmark set where the TM1000 Image Coprocessor appears almost useless if we just select enough non-standard video algorithms.

One approach to a verifiable benchmark set is to enumerate all variants of a system including the expected changes. In [44], such a solution was proposed for ASIP design, derived from a very similar solution for systems with concurrent processes but without variants [45]. Different tasks with timing requirements were serialized into a single large sequential process which was, then, synthesized such that all timing requirements of all tasks were met. Notably, synthesis turned out to be very sensitive to the order in which the variants (here: the different applications) were applied which underlines the challenges of a design process based on benchmarks. Similarly, the authors of [46] optimize different functions of an audio/video codec separately but define metrices to identify commonalties between functions which are then used for hardware optimization. Again, they use heuristics to select an appropriate variant order for co-synthesis.

Unfortunately, these solutions are not easily extendible to more complex systems with multiple concurrent processes. Also, with more complex systems, enumeration of potential system variants becomes prohibitive.

A more general approach is to explicitly describe mutually exclusive system variants in the specification. The process model in [47] uses boolean equations to describe sets of system parts which are executed together. This can be used to for scheduling of systems with different system modes (dynamic adaptation and reconfiguration) but equally well to describe mutually exclusive system variants. Originally developed for scheduling, this model can also include incomplete process descriptions where only data flow requirements or estimated timing with lower and upper bounds are provided. This can be used to summarize expected late changes in not yet fully specified parts. Such process models might become a step towards a more formal description of flexibility and adaptability requirements.

8. Selected Co-Design Problems

The classification of system specialization techniques has given structure to the confusing variety of seemingly unrelated embedded architectures. Understanding embedded target architectures is a prerequisite for useful and accurate modeling and estimation.

We should also find ways to apply this knowledge to system synthesis. On the one hand, the system specialization techniques which were derived from manual designs could be applied as optimization strategies of a synthesis and code generation process. Current research approaches include **instruction set definition** [48, 49], **instruction encoding** for code compression [50] and **memory optimization** [51, 52, 53]. The other techniques are less examined, in particular the system level techniques are largely unexplored. Just to highlight one of the problems, **global control and data flow optimization** would be a worthwhile topic which goes beyond the work in data dominated systems which has been mentioned above. Related to this problem are **communication channel selection and communication optimization. Component interface synthesis** is already an active research topic [54, 55, 19]. Specialization for **power minimization** would clearly be another topic of interest. All problems should be approached under **flexibility requirements.**

On the other hand, the classification of specialization techniques could be applied to support **component selection** and **component reuse**. In this case, we could characterize a component with respect to the specialization techniques and then compare if this specialization fits the application. Such an abstract approach could also include abstract flexibility requirements while synthesis would require a more formal definition of flexibility.

9. Conclusion

There is a large variety of target architectures for hardware/software co-design. This variety is the result of manual architecture optimization to a single application or to an application domain under various constraints and cost functions. The optimization leads to specialized systems. Target system optimization can therefore be regarded as a specialization process. We have identified a set of system specialization techniques for components and systems which can serve to describe and to classify the resulting target systems. A number of target system examples for control dominated and data dominated application domains with different models of computation have been analyzed based on this classification. We identified various degrees of specialization and discussed the conflict of specialization and flexibility. Finally, we analyzed the system design process with respect to the related issues of reuse and system flexibility system discussed several approaches to include flexibility requirements in the design process. We identified many open problems which require further research in target architectures.

10. References

1 Semiconductor Industry Association (1997) *The national technology roadmap for semiconductors.*

2 Konkurrenzlos Preisgünstig - Siemens (1998) Embedded DRAM für Consumer- Applikationen (in German). *Journal Markt&Technik,* **3**, 24-27.

3 Stankovic, J.A. and Ramamritham, K. (1988) *Tutorial on Hard Real-Time Systems.* IEEE.

4 Lala, P.K. (1985) *Fault Tolerant & Fault Testable Hardware Design.* Prentice Hall.

5 Pradhan, D.K. (1986) *Fault Tolerant Computing Theory and Techniques* **I** and **II**. Prentice-Hall.

6 Ott, H.W. (1988) *Noise Reduction Techniques in Electronic Systems.* Wiley&Sons.

7 Hennessy, J.L. and Patterson, D.A. (1996) *Computer Architecture - A Quantitative Approach*. 2nd edition. Morgan-Kaufmann.

8 Flynn, M.J. (1972) Some Computer Organizations and their Effectiveness. *IEEE Trans. on Comp.*, 948-960.

9 DECchip 21030 (1994) Reference Manual. Digital Equipment.

10 Österling, A. et al. (1997) The COSYMA System, in J. Staunstrup, W. Wolf (eds.), *Hardware/Software Co-Design*, Kluwer, 263-282.

11 TMS320C80 (MVP) (1995) Collection of User's Guides. Texas Instruments.

12 Vissers, K.A. (1995) Architecture and Programming of two Generations Video Signal Processors. *Microprocessing and Microprogramming*, 373-390.

13 Lee, E.A. and Messerschmidt, D.G. (1987) Static Scheduling of Synchronous Data Flow Graphs for Digital Signal Processing. *IEEE Trans. on CAD*, 24-35.

14 Lippens, P., van Meerbergen, J., van der Werf, A., Verhaegh, W., Mc-Sweeney, B., Hiusken, J., McArdle, O. (1991) PHIDEO: A Silicon Compiler for High Speed Algorithms. *Proc. EDAC 91*, 436-441.

15 Jerraya, A. et al. (1997) Languages for System Level Specification and Design, in J. Staunstrup, W. Wolf (eds.), *Hardware/Software Co-Design*, Kluwer, 235-261.

16 SAB 80C515 Users Manual. Siemens, 1992.

17 80C552 Product Specification. Philips, 1993.

18 SAB 80C515 Users Manual. Siemens, 1992.

19 Chou, P., Ortega, R., Borriello, G. (1995) Interface Co-Synthesis Techniques for Embedded Systems. *Proc. ICCAD 95*, 280-287.

20 Oppenheim, A.V. and Schafer, R.W. (1997) *Digital Signal Processing*. Prentice Hall.

21 Peleg, A. and Weiser, U. (1996) MMX technology extension to the Intel architecture. *IEEE Micro*, Aug. 96, 42-50.

22 Tremblay, M. and O'Connor, J.M. (1996) VIS speeds new media processing. *IEEE Micro*, Aug. 96, 10-20.

23 Lee, R. (1996) Subword parallelism with MAX-2. *IEEE Micro*, Aug. 96, 51-59.

24 SH-DSP Programming. Hitachi Micro Systems, 1997.

25 SH-DSP Overview. Hitachi Micro Systems, 1996.

26 Liem, C. and Paulin, P. (1997) Compilation Techniques for Embedded Processor Architectures, in J. Staunstrup, W. Wolf (eds.), Hardware/Software Co-Design, Kluwer, 149-192.

27 Rathnam, S. and Slavenburg, G. (1998) Processing the new world of interactive media. *IEEE Signal Processing Magazine*, March 98, 108-117.

28 TM 1000 preliminary data book. Philips, 1997.

29 Riemens, A.K., Schutten, R.J., and Vissers, K.A. (1997) High speed video de-interlacing with a programmable TriMedia VLIW core. *Proc. ICSPAT 97*, 1375-1380.

30 Gillingham, P. and Vogley, B. (1997) SLDRAM: high performance, open-standard memory. *IEEE Micro*, vol 16, no 6, Dec. 97, 29-39.

31 Crisp, R. (1997) DirectRambus technology: the next main memory standard. *IEEE Micro*, vol 16, no 6, Dec. 97, 18-28.

32 Jack, K. (1996) Video Demystified. HighText Publications.

33 Chen, D. and Rabaey, J. (1992) A Reconfigurable Multiprocessor IC for Rapid Prototyping of Algorithm Specific, High Speed DSP Data Paths. *IEEE Journal of Solid State Circuits*, vol 27, no 12, Dec. 92, 1895-1904.

34 Leijten, J., van Meerbergen, J., Timmer, A., Jess, J. (1997) PROPHID: A heterogeneous multiprocessor architecture for multimedia. *IEEE Proc. ICCD 97*, Austin, 164 -169.

35 Leijten, J., van Meerbergen, J., Timmer, A., Jess, J. (1998) Stream Communication between Real-Time Tasks in a High-Performance Multiprocessor. *IEEE Proc. DATE 98*, Paris, 125-131.

36 Lee, E. and Parks, Th. (1995) Data Flow Process Networks. *Proceedings of the IEEE*, May 95, 773-799.

37 Gerndt, R. and Ernst, R. (1997) An Event-Driven Multi-Threading Architecture for Embedded Systems. *IEEE Proc. Workshop on Hardware/Software Co-Design (Codes) 97*, Braunschweig, 29-33.

38 MC68332 Reference Manual. Motorola, 1991.

39 Hauck, S. (1998) The Roles of FPGA's in Reprogrammable Systems. *Proceedings of the IEEE*, vol 86, no 4, Apr. 98, 615-638.

40 Rabaey, J. (1997) Reconfigurable Computing: The Solution to Low Power Programmable DSP. *IEEE Proc. ICASSP*, Munich, April 1997.

41 Abnous, A., Seno, K., Ichikawa, Y., Wan, M., Rabaey, J. (1998) Evaluation of a Low-Power Reconfigurable DSP Architecture. *Proc. Reconfigurable Architectures Workshop*, Orlando, March 1998.

42 DeHon, A. (1996) *Reconfigurable Architectures for General-Purpose Computing*. PhD Thesis, MIT, 1996.

43 Wolf, W. *Co-Synthesis Algorithms*. This Book.

44 Kim, K., Karri, R., Potkonjak, M. (1997) Synthesis of application specific programmable processors. *Proc. DAC 97*, Anaheim, 353-358.

45 Potkonjak, M. and Wolf, W. (1995) Cost Optimization in ASIC Implementation of Periodic Hard-Real Time Systems using Behavioral Synthesis Techniques. *Proc. ICCAD 95*, San Jose, 446 - 451.

46 Kalavade, A. and Subrahmanyam, P. (1997) Hardware/Software Partitioning for Multi-Function Systems. *Proc. ICCAD 97*, 516-521.

47 Ziegenbein, D., Richter, K., Ernst, R., Teich, J., Thiele, L. (1998) Representation of Process Mode Correlation for Scheduling. To appear: *IEEE Proc. ICCAD 98.*

48 Wilberg, J. and Camposano, R. (1996) VLIW Processor Codesign for Video Processing. *Journal on Design Automation for Embedded Systems*, vol 2, 79-119.

49 Athanas, P.M. and Silverman, H.F. (1993) Processor Reconfiguration Through Instruction Set Metamorphosis. *IEEE Trans. on Computers*, March 93, 11-18.

50 Kozuch, M. and Wolfe, A. (1994) Compression of Embedded System Programs. *Proc. ICCD 94*, 270-277.

51 Mirinda, M., Catthoor, F., Janssen, M., De Man, H. (1996) ADOPT: Efficient Hardware Address Generation in Distributed Memory Architectures. *Proc. ISSS 96*, 20-25.

52 Tomiyana, H. and Yasuura, H. (1996) Optimal Code Placement of Embedded Software for Instruction Caches. *Proc. ED&TC 96*, 96-101.

53 Schmit, H. and Thomas, D. (1995) Address Generation for Memories Containing Multiple Arrays. *Proc. ICCAD 95*, 510-514.

54 Ismail, T.D., Abid, M., Jerraya, A. (1994) COSMOS: A Codesign approach for communicating systems. *Proc. Workshop on Hardware/Software Codesign (Codes) 94*, 17-24.

55 Van Rompaey, K., Verkest, D., Bolsens, J., De Man, H. (1996) CoWare: A Design Environment for Heterogeneous Hardware/Software Systems. *Proc. EURODAC 96.*

MODELS OF COMPUTATION FOR EMBEDDED SYSTEM DESIGN

LUCIANO LAVAGNO
Department of Electronics
Politecnico di Torino
C. Duca degli Abruzzi 24, Torino, Italy
lavagno@polito.it

ALBERTO SANGIOVANNI-VINCENTELLI
Department of EECS
University of California
Berkeley, CA 94709 USA
alberto@eecs.berkeley.edu

AND

ELLEN SENTOVICH
Cadence Berkeley Laboratories
2001 Addison St.
Berkeley, California USA
ellens@cadence.com

Abstract. In the near future, most objects of common use will contain electronics to augment their functionality, performance, and safety. Hence, time-to-market, safety, low-cost, and reliability will have to be addressed by any system design methodology. A fundamental aspect of system design is the specification process. We advocate using an unambiguous formalism to represent design specifications and design choices. This facilitates tremendously efficiency of specification, formal verification, and correct design refinement, optimization, and implementation. This formalism is often called *model of computation*. There are several models of computation that have been used, but there is a lack of consensus among researchers and practitioners on the "right" models to use. To the best of our knowledge, there has also been little effort in trying to compare rigorously these models of computation. In this paper, we review current models of computation and compare them within a framework that has been recently proposed. This analysis demonstrates both the need for heterogeneity to capture the

A. A. Jerraya and J. Mermet (eds.), System-Level Synthesis, 45–102.

richness of the application domains, and the need for unification for optimization and verification purposes. We describe in detail our CFSM model of computation, illustrating its suitability for design of reactive embedded systems and we conclude with some general considerations about the use of models of computations in future design systems.

1. Introduction

1.1. EMBEDDED SYSTEM DESIGN TODAY

An embedded system is a complex object containing a significant percentage of electronic devices (generally including at least one computer) that interacts with the real world (physical environment, human users, etc.) through sensing and actuating devices. A system is heterogeneous, i.e., is characterized by the co-existence of a large number of components of disparate type and function. For example, it may contain programmable components such as micro-processors and Digital Signal Processors, as well as analog components such as A/D and D/A converters, sensors, transmitters and receivers. In the past, the system design effort has focused on these hardware parts, leaving the software design to be done afterwards as an implementation step. However, today more than 70% of the development cost for complex systems such as automotive electronics and communication systems is attributable to software development. This percentage is increasing constantly. The challenge posed to the semiconductor industry is to provide a new generation of programmable parts and of supporting tools to help system designers develop software faster and correctly the first time.

Today much attention is devoted to the hardware-software co-design issue, i.e., to the concurrent development of Application Specific Integrated Circuits and standard hardware components, selection of programmable components, and development of the application software that will run on them. We believe that this approach in fact enters the design process too late to explore interesting design trade-offs.

1.2. OUR DESIGN METHODOLOGY GOALS

The computer-aided design process should begin at a very early stage. We believe that the real key to shortening design time and coping with complexity is to start the design process *before* the hardware-software partitioning. For this reason, we believe that the key problem is not so much hardware-software co-design, but the sequence consisting of specifying what the system is intended to do with no bias towards implementation, of the initial functional design, its analysis to determine whether the functional design satisfies the specification, the mapping of this design to a candidate architecture, and the subsequent performance evaluation. It is then clear that the key aspect of system design is indeed *function-architecture co-design*.

Our approach is a design methodology that is based on the use of *formal models* to describe the behavior of the system at a high level of abstraction, before a decision on its decomposition into hardware and software components is taken. Our approach also facilitates the use of existing parts. As the complexity of embedded systems increases, it is unthinkable to design new systems from scratch. Already hardware components are often standard parts that are acquired from silicon vendors, and software is often incrementally upgraded from previous versions of the same product. In the future, *design re-use* will be the key to profitability and market timing. In addition, the decreasing feature size of silicon manufacturing processes will make it possible to incorporate multiple microprocessors, complex peripherals, and even sensors and actuators on the same silicon substrate, which will force system developers and IC designers/manufacturers to deal with the problem of *exchanging Intellectual Property* in the form of designs instead of chips.

1.3. DESIGN STRATEGY

The overall design strategy that we envision is depicted in Figure 1. There are, of course, other ways to design systems in common use today. The top-down nature of the design methodology that our group has advocated throughout the years may not be agreed upon by the system design community where a mixed top-down, bottom-up approach is mostly used. In our methodology, however, we believe that this approach is captured by the presence of architectural and functional libraries that could be the result of a bottom-up assembly of basic components. We strongly emphasize that no matter how the design is carried out, a *rigorous framework is necessary to reduce design iterations and to improve design quality.*

1.3.1. *Design Conception to Design Description*

At the functional level, a behavior for the system to be implemented is selected and analyzed against a set of specifications. The definition of specification and behavior is often the subject of hot debate. For some, there is no difference between specification and behavior. For some, specification is the I/O relation of the system to realize together with a set of constraints to satisfy and of goals to achieve, and behavior is the algorithm that realizes the function to be implemented. For others, specification is the algorithm

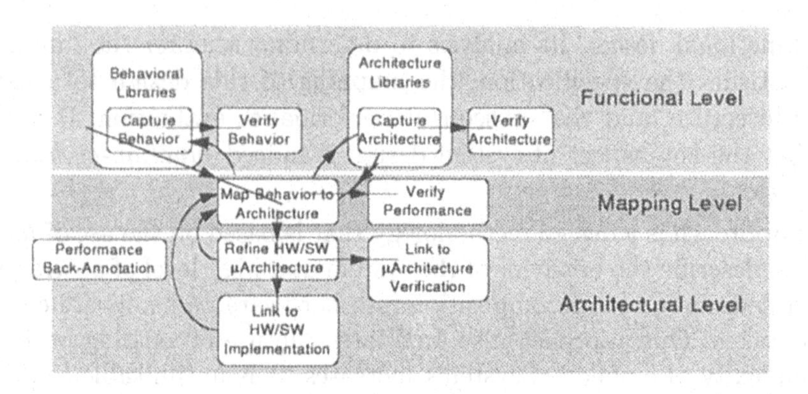

Figure 1. Proposed design strategy

itself. From a purist point of view, an algorithm is indeed the result of an implementation decision from a given set of specifications and we prefer to stick to this view in our design methodology. For example, if we specify the function that a system has to perform as "given a nonlinear function f over the set of reals, find x so that f(x)=0", then it is a design decision to chose the Newton-Raphson algorithm or a Gauss-Seidel nonlinear relaxation algorithm. On the other hand, for an MPEG encoder, the specification is the encoding of the compressed stream, and any implementation that creates it from a stream of images is "correct". In this second case, the first step of system design has already been decided upon and the designer has no freedom to alter the conceptual design.

1.3.2. *Algorithm Design*

Algorithm development is a key aspect of system design at the functional level. We believe that little has been done in this domain to help the designer to select an algorithm that satisfies the specifications. The techniques and environments for this step are often application dependent. We have experience in automotive engine control [7], where the algorithms have to have strong correctness properties due to the life critical aspects inherent in this application. In addition, the "plant" to be controlled (the combination of the engine and the drive-line) is a hybrid system consisting of continuous components (drive-line) and discrete ones (engine). To assess the properties of the algorithms, one must use control theory and sophisticated simulation techniques involving mixed differential equations-discrete event models. The understanding of the application domains yields a design methodology that integrates the application-specific view with general-purpose techniques that could be re-used in other domains of application. It is our strong belief that this step of system design carries the maximal leverage when combined with the design methodology proposed here.

1.3.3. *Algorithm Analysis*

The behavior of an algorithm is verified by performing a set of analysis steps. Analysis is a more general concept than simulation. For example, analysis may mean the formal proof that the algorithm selected always converges, that the computation performed satisfies a set of specifications, or that the computational complexity, measured in terms of number of operations, is bounded by a polynomial in the size of the input. In the view of design re-use, parts of the overall behavior may be taken from an existing library of algorithms. *Since it is the formal model that provides the framework for algorithm analysis, it is very important to decide which mathematical model to support in a design environment.*

1.4. ALGORITHM IMPLEMENTATION

Once the algorithm has been selected, there is an intermediate step before the selection of the architecture to support its implementation: its transformation into a set of functional components that are computationally tractable. This set of functional components have to be formally defined to ensure that the properties of the implementation of the algorithm can be assessed. To do so, the concept of models of computation is key. Most system designs use one or more of the following models of computation: Finite State Machines, Data Flow Networks, Discrete Event Systems, and Communicating Sequential Processes. A particular model of computation has mathematical properties that can be efficiently exploited to answer questions about system behavior without carrying out expensive verification tasks. An important issue here is how to compare and compose different models of computation.

Once the model(s) of computation have been selected, then we can safely proceed towards the implementation of the system by selecting the physical components (architecture) of the design.

1.5. OUR GOALS

The main goal of this paper is to review and compare the most important models of computations using a unifying theoretical framework introduced recently by Lee and Sangiovanni-Vincentelli [43]. We also believe that it is possible to optimize across model-of-computation boundaries to improve the performance of and reduce errors in the design at an early stage in the process.

There are many different views on how to accomplish this. There are two essential approaches: one is to develop encapsulation techniques for each pair of models that allow different models of computation to interact in a meaningful way, i.e., data produced by one object are presented to the other in a consistent way so that the object "understands" [17]. The other is to develop an encompassing framework where all the models of importance "reside" so that their combination, re-partition and communication happens in the same generic framework. While we realize that today heterogeneous models of computation are a necessity, we believe that the second approach will be possible and will provide the designer with a powerful mechanism to actually select the appropriate models of computation, (e.g., FSMs, Data-flow, Discrete-Event, that then become a lower level of abstraction with respect to the unified model) for the essential parts of the design.

At this level is also important to *orthogonalize concerns*, that is to separate different aspects of the design. In this regard, a natural dividing line is the separation between functionality and communication. That is, we view a design as composed of functional behavior (modules) and communication behavior (between modules), which are themselves further decomposed as we refine and analyze the design. It is our strong belief that communication is key in assembling systems on a chip from separate Hip's. Communication is a complex issue since the functionality of the components being interconnected must be preserved.

The separation between function and communication will be emphasized throughout the paper. In addition, a model of computation that encompasses the key aspects of Discrete Event, Data-Flow and Finite State Machine models will be presented in detail. This model, called network of Co-design Finite State Machines (CFSM), is the backbone of the POLIS system developed at the University of California at Berkeley, an environment for function-architecture co-design with particular emphasis on control-dominated applications and on software development. (See Figure 2 for a block diagram of the functionalities of the environment.) This model is also used as the basic semantic model for an industrial product of Cadence Design Systems, an environment for embedded system design including multi-media and telecommunication applications.

The paper is organized as follows. In Section 2, we present the mathematical machinery used to compare and describe the models of computation. In Section 3, we present and compare the most important models of computation. In Section 4, we introduce the CFSM model. In Section 5, we give some concluding remarks.

2. MOCs: Basic Concepts and the Tagged Signal Model

2.1. MODELING EMBEDDED SYSTEMS WITH MOCS

An MOC is composed of a description mechanism (syntax) and rules for computation of the behavior given the syntax (semantics). An MOC is

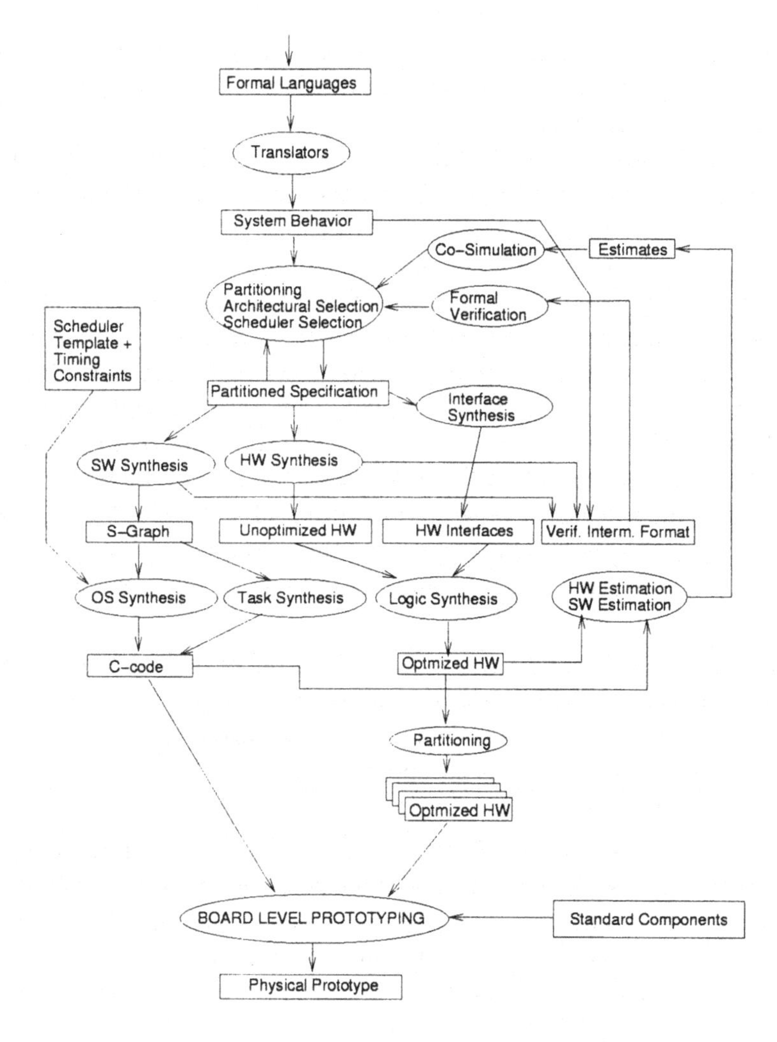

Figure 2. The POLIS design framework

chosen for describing a sub-behavior of a design based on its suitability: compactness of description, fidelity to design style, ability to synthesize and optimize the behavior to an appropriate implementation. For example, some MOCs are suitable for describing complicated data transfer functions and completely unsuitable for complex control, while others are designed with complex control in mind.

There are a number of basic ideas and primitives that are commonly used in formulating models of computation. Most MOCs permit distributed system description (a collection of communicating modules), and give rules dictating how each module computes (function) and how they transfer information between them (communication). Some of the primitives include combination Boolean functions and synchronous state machines for specifying function, and queues, buffers, schedulers for specifying communication. Function and communication are often not described completely separately for efficiency and optimization.

More precisely, MOCs are typically realized (implemented in practice) by a particular language and its semantics. We elaborate on the distinction between MOCs and languages in Section 2.2.

A *language* is a set of symbols, rules for combining them (its *syntax*), and rules for interpreting combinations of symbols (its *semantics*). Two approaches to semantics have evolved, *denotational* and *operational*. A language can have both (ideally they are consistent with one another, although in practice this can be difficult to achieve). Denotational semantics, first developed by Scott and Strachey [59], gives the meaning of the language in terms of relations. Operational semantics, which dates back to Turing machines, gives meaning of a language in terms of actions taken by some abstract machine, and is typically closer to the implementation.

Models of computation can be viewed based on the following characteristics:

- the kinds of relations that are possible in a denotational semantics
- how the abstract machine behaves in an operational semantics
- how individual behavior is specified and composed
- how hierarchy abstracts this composition
- communication style

A design (at all levels of the abstraction hierarchy from functional specification to final implementation) is generally represented as a set of components, which can be considered as isolated monolithic blocks, which interact with each other and with an environment that is not part of the design. The model of computation defines the behavior and interaction of these blocks.

We view MOCs at two levels of abstraction. At the higher level, we take the view of the tagged signal model (which we call here TSM) described in section 2.3. The TSM abstraction defines processes and their interaction using signals composed of partially ordered events, in turn composed of tags and values. We use processes to describe both functional behavior and communication behavior. This is a denotational view, though it is not associated with a particular language. We use this model to compare elements of different models of computation, styles of sequential behavior, concurrency, and communication at a high level.

At the lower level of abstraction, we take the view of general primitives for function and timing (used in the refinement of TSM processes), where each MOC constitutes a particular choice of these two. This is a more operational view. We give precise definitions for a number of terms, but these definitions will inevitably conflict with standard usage in some communities. We have discovered that, short of abandoning the use of most common terms, no terminology can be consistent with standard usage in all related communities. We attempt to avoid confusion by being precise, even at the risk of being pedantic. The basic primitive concepts are describe in Section 2.4. The primitive building blocks for specification and implementation are given in Section 2.5.

All these basic primitives and concepts are then used in Section 3 to classify and describe the main MOCs that appear in the literature.

2.2. LANGUAGES AND MODELS OF COMPUTATION

The distinction between a language and its underlying model of computation is important. The same model of computation can give rise to fairly different languages (e.g., the imperative Algol-like languages C, C++, Pascal, and FORTRAN). Some languages, such as VHDL and Verilog, support two or more models of computation[1].

The model of computation affects the *expressiveness* of a language — which behaviors can be described in the language, whereas the syntax affects compactness, modularity, and reusability. Thus, for example, object-oriented properties of imperative languages like C++ are more a matter of syntax than a model of computation.

The expressiveness of a language is an important issue. A language that is not expressive enough to specify a particular behavior is clearly unsuitable, while a language that is too expressive is often too complex for analysis and synthesis. For very expressive languages, many analysis and synthesis problems become undecidable: no algorithm will solve all problem instances in finite time.

A language in which a desired behavior cannot be represented succinctly is also problematic. The difficulty of solving analysis and synthesis problems is at least linear in the size of the problem description, and can be as bad as several times exponential, so choosing a language in which the description of the desired behavior of the system is compact can be critical.

A language may be very incomplete and/or very abstract. For example, it may specify only the interaction between computational modules, and not the computation performed by the modules. In this case, it provides an interface to a host language that specifies the computation, and is called a coordination language (examples include Linda [20], Granular Lucid [34], and Ptolemy domains [17]). Another language may specify only the causality constraints of the interactions without detailing the interac-

[1] They directly support the Imperative model within a process, and the Discrete Event model among processes. They can also support Extended Finite State Machines under suitable restrictions known as the "synthesizable subset".

tions themselves nor providing an interface to a host language. In this case, the language is used as a tool to prove properties of systems, as done, for example, in process calculi [33, 46] and Petri nets [50, 53]. In still more abstract modeling, components in the system are replaced with nondeterminate specifications that give constraints on the behavior, but not the behavior itself. Such abstraction provides useful simplifications that help formal verification.

2.3. THE TAGGED-SIGNAL MODEL

At the highest level of abstraction, we adopt the tagged-signal model (TSM) proposed by Lee and Sangiovanni-Vincentelli [42]. It is a formalism for describing aspects of models of computation for embedded system specification. It is denotational in the Scott and Strachey [59] sense, and it defines a semantic framework (of signals and processes) within which models of computation can be studied and compared. It is very abstract—describing a particular model of computation involves imposing further constraints that make it more concrete.

2.3.1. *Signals, tags and events*

The fundamental entity in the TSM is an event: a value/tag pair. Tags are often used to denote temporal behavior. A set of events (an abstract aggregation) is a signal. Processes are relations on signals, expressed as sets of n-tuples of signals. A particular model of computation is distinguished by the order it imposes on tags and the character of processes in the model. More formally, given a set of *values* V and a set of *tags* T, an *event* is a member of $T \times V$. A *signal* s is a set of events, and thus is a subset of $T \times V$. A *functional* (or deterministic) *signal* is a (possibly partial) function from T to V. The set of all signals is denoted S. A *tuple* of n signals is denoted $\mathbf{s}$, and the set of all such tuples is denoted S^n.

The different models of time that have been used to model embedded systems can be translated into different order relations on the set of tags T in the tagged-signal model. In a *timed system* T is totally ordered, i.e., there is a binary relation $<$ on members of T such that if $t_1, t_2 \in T$ and $t_1 \neq t_2$, then either $t_1 < t_2$ or $t_2 < t_1$. In an *untimed system*, T is only partially ordered.

2.3.2. *Processes*

A *process* P with n signals is a subset of the set of all n-tuples of signals, S^n for some n. A particular $\mathbf{s} \in S^n$ is said to *satisfy* the process if $\mathbf{s} \in P$. An $\mathbf{s}$ that satisfies a process is called a *behavior* of the process (intuitively, it is the generalization of a "simulation trace"). Thus a *process* is a set of possible *behaviors*, or a constraint on the set of "legal" signals.

Intuitively, processes in a system operate *concurrently*, and constraints imposed on their signal tags define *communication*[2] among them. The environment in which the system operates can be modeled with a process as well.

For many (but not all) applications, it is natural to partition the signals associated with a process into *inputs* and *outputs*. Intuitively, the process does not determine the values of the inputs, and does determine the values of the outputs. If $n = i + o$, then (S^i, S^o) is a partition of S^n. A process with i inputs and o outputs is a subset of $S^i \times S^o$. In other words, a process defines a *relation* between input signals and output signals. A $(i+o)$-tuple $\mathbf{s} \in S^{i+o}$ is said to *satisfy* P if $\mathbf{s} \in P$. It can be written $\mathbf{s} = (\mathbf{s}_1, \mathbf{s}_2)$, where $\mathbf{s}_1 \in S^i$ is an i-tuple of *input signals* for process P and $\mathbf{s}_2 \in S^o$ is an o-tuple of *output signals* for process P. If the input signals are given by $\mathbf{s}_1 \in S^i$, then the set $I = \{(\mathbf{s}_1, \mathbf{s}_2) \mid \mathbf{s}_2 \in S^o\}$ describes the inputs, and $I \cap P$ is the set of behaviors consistent with the input $\mathbf{s}_1$.

A process F is *functional* (or deterministic) with respect to an input/output partition if it is a single-valued, possibly partial, mapping from S^i to S^o. That is, if $(\mathbf{s}_1, \mathbf{s}_2) \in F$ and $(\mathbf{s}_1, \mathbf{s}_3) \in F$, then $\mathbf{s}_2 = \mathbf{s}_3$. In this case, we can write $\mathbf{s}_2 = F(\mathbf{s}_1)$, where $F : S^i \to S^o$ is a (possibly partial) function. Given the input signals, the output signals are determined (or there is unambiguously no behavior). A process is *completely specified* if it is a total function, that is, for all inputs in the input space, there is a unique behavior.

2.3.3. *Process composition*

Process composition in the TSM is defined by the *intersection* of the constraints each process imposes on each signal. To facilitate its definition, we assume that all the processes that are composed are defined on the same set of signals[3]. Hence a composition of a set of processes is also a process.

In the rest of this paper, we will use processes to model both *function* and *communication*. Generally, an MOC defines a flexible mechanism for modeling function, and a rigid mechanism (signal, queue, shared variable, ...; see Section 2.5) for modeling communication. On the other hand, the TSM must compare different MOCs and hence be flexible when modeling communication as well. It is, however, useful to distinguish between the two at least conceptually, since:

1. Functional processes are mostly concerned with the *value* component of their signals, and generally do not have much to do with the *tag* com-

[2]This is often called also synchronization, but we will try to avoid using the term in this sense because it is too overloaded.

[3]This can be obtained trivially, since a process can be extended to any new signal by simply not imposing any constraint on it.

ponent. In other terms, the constraints a functional process imposes on its input and output signals are generally complex with respect to values, but much simpler with respect to tags.

2. Communication processes are *solely* concerned with the *tag* component of their signals, while values are left untouched.

One of the most useful and important questions to ask when composing processes, is what properties of the isolated processes are *preserved by composition*. Here we focus only on two fundamental properties: functionality (unique output n-tuple for every input n-tuple) and complete specification (for every input n-tuple there exists a unique output n-tuple).

To analyze this aspect, we note that, given a formal model of the functional specifications and of the properties, three situations may arise:

1. The property is *inherent* for the model of the specification (i.e., it can be shown formally to hold for all specifications described using that model).
2. The property can be verified *syntactically* for a given specification (i.e., it can be shown to hold with a simple, usually polynomial-time, analysis of the specification).
3. The property must be verified *semantically* for a given specification (i.e., it can be shown to hold by executing, at least implicitly, the specification for all inputs that can occur).

For example, consider the functionality property. Any design described by a dataflow network (a formal model to be described later) is functional (also called deterministic or determinate in data-flow vernacular), and hence this property need not be checked for this model of computation. If the design is represented by a network of FSMs (for example, synchronous composition of Mealy Finite State Machines), even if the components are functional and completely specified, the result of the composition may be either incompletely specified (the composition has no solution) or non-functional (the composition has multiple solutions). These situations arise if and only if when a combinational feedback loop exists in the composition: with an odd number of Boolean inverters, there is no "solution" and the composition is incompletely specified, with an even number of inverters, there are multiple solutions and the composition is non-functional. A syntactical check on the composition to verify whether combinational loops exist can be carried out. If none exist, then the composition is functional and completely specified. With Petri nets, on the other hand, functionality is difficult to prove: it must be checked by exhaustive simulation.

2.3.4. *Examples*

Consider, as a motivating example introducing these several mechanisms to denote temporal behavior, two problems: one of analysis, modeling a time-invariant dynamical system on a computer, and one of design, the design of a two-elevator system controller.

Analysis Example The underlying mathematical model of a time-invariant dynamical system, a set of differential equations over continuous time, is not directly implementable on a digital computer, due to the double quantization of real numbers into finite bit strings, and of time into clock cycles. Hence a first translation is required, by means of an *integration rule*, from the differential equations to a set of *difference equations*, that are used to compute the values of each signal with a given tag from the values of some other signals with previous and/or current tags.

If it is possible to identify several strongly connected components in the dependency graph[4], then the system is *decoupled.* It becomes then possible to go from the total order of tags implicit in physical time to a *partial order* imposed by the depth-first ordering of the components. This partial ordering gives us some freedom in implementing the integration rule on a computer. We could, for example, play with scheduling by embedding the partial order into the total order among clock cycles. It is often convenient, for example, to evaluate a component completely, for all tags, before evaluating components that depend on it. It is also possible to spread the computation among multiple processors.

In the end, time comes back into the picture, but the *double transformation*, from total to partial order, and back to total order again, is essential to

1. *prove properties* about the implementation (such as stability of the integration method, or a bound on the maximum execution time),
2. *optimize* the implementation with respect to a given cost function (e.g., size of the buffers required to hold intermediate signals versus execution time, or satisfaction of a constraint on the maximum execution time).

Design Example. One of the key motivations for the tagged signal model was to avoid over-specification of designs. For a two-elevator system controller, a simplistic set of specifications can be expressed as follows: respond to all requests in the exact order they are received with the criterion that the maximum delay from the time a request is received and the time the elevator service is offered, is minimized. It is clear that the two elevators are concurrent subsystems and that their operation can be controlled with no need to "synchronize" their operation. It is determined by analyzing the order of events not the exact time of occurrence. However, if no assumption is

[4]A directed graph with a node for each signal, and an edge between two signals whenever the equation for the latter depends on the former.

made about the way requests are made, then we may end up in a dead-lock situation due to the nature of the specification. In fact, if three requests are made at the same instant of time, then the response cannot follow the specification, since there are only two elevators available. One solution to this problem is to assume that no two requests may happen at the exact same time; then the specification can be met for the two elevator system. Another solution is to arbitrarily assign priorities among requests that happen at the same time: the specification is changed to reflect these priorities instead of those implied by the order of occurrence. The most important aspect in design is to capture the intent of the designer by abstracting away the non-essential aspects of the system. This example illustrates that it is essential to classify MOCs by their treatment of events with the same tag. This aspect is strictly related to the notion of synchrony and asynchrony as we will see later.

Once the control algorithm has been developed, then its implementation needs to be carried out. If the algorithm is implemented in software running on a single processor, then all events processed are totally ordered and the order is determined by the intrinsic order coming from the specifications (order of occurrence of the requests) and by the existence of limited resources. Even if the partial order dictated by the algorithm exposes some potential parallelism, the presence of a single processor forces sequential execution determined by a scheduling algorithm that decides in which order the operations are executed. Hence, in the end, we need to map an abstract design into the physical world characterized by real time and limited resources that imposes a global ordering on events.

2.4. COMPARING MODELS OF COMPUTATION

A TSM process is, according to the definition given, a partial mapping from input signals to output signals. In order to consider more concrete mappings, we introduce some primitive concepts on which they are based.

System behavior, as we have previously stated, is composed of *functional behavior* and *communication behavior*, each represented by TSM processes. A process in turn is composed of *functional behavior* and *timing behavior*. Function is how things happen, or in the TSM, how events are related (how inputs are used to compute outputs) "around" a particular tag. Time is the order in which things happen, or in the TSM, the assignment of a tag to each event. The distinction between function and time is not this clean in every context. For example, a state in a finite state machine cannot be labeled as belonging exclusively to the function or time component of the behavior of the machine, but is rather based on the history of both. Nonetheless, the division between function and time, particularly at a primitive level, is useful in the conception and understanding of MOCs.

System operation can be viewed as a series of process computations, sometimes called *firings*. We will use function, time, computation (firing) to describe MOCs and their primitives.

In the sections that follow, we consider the fundamental concepts which are used to refine our processes. For functional processes, we will first consider state-less processes in which only inputs with a given tag concur to form outputs with the same tag. We then introduce the notion of state in the context of process networks. For communication processes, we then consider the primitives for concurrency and inter-process communication. Finally, we give the basic building blocks used to realize these concepts in practice. It is from these that today's most prevalent models of computation are built.

2.4.1. *Process function*

In the control-dominated arena, since the pioneering work of Shannon, Boolean functions have been used as a representation of both a system specification and its implementation in hardware (relay networks in Shannon's time, CMOS gates now). Several formally equivalent (but often with different levels of convenience in practice) representations for binary- and multi-valued boolean functions have been proposed, such as:

- truth table,
- Boolean network [57], which is a Directed Acyclic Graph (DAG), with a truth table associated with each node and edges carrying Boolean variable values,
- Binary Decision Diagram [15], that is also a DAG with one level of nodes for each input variable, and each node acting as a "multiplexer" between the function values associated with every variable values.

In the data-dominated arena, Data Flow actors play the role of processes and represent functions from simple state-less arithmetic operations such as addition and multiplication, to higher-level "combinational" transformations, such as Fast Fourier Transform.

2.4.2. *Process State*

Most models of computation include components with state, where behavior is given as a sequence of state transitions. State in a process network can always be simply implemented by means of feedback. An output and an input signal can be connected together, and thus provide a connection between process inputs and outputs beyond the tag barrier. However, we can also consider a notion of state *within* a process, since this can be useful in order to "hide" the implementation of the state information.

We can formalize this notion by considering a process F that is functional with respect to partition (S^i, S^o). Let us assume for the moment that F belongs to a timed system, in which tags are totally ordered[5]. Then for any tuple of signals $\mathbf{s}$, we can define $\mathbf{s}_{>t}$ to be a tuple of the (possibly empty) subset of the events in $\mathbf{s}$ with tags greater than t.

Two input signal tuples $\mathbf{r}, \mathbf{s} \in S^i$ are in relation E_t^F (denoted $(r^i, s^i) \in E_t^F$) if $\mathbf{r}_{>t} = \mathbf{s}_{>t}$ implies $F(\mathbf{r})_{>t} = F(\mathbf{s})_{>t}$. This definition intuitively means that process F cannot distinguish between the "histories" of $\mathbf{r}$ and $\mathbf{s}$ prior to time t. Thus, if the inputs are identical after time t, then the outputs will also be identical.

E_t^F is obviously an equivalence relation, partitioning the set of input signal tuples into equivalence classes for each t. Following a long tradition, we call these equivalence classes the *states* of F. In the hardware community, components with only one state for each t are called *combinational*, while components with more than one state for some t are called *sequential*. Note however that the term "sequential" is used in very different ways in other communities.

2.4.3. *Concurrency and Communication*

The sequential or combinational behavior just described is related to individual processes, and embedded systems will typically contain several coordinated concurrent processes. At the very least, such systems interact with an environment that evolves independently, at its own speed. It is also common to partition the overall model into tasks that also evolve more or less independently, occasionally (or frequently) interacting with one another. This interaction implies a need for coordinated communication.

Communication between processes can be *explicit* or *implicit*. Explicit communication implies forcing an order on the events, and this is typically realized by designating a *sender* process which informs one or more *receiver* processes about some part of its state. Implicit communication implies the sharing of tags (i.e., of a common time scale), which forces a common partial order of events, and a common notion of state. The problem with this form of communication is that it must be *physically* implemented via shared signals (e.g., a common reference clock), whose distribution may be difficult in practice.

Basic Time Time plays a larger role in embedded systems than in classical computation. In classical transformational systems, the correct result is the primary concern—when it arrives is less important (although *whether* it arrives, the termination question, *is* important). By contrast, embedded

[5] A definition of state for untimed systems is also possible, but it is much more involved.

systems are usually real-time systems, where the time at which a computation takes place is very important.

As mentioned previously, different models of time become different order relations on the set of tags T in the tagged-signal model. Recall that in a *timed system* T is totally ordered, while in an *untimed system* T is only partially ordered. Implicit communication generally requires totally ordered tags, usually identified with physical time.

The tags in a *metric-time system* have the notion of a "distance" between them, much like physical time. Formally, there exists a partial function $d : T \times T \to \mathbf{R}$ mapping pairs of tags to real numbers such that $d(t_1, t_2) = 0 \Leftrightarrow t_1 = t_2$, $d(t_1, t_2) = d(t_2, t_1)$ and $d(t_1, t_2) + d(t_2, t_3) >= d(t_1, t_3)$.

Two events are *synchronous* if they have the same tag (the distance between them is 0). Two signals are synchronous if each event in one signal is synchronous with an event in the other signal and vice versa.

Treatment of Time in Systems A *discrete-event system* is a timed system where the tags in each signal are order-isomorphic with the natural numbers [42]. Intuitively, this means that any pair of ordered tags has a finite number of intervening tags. This is the basis of the underlying MOC of the Verilog and VHDL hardware description languages [62, 49].

A *synchronous system* is one in which every signal in the system is synchronous with every other signal in the system.

A *discrete-time system* is a synchronous discrete-event system.

An *asynchronous system* is a system in which no two events can have the same tag. If tags are totally ordered, the system is *asynchronous interleaved*, while if tags are partially ordered, the system is *asynchronous concurrent*. For asynchronous systems concurrency and interleaving are, to a large extent, interchangeable, since interleaving can be obtained from concurrency by embedding the partial order into a total order, and concurrency can be reconstructed from interleaving by identifying "untimed causality" [48].

Note that time is a continuous quantity. Hence real systems are asynchronous by nature. Synchronicity is only a (very) convenient abstraction, that may be expensive to implement due to the need to share tags, and hence, as discussed above, to share a reference "clock" signal.

Synchronous/reactive languages (see e.g. [28]) deserve special mention. They have an underlying synchronous model in which the set of tags in any behavior of the system implies a global "clock" for the system. However, to make this MOC synchronous in the sense of the TSM, we need to assume that every signal conceptually has an event at every tag. In some synchronous/reactive designs this may not be the case but if we define the events in the process to include a value denoting the absence of an event, then all synchronous/reactive models can be defined as synchronous in our framework. At each clock tick, each process maps input values to output values. Note that if we include the absent value for events, then discrete-event systems are also synchronous.

The main differences are:

- in the granularity of tags: intuitively, synchronous models should be used for systems in which there are fewer tags, and
- in the number of events that have the absent value at any tag: intuitively, synchronous models should be used for systems in which only few events have absent values.

Particular attention has to be devoted to events with values at the same tag and that have cyclic dependencies (*"combinational cycles"*). The existence of such dependencies implies that the input-output relation is described implicitly as the solution of an algebraic set of equations. This set of equations may have either a single solution for each input value, in which case the process is completely specified, no solution for some input value, in which case the process is functional but not completely specified, or multiple solutions for some input value, in which case the process is not functional. This is the source of endless problems in systems described by VHDL or Verilog and difficulties in synchronous/reactive languages. A possibility when facing a cyclic dependency is to leave the result unspecified, resulting in nondeterminacy or, worse, infinite computation within one tick according to the particular input values (VHDL, Verilog and some variants of StateCharts belong to this class [65]). A better approach is to use fixed-point semantics, where the behavior of the system is defined as a set of events that satisfy all processes [11]. Given this approach to the problem, there are procedures that can determine the existence of single or multiple fixed points in finite time, thus avoiding nasty inconsistencies and difficulties.

Asynchronous systems do not suffer from this problem since there cannot be cyclic dependencies at the same tag given that only one event can have a value at any given tag. Note that often asynchronous systems are confused with discrete-event systems and thus it is not infrequent to find assertions in the literature that asynchronous systems may have inconsistent or multiple solutions when indeed this is never the case!

Implementation of Concurrency and Communication Concurrency in physical implementations of systems implies a combination of *parallelism*, which employs physically distinct computational resources, and *interleaving*, which means sharing of a common physical resource. Mechanisms for achieving interleaving, generally called *schedulers*, vary widely, ranging from operating systems that manage context switches to fully-static interleaving in which multiple concurrent processes are converted (compiled) into a single process. We focus here on the mechanisms used to manage communication between concurrent processes.

Parallel physical systems naturally share a common notion of time, according to the laws of physics. The time at which an event in one subsystem occurs has a natural ordering relationship with the time at which an event occurs in another subsystem. Physically interleaved systems also share a natural common notion of time: one event happens before another and the time between them can be computed (of course, accuracy is an issue).

Logical systems, on the other hand, need a mechanism to explicitly share a notion of time (*communicate*). Consider two imperative programs interleaved on a single processor under the control of a time-sharing operating system. Interleaving creates a natural ordering between events in the two processes, but this ordering is generally unreliable, because it heavily depends on scheduling policy, system load and so on. Some explicit communication mechanism is required for the two programs to cooperate. One way of implementing this could be by forcing both to operate based on a global notion of time, which in turn forces a total order on events. This can be extremely expensive. In practice, this communication is done explicitly, where the total order is replaced by a partial order. Returning to the example of two processes running under a time-sharing operating system, we take precautions to ensure an ordering of two events only if the ordering of these two events matters. We can do this by communicating through common signals, and forcing one process to wait for a signal from the other, which forces the scheduler to interleave the processes in a particular way.

A variety of mechanisms for managing the order of events, and hence for communicating information between processes, exists. We will now examine and classify them according to the tagged-signal model, by using "special-purpose" processes to model communication. Using processes to model communication (rather than considering it as "primitives" of the tagged-signal model) makes it easier to compare different MOCs, and also allows one to consider *refining* these communication processes when going from specification to implementation [56].

Recall that the *communication* primitive in the TSM is the event, which is a two-component entity whose value is related to function and whose tag is related to time. That is, communication is implemented by two operations:

1. the transfer of values between processes (function; TSM event value),
2. the determination of the relationship in time between two processes (time; TSM event tag).

Unfortunately, often the term "communication" (or data transfer) is used for the former, and the term "synchronization" is used for the latter. We feel, however, that the two are intrinsically connected in embedded systems: both tag and value carry information *about a* communication. *Thus, communication and synchronization, as mentioned before, are terms which cannot really be distinguished in this sense.*

2.5. BASIC COMMUNICATION PRIMITIVES

In this section, we define some of the communication primitives that have been described in the literature, following the classification developed in the previous sections.

Unsynchronized In an unsynchronized communication, a producer of information and a consumer of the information are not coordinated. There is some connection between them (e.g., a buffer) but there is no guarantee that the consumer reads "valid" information produced by the producer, and no guarantee that the producer will not overwrite previously produced data before the consumer reads the data. In the tagged-signal model, the repository for the data is modeled as a process, and the reading and writing actions are modeled as events without any enforced ordering of their tags.

Read-modify-write Commonly used for accessing shared data structures in software, this strategy locks a data structure during a data access (read, write, or read-modify-write), preventing any other accesses. In other words, the actions of reading, modifying, and writing are atomic (indivisible, and thus uninterruptible). In the tagged-signal model, the repository for the data is modeled as a process where events associated with this process are totally ordered (resulting in a partially ordered model at the global level). The read-modify-write action is modeled as a single event.

Unbounded FIFO buffered This is a point-to-point communication strategy, where a producer generates (writes) a sequence of data tokens and a consumer consumes (reads) these tokens, but only after they have been generated (i.e., only if they are valid). In the tagged-signal model, this is a simple connection where the signal on the connection is constrained to have totally ordered tags. The tags model the ordering imposed by the FIFO model. If the consumer process has unbounded FIFOs on all inputs, then all inputs have a total order imposed upon them by this communication choice. This model captures essential properties of both Kahn process networks and dataflow [35].

Bounded FIFO buffered In this case (we discuss only the point-to-point case for the sake of simplicity), the data repository is modeled as a

process that imposes ordering constraints on its inputs (which come from the producer) and the outputs (which go to the consumer). Each of the input and output signals are internally totally ordered, while their combination is partially ordered. The simplest case is where the size of the buffer is one, in which case the input and output events must be perfectly interleaved (i.e., that each output event lies between two input events). Larger buffers impose a maximum difference (often called *synchronic distance* [51]) between the number of input or output events occurring in succession.
Note that some implementations of this communication mechanism may not really block the writing process when the buffer is full, thus requiring some higher level of flow control to ensure that this never happens, or that it does not cause any harm.

Petri net places This is a multi-partner communication strategy, where several producers generate tokens and several consumers consume these tokens [51]. In the tagged-signal model, this is modeled as a process that keeps track of the tags of its input (from producers) and output (to consumers) signals. As in the previous case, each signal has totally ordered events, and the process makes sure that the number of input events is always greater than or equal to that of output events.

Rendezvous In the simplest form of rendezvous, which is embodied in the underlying MOC of the Occam and Lotos [64] languages, a single writing process and a single reading process must simultaneously be at the point in their control flow where the write and the read occur. It is a convenient communication mechanism, because it has the semantics of a single assignment, in which the writer provides the right-hand side, and the reader provides the left-hand side. In the tagged-signal model, this is imposed by events with identical tags [42]. Lotos offers, in addition, multiple rendezvous, in which one among multiple possible communications is *non-deterministically* selected. Multiple rendezvous is more flexible than single rendezvous, because it allows the designer to specify more easily several "expected" communication ports at any given time, but it is very difficult and expensive to implement correctly.

Of course, various combinations of the above models are possible. For example, in a model that partially uses the unsynchronized communication scheme, a consumer of data may be required to wait until the first time a producer produces data, after which the communication is unsynchronized.

The essential features of the concurrency and communication styles described above are presented in Table 1. These are distinguished by the number of transmitters and receivers (e.g., broadcast versus point-to-point communication), the size of the communication buffer, whether the transmitting or receiving process may continue after an unsuccessful commu-

	Transmitters	Receivers	Buffer Size	Blocking Reads	Blocking Writes	Single Reads
Unsynchronized	many	many	one	no	no	no
Read-Modify-Write	many	many	one	yes	yes	no
Unbounded FIFO	one	one	unbounded	yes	no	yes
Bounded FIFO	one	one	bounded	maybe	maybe	yes
Petri net place	many	many	unbounded	no	no	yes
Single Rendezvous	one	one	one	yes	yes	yes
Multiple Rendezvous	many	many	one	no	no	yes

TABLE 1. A comparison of concurrency and communication schemes.

nication attempt (blocking reads and writes), and whether the result of each write can be read at most once (single reads). Note that, strictly speaking, the blocking/nonblocking read and write aspects are part of the "functional" processes, and not of the "communication" processes. However, these communication schemes also specify that aspect, and hence we chose to include in the table. A "maybe" entry means that MOCs considering both the "yes" and "no" answer have been proposed in the literature.

3. Common Models of Computation

We are now ready to use the scheme developed in the previous Section to classify and analyze several models of computation that have been used to describe embedded systems. We will consider issues such as ease of modeling, efficiency of analysis (simulation or formal verification), automated synthesizability, and optimization space versus over-specification.

We assume a background knowledge of basic, non-concurrent MOCs such as Finite Automata, Turing Machines, and Algebraic State Machines, and we focus on the timing, concurrency and communication aspects instead.

3.1. DISCRETE-EVENT

Time is an integral part of a discrete-event model of computation. Events usually carry a totally-ordered time stamp indicating the time at which the event occurs. A DE simulator usually maintains a global event queue that sorts events by time stamp.

Digital hardware is often simulated using a discrete-event approach. The

Verilog language [62], for example, was designed as an input language for a discrete-event simulator. The VHDL language [49] also has an underlying discrete-event model of computation.

Discrete-event modeling can be expensive—sorting time stamps can be time-consuming. Moreover, ironically, although discrete-event is ideally suited to modeling distributed systems, it is very challenging to build a distributed discrete-event simulator. The global ordering of events requires tight coordination between parts of the simulation, rendering distributed execution difficult.

Discrete-event simulation is most efficient for large systems with large, frequently idle or autonomously operating sections. Under discrete-event simulation, only the changes in the system need to be processed, rather than the whole system. As the activity of a system increases, the discrete-event paradigm becomes less efficient because of the overhead inherent in processing time stamps.

Simultaneous events, especially those arising from zero-delay feedback loops, present a challenge for discrete-event models of computation. In such a situation, events may need to be ordered, but are not.

Consider the discrete-event system shown in Figure 3. Process B has zero delay, meaning that its output has the same time stamp as its input. If process A produces events with the same time stamp on each output, there is ambiguity about whether B or C should be invoked first, as shown in Figure 3(a).

Suppose B is invoked first, as shown in Figure 3(b). Now, depending on the simulator, C might be invoked once, observing both input events in one invocation, or it might be invoked twice, processing the events one at a time. In the latter case, there is no clear way to determine which event should be processed first.

The problem could be solved by requiring the user to provide a delay for each process, but this is not convenient in general. Hence various simulators have resorted to various heuristic techniques:

- The VHDL simulation semantics [49] uses a *synchronous* model (with unit delay, called "delta step") in order to provide a two-level structure of time and thus solve non-determinism within a given "real time" instant. Each instant of time (level 1) is broken into (a potentially infinite number of) totally ordered delta steps (level 2). A "zero-delay" process in this model actually has delta steps, or ordered progress towards a solution though no real time elapses. For example, if Process B contains a delta step between input and output, firing A followed by B would result in the situation in Figure 3(c). The next firing of C will see the event from A only; the firing after that will see the (delay-ordered) event from B.

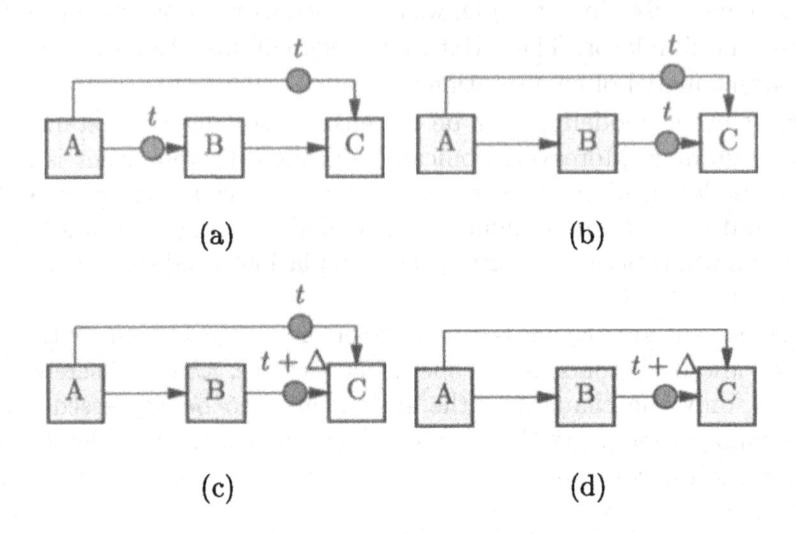

Figure 3. Simultaneous events in a discrete-event system. (a) Process A produces events with the same time stamp. Should B or C be fired next? (b) Zero-delay process B has fired. How many times should C be fired? (c) Delta-delay process B has fired; C will consume A's output next. (d) C has fired once; it will fire again to consume B's output.

- The Discrete Event domain in Ptolemy [17] uses a synchronous model, but with mostly zero delay and only enough delta steps to eliminate all zero-delay cycles.
- The BONES simulator by Cadence uses an *asynchronous* model.

Adding a feedback loop from Process C to A in Figure 3 would create a problem if events circulate through the loop without any increment in time stamp. The same problem occurs in synchronous languages, where such loops are called causality loops. No precedence analysis can resolve the ambiguity. In synchronous languages, the compiler may simply fail to compile such a program. Discrete-event simulators attempt to identify such cases and report them to the user.

We wish to stress that delta steps do not have a meaning of time (though they are often called delta "delay"). They are just a clever mechanism to implement a fixed point computation used to compute the behavior of the system at a point in time. Fixed point iteration can also be used in the synchronous/reactive model to define its semantics and make it determinate. Hence "delta steps" can also be thought of as an "iteration index". Moreover, VHDL uses an event model that is not monotonic, and hence the fixed point may never be reached, as discussed above. On the other hand, synchronous language use a ternary logic model, in which fixed point convergence in ensured in a finite number of steps [16].

The reason why DE is a popular MOC in practice is that it has been implemented efficiently in a number of event-driven simulators, and it is quite convenient to evaluate the performance of very large and complex systems. By imposing little restriction on the modeling style, it makes simulation simple and synthesis as well as formal verification hard.

3.2. DATAFLOW PROCESS NETWORKS

In dataflow, a program is specified by a directed graph where the nodes (called *actors*) represent computations and the arcs represent totally ordered sequences (called *streams*) of events (called *tokens*). In figure 4(a), the large circles represent actors, the small circle represents a token and the lines represent streams. The graphs are often represented visually and are typically hierarchical, in that a node in a graph may represent another directed graph. The nodes in the graph can be either language primitives or subprograms specified in another language, such as C or FORTRAN. In the latter case, we are actually mixing two models of computation, where dataflow serves as the coordination language for subprograms written in an imperative host language.

Dataflow is a special case of Kahn process networks [35, 41]. In a Kahn process network, communication is by unbounded FIFO buffering, and processes are constrained to be continuous mappings from input streams to output streams. "Continuous" in this usage is a topological property that ensures that the program is determinate [35]. Intuitively, it implies a form of causality without time; specifically, a process can use partial information about its input streams to produce partial information about its output streams. Adding more tokens to the input stream will never result in having to change or remove tokens on the output stream that have already been produced. One way to ensure continuity is with blocking reads, where any access to an input stream results in suspension of the process if there are no tokens. One consequence of blocking reads is that a process cannot test an input channel for the availability of data and then branch conditionally to a point where it will read a different input.

In dataflow, each process is decomposed into a sequence of *firings*, indivisible quanta of computation. Each firing consumes and produces tokens. Dividing processes into firings avoids the multi-tasking overhead of context switching in direct implementations of Kahn process networks. In fact, in many of the signal processing environments, a major objective is to statically (at compile time) schedule the actor firings, achieving an interleaved implementation of the concurrent model of computation. The firings are organized into a list (for one processor) or set of lists (for multiple processors). Figure 4(a) shows a dataflow graph, and Figure 4(b) shows a single processor schedule for it. This schedule is a list of firings that can be repeated indefinitely. One cycle through the schedule should return the graph to its original state (here, state is defined as the number of tokens on each arc). This is not always possible, but when it is, considerable simplification results [12]. In many existing environments, what happens within a firing can only be specified in a host language with imperative semantics, such as C or C++.

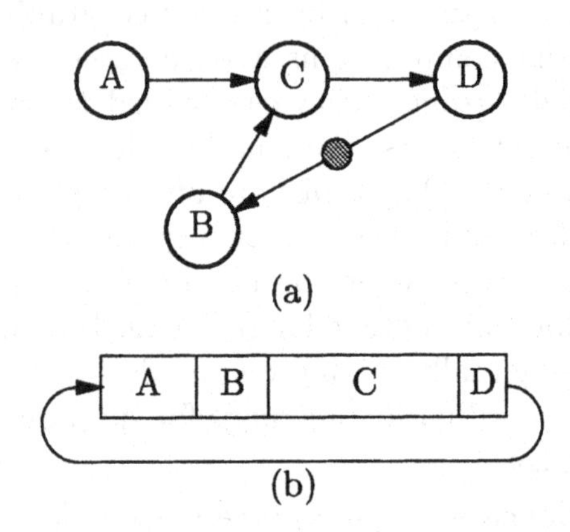

Figure 4. (a) A dataflow process network (b) A single-processor static schedule for it

A useful formal device is to constrain the operation of a firing to be functional, i.e., a simple, stateless mapping from input values to output values. Note, however, that this does not constrain the process to be stateless, since it can maintain state in a self-loop: an output that is connected back to one of its inputs. An initial token on this self-loop provides the initial value for the state.

Many possibilities have been explored for precise semantics of dataflow coordination languages, including Karp and Miller's computation graphs [37], Lee and Messerschmitt's synchronous dataflow graphs [40], Lauwereins *et al.*'s cyclo-static dataflow model [39, 13], Kaplan *et al.*'s Processing Graph Method (PGM) [36], Granular Lucid [34], and others [1, 20, 22, 60]. Many of these limit expressiveness in exchange for formal properties (e.g., provable liveness and bounded memory).

Synchronous dataflow (SDF) and cyclo-static dataflow require processes to consume and produce a fixed number of tokens for each firing. Both have the useful property that a finite static schedule can always be found that will return the graph to its original state. This admits extremely efficient implementations [12]. For more general dataflow models, it is undecidable whether such a schedule exists [18].

A looser model of dataflow is the tagged-token model, in which the partial order of tokens is explicitly carried with the tokens [3]. A significant advantage of this model is that while it logically preserves the FIFO semantics of the channels, it permits out-of-order execution.

Some examples of graphical dataflow programming environments intended for signal processing (including image processing) are Khoros [52], and Ptolemy [17].

3.3. PETRI NETS

Petri nets [50, 53] are, in their basic form, an infinite state model (just like dataflow) for which, however, most properties are decidable in finite time and memory. They are interesting as an uninterpreted model for several very different classes of problems, including some relevant to embedded system design (e.g., process control, asynchronous communication, and scheduling).

Moreover, a large user community has developed an impressive body of theoretical results and practical design aids and methods based on Petri nets. In particular, partial order-based verification methods (e.g. [63], [27], [45]) are one possible answer to the state explosion problem plaguing Finite State Machine-based verification techniques.

A Petri net (PN) is a directed bipartite graph $N = \{P, T, F\}$. P is a set of places holding the distributed state (via tokens) of the system. T is a set of transitions, denoting the activity of the system. $F \subseteq P \times T \cup T \times P$ is the flow relation, from places to transitions and vice-versa. Nodes linked by F are said to be in a predecessor/successor relationship.

Transitions are often labeled with statements in a host language, just as in the case of DF actors. The state of the PN is the *marking* of the places, that is a non-negative integer valuation ("token assignment") of each place. The dynamic evolution of the PN is determined by the firing process of transitions. A transition may fire whenever all its predecessor places are marked, and if it fires, it decrements the marking (removes a token) of each predecessor and increments the marking of each successor (adds a token).

PNs are interesting in general, and in particular in embedded system design, because they are a very general model of control, potentially with infinite state, yet very powerful analysis techniques, both exact and approximate, have been defined for them.

In particular, the firing rule of a PN bears a strong connection with linear algebra. If we represent the graph of the flow relation (given arbitrary orderings of the sets T and P) as an incidence matrix I, and if we represent the current marking as an integer vector M, we can model the effect of a

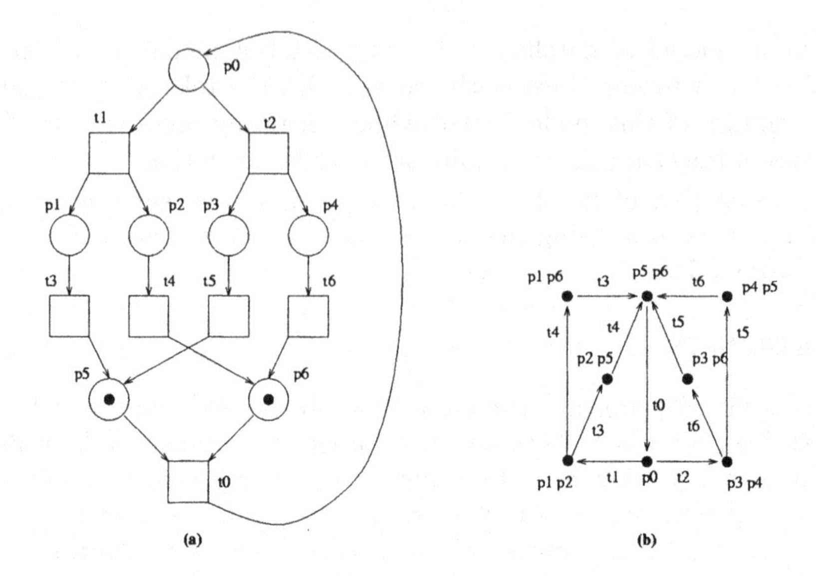

Figure 5. Example of Free Choice Petri net and its Reachability Graph

sequence of transitions σ starting from M as follows. Let us denote by f^σ the "firing vector" of σ, that is a vector whose i-th position contains the number of times the i-th transition appears in σ. The marking M' reached after σ is given by

$$M' = If^\sigma + M$$

For example, consider the PN in Figure 5.(a), whose set of reachable markings is shown in Figure 5.(b). Its incidence matrix (one row for each place and one column for each transition) is:

$$\left|\begin{array}{ccccccc} 1 & -1 & -1 & 0 & 0 & 0 & 0 \\ 0 & 1 & 0 & -1 & 0 & 0 & 0 \\ 0 & 1 & 0 & 0 & -1 & 0 & 0 \\ 0 & 0 & 1 & 0 & 0 & -1 & 0 \\ 0 & 0 & 1 & 0 & 0 & 0 & -1 \\ -1 & 0 & 0 & 1 & 0 & 1 & 0 \\ -1 & 0 & 0 & 0 & 1 & 0 & 1 \end{array}\right|$$

The first line corresponds to place $p0$, and has a 1 in position 0, because $t0$ adds 1 token to place $p0$, and -1 in positions 1 and 2 because $t1$ and $t2$ remove one token from it.

Consider now firing sequence $\sigma = t0, t1, t3$ whose firing vector (transposed) is $f^\sigma = |1101000|^t$. The marking M' reached from the initial marking

$M = |0000011|^t$ after firing σ is:

$$\begin{vmatrix} 0 \\ 0 \\ 1 \\ 0 \\ 0 \\ 1 \\ 0 \end{vmatrix} = \begin{vmatrix} 1 & -1 & -1 & 0 & 0 & 0 & 0 \\ 0 & 1 & 0 & -1 & 0 & 0 & 0 \\ 0 & 1 & 0 & 0 & -1 & 0 & 0 \\ 0 & 0 & 1 & 0 & 0 & -1 & 0 \\ 0 & 0 & 1 & 0 & 0 & 0 & -1 \\ -1 & 0 & 0 & 1 & 0 & 1 & 0 \\ -1 & 0 & 0 & 0 & 1 & 0 & 1 \end{vmatrix} \begin{vmatrix} 1 \\ 1 \\ 0 \\ 1 \\ 0 \\ 0 \\ 0 \end{vmatrix} + \begin{vmatrix} 0 \\ 0 \\ 0 \\ 0 \\ 0 \\ 1 \\ 1 \end{vmatrix}$$

that corresponds to $p2, p5$ being marked, as expected.

As another example, consider the initial marking with two tokens each in p_5, p_6 and the firing sequence $\sigma = t0, t0, t1, t3$. In that case,
$M = |0000022|^t$,
$f^\sigma = |2110000|^t$, and
$M' = |0111100|^t$.

This equation provides an interesting characterization of sequences of transitions that, when fired from a marking M, return the net to same M. These sequences, also called *T-invariants*, must be solutions to

$$0 = If^\sigma$$

This is only a *necessary* condition, of course, since the sequences must also be fireable from M (some intermediate step may yield a negative marking), but it is useful, e.g., when proving liveness conditions (e.g., showing that some transition can fire infinitely often) or schedulability properties [47].

For example, in Figure 5 firing sequence $t0, t1, t3, t4$ is a T-invariant $|1101100|^t$ that happens to be fireable from the initial marking. The reader can check that this invariant is indeed a solution of the equation shown above.

By duality (a very useful concept in Petri nets, based on exchanging the roles of places and transitions), one can also identify sets of places whose total cumulative marking cannot be changed by any firing sequence of the net. These sets, also called *S-invariants* can be used to establish the *unreachability* of a given marking, if it cannot be expressed as a linear combination of a basis of S-invariants [25]. Hence they can be very useful in proving (but not disproving) safety properties (e.g., the fact that some "dangerous" marking cannot be reached).

Invariant-based techniques become *necessary and sufficient* for a restricted (but expressive) class of PNs called *free-choice nets* [24], in which a multi-successor place must be the only predecessor of its successors. The net of Figure 5.(a) is a free-choice net, since the only multi-successor place ($p0$) has only single-predecessor successors ($t1, t2$).

In addition, reachability-based techniques for analysis, based on building the complete state space (or deciding in finite time that it is actually infinite), can also be used to prove properties of a given PN.

The basic PN model is interesting but somewhat limited in expressive power[6]. For this reasons, people have extended it in various ways, such as adding *colors* to tokens. Colored PNs are similar to Dataflow networks (with places playing the role of FIFOs and transitions playing the role of actors), but allow multiple predecessors and successors for a place/FIFO. In this way, they lose one of the most interesting properties of DF networks, *determinacy*, and of course gain something in terms of compactness and expressiveness[7].

Time can also play an explicit role in PNs. Time has been associated with transitions and places, in various combinations and forms ([55]). Generally speaking, time is associated with tokens, that carry a time stamp, and time stamps determine when transitions may fire (and thus create new tokens with new time stamps). The problem with timed PNs is, as usual with real-time MOCs, that they suffer from a particularly serious combinatorial explosion problem when reducing the originally infinite timed state space to a finite set of equivalence classes, as discussed more in detail in Section 3.8.

3.4. SYNCHRONOUS/REACTIVE

In a synchronous model of computation, all events are synchronous, i.e., all signals have events with identical tags. The tags are totally ordered, and globally available. Unlike the discrete-event model, all signals have events at all clock ticks, simplifying the simulator by requiring no sorting. Simulators that exploit this simplification are called cycle-based or cycle-driven simulators. Processing all events at a given clock tick constitutes a cycle. Within a cycle, the order in which events are processed may be determined by data precedences, which define the delta steps. These precedences are not allowed to be cyclic, and typically impose a partial order (leaving some arbitrary ordering decisions to the scheduler). Cycle-based models are excellent for clocked synchronous circuits, and have also been applied successfully at the system level in certain signal processing applications.

A cycle-based model is inefficient for modeling systems where events do not occur at the same rate in all signals. While conceptually such systems can be modeled (using, for example, special tokens to indicate the absence of

[6]It is more powerful than regular grammars, is incomparable with context-free grammars, and is less powerful than Turing machines.

[7]The formal power is the same, being that of Turing machines for both general CPN and general DF.

an event), the cost of processing such tokens is considerable. Fortunately, the cycle-based model is easily generalized to multirate systems. In this case, every nth event in one signal aligns with the events in another.

A multirate cycle-based model is still somewhat limited. It is an excellent model for synchronous signal processing systems where sample rates are related by constant rational multiples, but in situations where the alignment of events in different signals is irregular, it can be inefficient.

The more general synchronous/reactive model is embodied in the so-called synchronous languages [8]. Esterel [14] is a textual imperative language with sequential and concurrent statements that describe hierarchically-arranged processes. Lustre [29] is a textual declarative language with a dataflow flavor and a mechanism for multirate clocking. Signal [9] is a textual relational language, also with a dataflow flavor and a more powerful clocking system. Argos [44], a derivative of Harel's Statecharts [30], is a graphical language for describing hierarchical finite state machines (described more in detail in the next section). Halbwachs [28] gives a good summary of this group of languages.

The synchronous/reactive languages describe systems as a set of concurrently-executing synchronized modules. These modules communicate through signals that are either present or absent in each clock tick. The presence of a signal is called an event, and often carries a value, such as an integer.

Most of these languages are static in the sense that they cannot request additional storage nor create additional processes while running. This makes them well-suited for bounded and speed-critical embedded applications, since their behavior can be extensively analyzed at compile time. This static property makes a synchronous program finite-state, greatly facilitating formal verification.

Verifying that a synchronous program is causal (non-contradictory and deterministic) is a fundamental challenge with these languages. Since computation in these languages is delay-free and arbitrary interconnection of processes is possible, it is possible to specify a program that has either no interpretation (a contradiction where there is no consistent value for some signal) or multiple interpretations (some signal has more than one consistent value). Both situations are undesirable, and usually indicate a design error. A conservative approach that checks for causality problems structurally flags an unacceptably large number of programs as incorrect because most will manifest themselves only in unreachable program states. The alternative, to check for a causality problem in any reachable state, can be expensive since it requires an exhaustive check of the state space of the program.

In addition to the ability to translate these languages into finite-state descriptions, it is possible to compile these languages directly into hardware.

Techniques for translating both Esterel [10] and Lustre [54] into hardware have been proposed. The result is a logic network consisting of gates and flip-flops that can be optimized using traditional logic synthesis tools. To execute such a system in software, the resulting network is simply simulated. The technique is also the basis to perform more efficiently causality checks, by means of implicit state space traversal techniques [58].

3.5. COMMUNICATING SYNCHRONOUS FINITE STATE MACHINES

Finite State Machines (FSMs) are an attractive model for embedded systems. The amount of memory required by such a model is always decidable, and is often an explicit part of its specification. Halting and performance questions are always decidable since each state can, in theory, be examined in finite time. In practice, however, this may be prohibitively expensive, and thus formal verification techniques based on interacting FSMs require various forms of (non-trivial and non-automatable) abstraction in order to be kept manageable [38, 45].

A traditional FSM consists of:

- a set of input symbols (the Cartesian product of the sets of values of the input signals),
- a set of output symbols (the Cartesian product of the sets of values of the output signals),
- a finite set of states with a distinguished initial state,
- an output function mapping input symbols and states to output symbols, and
- a next-state function mapping input symbols and states to (next) states.

The input to such a machine is a sequence of input symbols, and the output is a sequence of output symbols. The model is *synchronous* (i.e., all signals have the same tags), and hence input and output symbols are well defined (they correspond to the set of events with a given tag). It is also semantically identical to that of previous section. However, there are enough syntactic differences to warrant a separate treatment (see [28, 11] for a discussion of possible mappings between the two).

Traditional FSMs are good for modeling sequential behavior, but are problematic for modeling system with concurrency or large memories, because of the state explosion problem. Every global state of a concurrent system must be represented individually, even when interleaving of independent actions may give rise to an exponential number of states. Similarly, a memory has as many states as the number of values that can be stored at each location *raised to the power* of the number of locations. The number of states alone is not always a good indication of complexity, but it often has a strong correlation.

Harel advocated the use of three major mechanisms that reduce the size (and hence the visual complexity) of finite automata for modeling practical systems [31]. The first one is hierarchy, in which a state can represent an enclosed state machine. That is, being in a particular state a has the interpretation that the state machine enclosed by a is active. Equivalently, being in state a means that the machine is in one of the states enclosed by a. Under the latter interpretation, the states of a are called "or states." Or states can exponentially reduce the complexity (the number of states) required to represent a system. They compactly describe the notion of *preemption* (a high-priority event suspending or "killing" a lower priority task), that is fundamental in embedded control applications.

The second mechanism is concurrency. Two or more state machines are viewed as being simultaneously active. Since the system is in one state of each parallel state machine simultaneously, these are sometimes called "and states." They also provide a potential exponential reduction in the size of the system representation.

The third mechanism is non-determinism. While often non-determinism is simply the result of an imprecise (maybe erroneous) specification, it can be an extremely powerful mechanism to reduce the complexity of a system model by *abstraction*. This abstraction can either be due to the fact that the exact functionality must still be defined, or that it is irrelevant to the properties currently considered of interest. E.g., during verification of a given system component, other components can be modeled as non-deterministic entities to compactly constrain the overall behavior. A system component can also be described non-deterministically to permit some optimization during the implementation phase. Non-determinism can also provide an exponential reduction in complexity. Note that non-determinism can be divided into and-non-determinism and or-non-determinism. In the first, the resolution of the non-determinism executes all possibilities, while in the second, resolution chooses just one. And-non-determinism is equivalent to hierarchy.

These three mechanisms have been shown in [26] to cooperate synergistically and orthogonally, to provide a potential triple exponential reduction in the size of the representation with respect to a single, flat deterministic FSM[8].

[8]The exact claim in [26] was that and-non-determinism (in which all non-deterministic choices must be successful), rather than hierarchical states, was the third source of exponential reduction together with "or" type non-determinism and concurrency. Hierarchical states, on the other hand, were shown in that paper to be able to simulate "and" non-determinism with only a polynomial increase in size.

Harel's Statecharts model uses a synchronous concurrency model (also called synchronous composition). The set of tags is a totally ordered countable set that denotes a global "clock" for the system. The events on signals are either produced by state transitions or inputs. Events at a tick of the clock can trigger state transitions in other parallel state machines at the same clock. Unfortunately, Harel left open some questions about the semantics of causality loops and chains of instantaneous (same tick) events, triggering a flurry of activity in the community that has resulted in at least twenty variants of Statecharts [65].

A model that is closely related to FSMs is Finite Automata. FAs emphasize the acceptance or rejection of a sequence of inputs rather than the sequence of output symbols produced in response to a sequence of input symbols. Most notions, such as composition and so on, can be naturally extended from one model to the other. FAs without accepting conditions are also called Labeled Transition Systems in the literature.

3.6. PROCESS ALGEBRAE

Synchronous FSMs, as described above, have a clear and deterministic composition mechanism that makes them relatively easy to understand, synthesize and verify. Of course, there is also a significant drawback: deciding when composition is well-defined (loosely speaking, there are no combinational loops) has a high computational complexity.

Moreover, for many applications, the tight coordination implied by the synchronous model is inappropriate. In particular, it is very difficult to keep a tight synchronization between heterogeneous components of an embedded system, since the pace of a synchronous system is dictated by its slowest component. In response to this, a number of more loosely coupled asynchronous FSM models have evolved, including CSP [33], CCS [46], behavioral FSMs [61], SDL process networks [61], and codesign FSMs [21].

In this section we focus on process algebraic models that constitute the semantical foundation of the Occam and Lotos [64] languages[9]: Communicating Sequential Processes [33] and the related Calculus of Communicating Systems [46]. In the following we discuss only the control aspect of CSP and CCS, and ignore the fact that their processes can also manipulate data via assignments, tests and so on. We also do not consider recursion, that can be defined in the process algebra but has limited interest (except for tail recursion, that defines looping) in the context of embedded systems.

The behavior of each process is modeled by a Labeled Transition System (only finite LTSs are of interest in embedded system design, for obvious

[9] Ada also uses rendezvous, although the implementation is stylistically quite different, using remote procedure calls rather than more elementary communication primitives.

reasons). Arcs in the transition system are labeled with signal names, and the state transition activity imposes a total order on the signals of each process. Communication is based on rendezvous. That is, two LTSs may share a signal, thus imposing that all the events of that signal must occur in both processes ("at the same time", if we interpret tags as time). Finally, process algebrae generally imply a completely *interleaved* view of concurrent actions, meaning that *no two events may have the same tag*. Concurrent (i.e., independent) events occur in all possible interleaving in the LTS.

No two events may have have the same tag, and hence process algebrae are an inherently asynchronous model. Note that a single LTS is an *interleaved* asynchronous model, while multiple LTSs communicating via rendezvous (and, equivalently, Petri nets in which at most one token can reside in each place in each reachable marking) are a *partially ordered* asynchronous model. As mentioned above, the rich theory of *regions* [48, 23] can be used to freely move between the two classes of models.

The result of process *composition* using this communication mechanism is another LTS, thus resulting in a hierarchical compositional model[10]. Compositionality is very important for proving properties of the system in a hierarchical fashion. This property is also true of communicating *synchronous* Finite State Machines, but not of dataflow networks (i.e., a dataflow network is different from an actor).

Let us consider a simple case of an interface with error detection. The LTSs specifying the protocols followed by the two partners are shown in Figure 6.(a-b).

1. The sender has two states, first sending a *request* on signal R, then waiting for either an *acknowledgment* of correct reception on signal A, or an *error* indication on signal E.
2. The receiver has a similar behavior, but in case of error, it requires one *internal action* (labeled τ) to resynchronize, and hence it has a third state.

The composed LTS using the rendezvous mechanism is shown in Figure 6.(c). Note how the state space of the composition is the product of the two state spaces, and the two LTSs synchronize on common edge labels.

For the sake of comparison, Figure 6.(d) shows the *synchronous* composition of the same two LTSs. Note how in case of error, the receiver waits for one clock tick, and hence becomes de-synchronized with the transmitter, thus leading to a deadlock[11].

[10] Compositionality means that two or more communicating processes can be viewed as a single process, that can in turn be used as a unit and composed with others.

[11] Of course, the fact that synchronous composition deadlocks while asynchronous composition does not is just a coincidence. It is easy to construct an example where the

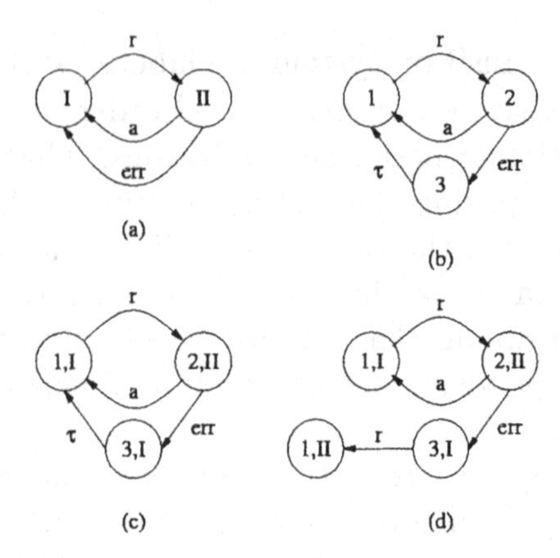

Figure 6. Example of Labeled Transition Systems and rendezvous communication

Rendezvous-based models of computation are sometimes called *synchronous* in the literature. However, by the definition we have given, they are not synchronous. Events are partially ordered, not totally ordered, with rendezvous points imposing the partial ordering constraints.

3.7. SDL PROCESS NETWORKS

SDL [61] is a language for specification, simulation and design of telecommunication protocols. Its underlying semantical model[12] is based on a process network. Each process is an FSM, and communication is via one unbounded FIFO queue per process. If we ignore the ability of a process to manipulate its input queue, the MOC is roughly equivalent to DE, with the restriction that the FSM can only read one event at a time.

SDL networks have a basic implementability problem, since both the size of the queues and the topology of the network can change at run time. (Processes can be created on the fly, and signals can be routed dynamically based on process identifiers.) Hence they either require a software implementation based on a Real-Time Operating System with dynamic memory allocation and task instantiation, or require the designer to pre-size queues and pre-instantiate all processes.

converse can happen.

[12] As usual, we focus on the control and communication aspects over the data computations, which are commonly specified with an imperative host language, in addition to a more formal and less practical treatment based on Abstract Data Types.

3.8. TIMED AUTOMATA

Synchronous and asynchronous Finite State Machines cannot reason easily about time, since in the best case (the synchronous one) time must be represented by counting clock ticks. This may cause a state explosion, and has been proven to be an inadequate abstraction of reality unless special care is taken [19].

For this reason, Alur and Dill [2] have proposed explicitly introducing time as a continuous quantity in the Timed Automata MOC. A Timed Automaton (TA) is a special case of hybrid systems, which are described in the next section. It is sufficiently restricted, so that most properties become decidable. A TA is a Finite Automaton (FA) plus a set of *clocks*. The state of the TA is the state of the FA together with a real valuation of the clocks. A transition of the TA is labeled with a symbol (from the FA alphabet) and a Boolean formula over atomic propositions comparing clocks with integers. The transition can also reset some clocks to zero.

While the state space of a TA is clearly infinite, a key result by Alur and Dill shows that it admits a finite state representation, by means of a partition into equivalence classes. Basically, [2] showed that the exact value of a clock does not matter after it grows beyond the largest constant with which it can be compared in any transition label. This imposes an equivalence relation on those portions of the state space that grow towards infinity. Moreover, since comparisons involve only integers, one can also partition the remaining part of the space into a *finite set of equivalence classes* (called regions), that admit a normal form representation (computed via an all-pair shortest path algorithm).

This is a very significant contribution, however it has shown only limited practical applicability so far because the state explosion problem is even more severe than in the communicating FSM case. Good generally applicable abstraction techniques are only beginning to be developed for TAs.

3.9. HYBRID SYSTEMS

A hybrid system is a Finite Automaton in which each state is associated with a set of differential equations, and transitions occur when inequalities over the continuous variables of the differential equations are satisfied. Hybrid systems are a powerful mechanism for modeling non-linear dynamic systems, and thus are becoming an essential tool in control theory. However, they are clearly Turing-equivalent, and hence too powerful, in almost all of their incarnations, with the notable exceptions of Timed Automata described above. It is likely that they will play an ever increasing role in embedded system design due to the growing need to raise the level of abstraction, but it is difficult to give them a complete and fair treatment in this brief overview, and we refer the interested reader to [32].

In the TSM, there are two possible views of hybrid systems (and hence of TAs).

1. A hybrid system (FA plus differential equations) can be modeled as a single TSM process. This provides an easy mechanism for composing hybrid systems. Signal tags in this case are order-isomorphic with the real numbers, but tags in which a transition of the automaton can occur can be only discrete.
2. A hybrid system can be modeled as a set of TSM processes. In this case we have two components for each hybrid system:
   - one process, whose signal tags can only be discrete, represents the automaton, and *multiplexes* the hybrid system outputs between
   - a set of processes, each behaving as a set of differential equations.

4. Codesign Finite State Machines

Codesign Finite State Machines (CFSMs) are the underlying MOC of the POLIS embedded system design environment [21, 6]. We describe them at length because, as we will argue later, they combine interesting aspects from several other MOCs, while preserving both formality and efficiency in implementation. As we pointed out above, one of the most important properties of an MOC is synchronicity or asynchronicity. We wish to summarize our views on this topic to motivate the introduction of yet another MOC.

4.1. SYNCHRONY AND ASYNCHRONY

Synchrony and asynchrony represent two fundamentally different views of time. That is, synchrony uses the notions of zero and infinite time, while asynchrony uses non-zero finite (and typically bounded) time. Both synchrony and asynchrony have appeared a number of times in our previous descriptions of various models of computation. In this section, we summarize our previous presentations of synchrony and asynchrony, and consider the differences in the behaviors produced under each model.

As usual, we consider a system of processes interacting through events.

4.1.1. *Synchrony*

Basic operation: At each clock tick (i.e., tag of its signals), each module reads inputs, computes, and produces outputs simultaneously. That is, all the synchronous events (both inputs and outputs) happen simultaneously, implying zero-delay calculations. In between clock ticks, an "infinite" time passes. Of course, no calculations happen in zero time in practice, nor does one wait an infinite amount of time between ticks (it is normally finite but unspecified). In practice, the computation times are much smaller than the clock rate, and thus can be considered to be zero with respect to the reaction times of the environment. The very desirable feature of designs implemented as synchronous systems with no cyclic dependencies among values of events with the same tags, is that the behavior of the implementation is not dependent upon the timing of the signals, thus simplifying tremendously the verification task.

Triggering and Ordering: All modules are triggered to compute at every clock tick. At a tick, there is no ordering of reading of inputs, computations, or writing of outputs. However, an ordering can be imposed in addition with the concept of delta steps (delays). A delta step (delay), as previously mentioned, is the (zero) time that passes between events at the same clock tick and which serves simply to order the events.

System solution: The system solution is the output reaction to a set of inputs. A well-designed synchronous system will have a unique solution (assignment to all signals) at each clock tick, though the corresponding models of computation, as well as many synchronous languages or specification methods, allow the designer to specify systems that do not have this property (see, e.g, [65]). We recall that the presence of cyclic dependencies among values of events with the same tag are responsible for this difficulty. It is the domain of the language and its semantical interpretation to verify whether a unique solution exists. Synchronous systems that have a unique solution, have a "single" finite state machine equivalent even though they consist of several interconnected components, and thus can be analyzed and verified with efficient techniques.

Implementation cost: Adherence to the synchronous assumption, that is, a process computes in negligible time compared to its environment, is a property that must be verified or enforced on the final design, and which may be expensive to implement. The assumption is checked on the final implementation. For hardware, one must ensure that the clock period is higher than the maximum possible computation time for a synchronous block; this may imply a clock rate that is much slower than might otherwise be achieved. For software, one must ensure that an invoked process is allowed to complete before another process or the operating system changes its inputs.

4.1.2. *Asynchrony*

Basic operation: Asynchronous events always have a non-zero amount of time between them: it is impossible to specify that two events happen simultaneously in a truly asynchronous system (as in real life ...). An individual process can run whenever it has a change on its inputs, and it may take an arbitrary time (that is typically bounded) to complete its computation.

Triggering and Ordering: A module is only triggered to run (and always triggered to run) when it has inputs that have changed. However, among the triggered modules, there is no a priori ordering of processes. One may later be imposed by a scheduling algorithm, but this is part of the implementation choice.

System solution: There is strong dependency of the solution from input signals and their timing. Thus, asynchronous systems are much more difficult to analyze. In addition, in a practical implementation or a model thereof, some events may *appear* to happen simultaneously. In practice it may be difficult and expensive to maintain the total ordering. If the actual order of these seemingly simultaneous events is not preserved, any order may be used possibly resulting in multiple behaviors. This is no longer an asynchronous model but a discrete-event model that has no guarantee of uniqueness of the solution because of the possible cyclic dependency of values of events with the same tags. It is this practical aspect that has misled many when assessing the properties of asynchronous systems, loading asynchronous systems with problems that are typical of discrete-event systems.

Implementation cost: Asynchronous implementations are usually chosen when the cost, particularly in terms of computation time, is too high for a synchronous solution. The flexibility provided by an asynchronous implementation implies that different parts of the same system (or the same system under different inputs) can operate at quite different rates, only communicating at particular check-points in the computation. For system design, it is usually imperative to have an asynchronous model at the highest level of communication. On the other hand, analysis of the behavior of designs implemented as asynchronous systems has to take into consideration the timing of the signals and, hence, is much more complex than the analysis of synchronous systems. This is the reason why much research on asynchronous system has been devoted to implementations that are more or less insensitive to "internal" delays, thus retaining the most desirable property of synchronous systems without paying the full penalty implied in a synchronous implementation.

4.1.3. *Combining Synchrony and Asynchrony*

An ideal MOC for system design should combine the advantages of verifiability in synchrony and flexibility in asynchrony in a globally asynchronous, locally synchronous (GALS) model. It is important to be explicit about where the boundary is between synchrony and asynchrony, because the behavior of the two, clearly, is very different. The differences can be illustrated simply in terms of event buffering and timing of event reading/writing.

In an asynchronous implementation, there is typically a need for an explicit buffering mechanism for the events, since it is not known when a module will run and hence read its inputs, and since different modules will run at different times and use the same input at different times. For synchrony, inputs are all read once at the beginning of a computation, so one global copy of each event value suffices and this one copy is cleared at the end of each tick. Thus, a synchronous communication transmits all events simultaneously and in zero time with no buffering; every module is guaranteed to see the same set of events at each clock tick. An asynchronous communication transmits events when ready and through buffers; each module sees its own stream of inputs which depends on the global scheduling.

Many programs will behave the same for an asynchronous or synchronous implementation, and such systems are typically more tolerant to implementation fluctuations. One can program in a style that is more robust with respect to these differences, by, for example

- Never assuming or waiting for simultaneous (synchronous) events. Since simultaneity is nearly impossible to guarantee, it is more robust to wait for the occurrence of two events rather than the *simultaneous* occurrence of them.
- Never programming with a global timing, e.g. a global clock tick, in mind. Synchronous languages have mechanisms for allowing a clock tick to pass, and thus for counting clock ticks and waiting for a certain amount of this artificial time to pass. Asynchronous systems of course do not have such a specific notion of time, so the same style of programming with an asynchronous model interpretation (in which a clock tick usually forces an ordering rather than referring to a time) will produce different behavior.

One may certainly use these programming techniques within a synchronous portion of the design. At the system level, however, a time-intolerant style of programming and thinking about the behavior of a design should be employed.

Our CFSM model reflects these views and was strongly motivated by the need of combining synchronous and asynchronous behavior where it made most sense.

4.2. CFSM OVERVIEW

Each CFSM is an extended FSM, where the extensions add support for data handling and asynchronous communication. In particular, a CFSM has

- a *finite state machine part* that contains a set of inputs, outputs, and states, a transition relation, and an output relation.
- a *data computation part* in the form of references in the transition relation to external, instantaneous (combinational) functions.
- a *locally synchronous behavior*: each CFSM executes a transition by producing a single output reaction based on a single, snap-shot input assignment in zero time. This is synchronous from *its own perspective.*
- a *globally asynchronous behavior*: each CFSM reads inputs, executes a transition, and produces outputs in an unbounded but finite amount of time as seen by the rest of the system. This is asynchronous interaction from the *system perspective.*

This semantics, along with a scheduling mechanism to coordinate the CFSMs, provides a GALS communication model: Globally (at the system level) Asynchronous and Locally (at the CFSM level) Synchronous.

4.3. COMMUNICATION PRIMITIVES

4.3.1. *Signals*

CFSMs, as TSM processes, communicate through *signals*, which carry information in the form of *events*. They may function as *inputs*, *outputs*, or *state signals.*

A signal is communicated between two CFSMs via a *connection* (single-input, single-output communication process) that has an associated *input buffer* (or 1-place buffer), which contains one memory element for the event (event buffer) and one for the data (data buffer).

The event is *emitted* (produced) by a *sender* CFSM setting the event buffer to 1. It may be *detected* and *consumed* by a *receiver* CFSM. It is detected by reading the event buffer; it is consumed by setting the buffer to 0.

A signal is therefore *present* if it has been emitted and not yet consumed. In the tagged signal model, this means that the input signal of the connection has had an event with a tag larger than the largest tag of the output signal.

The data may be *written* by a sender and *read* by a receiver. Reading and writing is done on the data buffer on the connection between the sender and the receiver.

A control signal carries only event information, i.e., it may only be emitted and detected/consumed and its value is irrelevant. A data signal carries only data information, i.e., it may only be read and written.

An input signal can only be detected/consumed and read (depending on its status as a signal, control signal, or data signal). A (possibly incomplete) set of values for the input signals of a CFSM is termed *input assignment*, a set of input values read by a CFSM at a particular time is termed *captured input assignment*, and an input assignment with at least one present event is termed *input stimulus.*

An output signal can only be emitted and written. A (possibly incomplete) set of values for the outputs of a CFSM is termed *output reaction.*

A state signal is an internal input/output data signal; it may be written and subsequently read by its CFSM. A *state* is a set of values for the state signals. A set of states may be given by a subset of state values. States are implicitly represented by the state signals and hence may be encoded or symbolic. The state signals could be considered part of the input and output signal sets, and it is only for the exposition that they are separated: discussion of scheduling and runnable CFSMs is facilitated by identifying an input assignment that triggers the CFSM separately from its state.

Where the type is unimportant, we may refer to any of the basic signal types (signal, control signal, data signal, input signal, output signal, state signal) simply as signal.

As will be seen in the behavior sections to follow, CFSMs *initiate communication through events.* The input events of a CFSM determine when it may react. That is, the model forbids a CFSM to react unless it has at least one input event present (except for the initial reaction, described in the functional behavior section). Without this restriction, a global clock would be required to execute the CFSMs at regular intervals, and this clock would in fact be a triggering input for all CFSMs. This would clearly imply a more costly implementation. A CFSM can trigger itself by emitting an output and detecting that same signal in the next execution. A CFSM with at least one present input event is termed *runnable.*

For CFSM A to send a signal S to B, A writes the data of S and then emits its event. This order ensures that the data is ready when the event is communicated. B is scheduled, sees the event (which is its stimulus), reads the corresponding data, and reacts. Pure data signals will only be read and written by a CFSM that has already been triggered by the presence of another input event.

4.3.2. CFSM *Networks*

A *net* is a set of connections on the same output signal, i.e., it is associated with a single sender and at least one receiver (in the TSM, it is a set of connection processes with the same input). There is an input buffer (TSM connection) for each receiver on a net, hence the communication mechanism is *multi-cast*: a sender communicates a signal to several receivers with a single emission, and each receiver has a private copy of the communicated signal. Each CFSM can thus independently detect/consume and read its inputs.

A *network* is a set of CFSMs and nets. The behavior of the network (and even of a single CFSM) depends on both the individual behavior, and that of the global system. In the mathematical model, the system is composed of CFSMs and a scheduling mechanism coordinating them. It can be implemented as:

- a set of CFSMs in software (e.g., C), a compiler, an operating system, and a microprocessor (the *software domain*),
- a set of CFSMs in hardware (e.g., gates mapped to an FPGA), a hardware initialization scheme, and a clocking scheme (the *hardware domain*), and
- the *interface* between them (e.g., a polling or interrupt scheme to pass events from hardware CFSMs to software ones via the RTOS, a memory-mapped scheme to pass events from software to hardware).

Thus the scheduling mechanism in the model may take several forms in the implementation: a simple RTOS scheduler for software on a single processor and concurrent execution for hardware, or a set of RTOSs on a heterogeneous multi-processor for software and a set of scheduling FSMs for hardware.

The CFSM model does not require any coordination between these schedulers in order to guarantee correct behavior, apart from an implementation of the event delivery mechanism (the interface). Explicit or implicit coordination is required only in order to satisfy timing constraints, which in turn may guarantee an ordering of events and/or a particular functional behavior.

4.4. TIMING BEHAVIOR

In the CFSM model, a global "scheduler" controls the interaction of the CFSMs, and invokes each appropriately during execution of the design. The system output will depend on the functional and timing behavior of the individual CFSMs, and the functional and timing behavior of their ensemble.

The scheduler operates by continually deciding which CFSMs can be run, and calling them to be executed. Each CFSM is either *idle* (waiting for input events or waiting to be run by the scheduler), or *executing* (generating a single reaction). During an *execution*, a CFSM reads its inputs, performs a computation, and possibly changes state and writes its outputs.

The mathematical model places few restrictions on the timing of an execution. Each CFSM execution can be associated with a single *transition point*, t_i, in time. The model dictates that it is at this point that the CFSM begins reacting: reading inputs, computing, changing state, and writing outputs. Since the reaction time is unbounded, one cannot say exactly at which time a particular input (event or data) is read, at which time that input had previously been written, or at which time a particular output is written. There are, however, some restrictions. For each execution, each input signal is read at most once, each input event is cleared at every execution, and there is a partial order on the reading and writing of signals. Since the data value of a signal (with an event and data part) only has meaning when that signal is present, the model dictates that the event is read before the data. Similarly for the outputs, the data is written before the event, so that it is valid at the time the event is emitted.

This means that for transition point t_i,

- an input may be read at any time between t_i and t_{i+1} (but not later, because that would correspond to transition point t_{i+1}),
- the event that is read may have occurred at any time between t_{i-1} and t_{i+1},
- the data that is read may have been written at any time between t_0 and t_{i+1}, and
- the outputs are written at some time between t_i and t_{i+1}.

After reading an input, its value may be changed by the sender before t_{i+1}, but the receiver reacts to the captured input and the new value is not read until the next reaction.

This flexibility in timing can have non-intuitive behavior.

Example: Event/data separation. Suppose a sender S writes data value v1 at t_1 for signal X and emits it at t_2. Let this emission be e1, so the pair is (e1, v1). This is illustrated in Figure 7. A receiver R at t_i sometime later reads the event, but takes longer than expected to read its corresponding value v1. S communicates X again: value v2 then emission of X, for pair (e2, v2). R now reads the value for X, and reads v2. The captured input for the R is thus the pair (e1, v2), which was not the pair intended by S. Furthermore, data v1 has been lost forever, even though it was sent with an event to signal its presence.

Problems such as this can easily be resolved by requiring the appropriate level of *atomicity* in the model and in the implementation, i.e., by re-

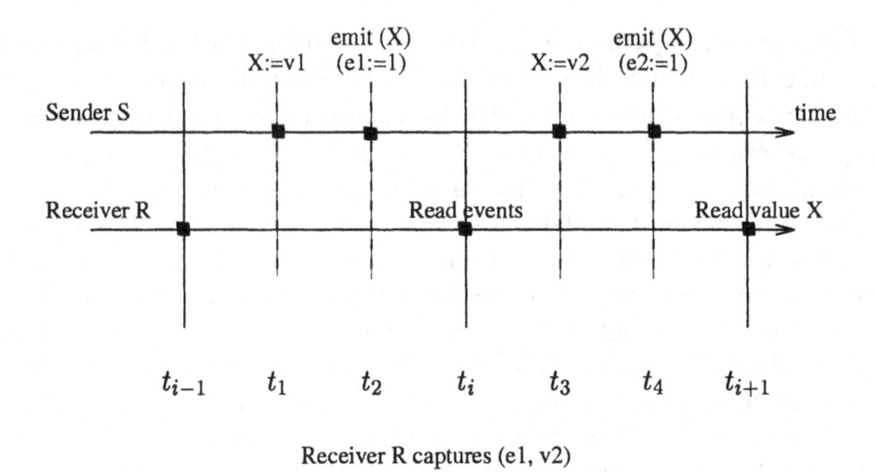

Figure 7. Event/data separation.

stricting some parts of the communication to take place simultaneously and instantaneously – as a single entity. In the CFSM model, the requirement is simply that the input events are read atomically. At each t_i, a CFSM reads its input events without interruption, and without those events being overwritten by the sending CFSMs. This is easily implemented in software by reading a bit-vector of input events in one instruction, and in hardware by clocking all CFSMs together, with a separate read phase and a compute/write phase (per clock cycle).

Atomicity of input event reading implies an implementation that retains much of the flexibility required for efficiency, while mitigating the worst of the synchronicity problems. It should be clear that allowing input event and data reading and output event and data writing to happen at completely arbitrary times leads to behavior with very difficult to predict and prescribe results. Given the event-based communication of CFSMs, atomicity of input event reading is a natural means of ensuring some predictability: a receiver CFSM is guaranteed to see a snapshot of input events that are simultaneously present at some point in real time. Additional constraints, if necessary, can be imposed to ensure that the values subsequently read are meaningful. These constraints will vary considerably depending on the implementation chosen and the design constraints.

Example: no atomicity of data reading. Consider a sender S and two receivers R1 and R2, as illustrated in figure 8. S is sending the value of signals X and Y. Both X and Y are currently 4 and are changing to 5. R1 reads X at t_1 and Y at t_6. R2 reads X at t_3 and Y at t_4. S changes X at t_2 and Y at t_5. R1 therefore captures $X = 4$ and $Y = 5$ while R2 captures $X = 5$ and $Y = 4$. Not only do they capture different input assignments,

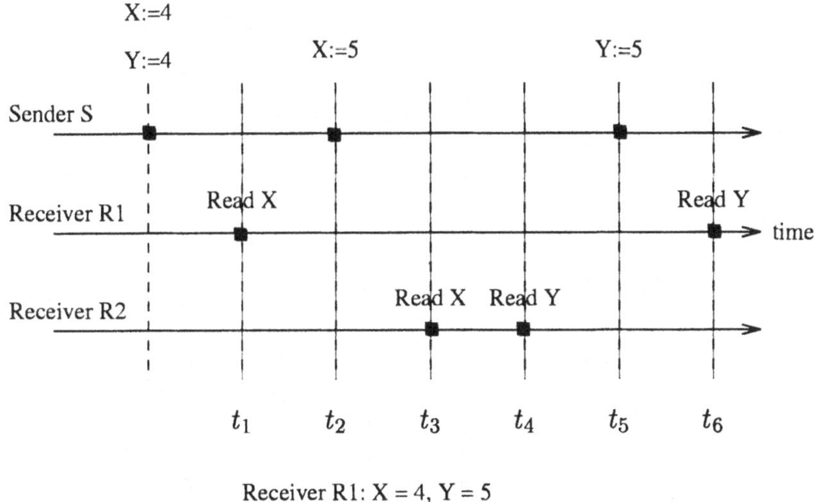

Figure 8. No atomicity of data value reading.

but R1 captures a set of values that never occurs simultaneously. Note that if X and Y were (control/data) signals, they would be sent with events as well, and they would both be changed to 5 before the events were emitted and hence before the receivers can read them. However, those new values can be overwritten by the sender if the receiver doesn't read fast enough, leading to the separation of the event/data pair as illustrated in the previous example.

Example: atomicity of event reading. Now consider a system with a sender S and three receivers R1, R2, and R3 as illustrated in Figure 9. S emits control signals X and Y at times t_2 and t_4. R1 reads at t_1 and sees both absent. R2 reads at t_3 and sees only X present. R3 reads at t_5 and sees both X and Y present. Though each has a different captured input assignment, each sees an input assignment that occurs at some point in real time.

4.5. FUNCTIONAL BEHAVIOR

The functional behavior of a CFSM at each execution is determined by the specified *transition relation* (TR). This relation is a set of tuples (**input_set**, **previous_state**, **output_set**, **next_state**) where **input_set** is an input assignment, **previous_state** is a state, **output_set** is an output assignment, and **next_state** the next state. Each tuple of the TR represents a specified *transition* of the machine, and the set of tuples is the specified behavior of the machine. A transition in which the **input_set** includes an

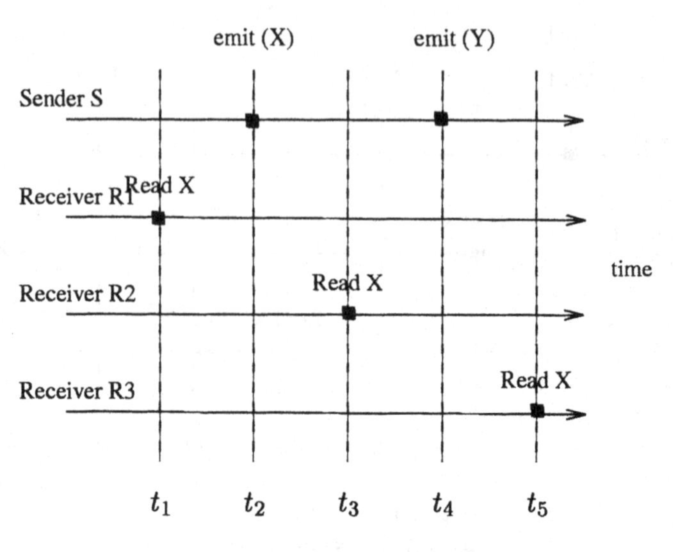

Figure 9. Atomicity of event reading.

input stimulus is termed a *valid transition.*

At each execution, a CFSM

1. Reads an input assignment.
2. Looks for a transition **Transition** = (**input_set**, **previous_state**, **output_set next_state**) such that the read input includes **input_set** and the present state of the CFSM matches **present_state** (hence the absence of an input from **input_set** means "don't care about that input").
3. If **Transition** is found, it is executed by
   (a) consuming the inputs (setting input event buffers to 0)
   (b) making the state transition to **next_state**
   (c) writing the new output events in **output_set** (hence absence from **output_set** means "don't emit/don't modify").
4. If **Transition** is not found, the CFSM consumes no inputs, makes no state change, and writes no outputs.

The last case, in which no matching transition is found, is known as the *empty execution.* If this can happen for some input stimulus, the transition relation is *incomplete*; otherwise it is *complete*. For software, this is precisely the same behavior that would be produced if this CFSM had not been scheduled by the RTOS. If several transitions match, the CFSM is non-deterministic and the execution can perform any of the matching transitions. For the implementation, all CFSMs must be deterministic in order to simulate and synthesize the behavior. Non-determinism can be used at the initial stages of a design in order to model partial specifications, later to be refined into deterministic CFSMs.

A *trivial transition* is one in which no output events are emitted, no output values are changed, and no state change is effected, but inputs are consumed. It effectively discards the current input assignment and waits for a new one. Trivial transitions are specified in the TR like any other transition, with **output_set** empty (or leaving state variables unchanged).

Each state variable may have a designated set of *reset values* (or initial values) that are specified with the transition relation. A set of reset values, one for each state variable, is a *reset state* (or initial state). If there are several reset values for a state variable, there are several reset states. This represents a non-deterministic starting condition which must be resolved before synthesis or simulation can be performed.

The *initial transition(s)* is a special transition(s) where the present state part is equal to the reset state. Moreover, this transition is allowed to not have any input events present in the input assignment. (Recall that for all other transitions, at least one input event is required to trigger the CFSM.) The initial transition(s) may be specified for a CFSM, but is not required. There may be several possible initial transitions depending on the initial state(s) and the values of the corresponding input assignments. If there is non-determinism, again, it must be resolved before synthesis and simulation.

4.6. CFSMS AND PROCESS NETWORKS

We can now classify CFSMs along the same lines that were used for the other formal models.

CFSMs are an asynchronous Extended FSM model, that is different from CSP and CCS because communication is not via rendezvous but via bounded (1-deep) non-blocking buffers, and different from SDL since queues are bounded and the process network topology is fixed.

Moreover, each CFSM can be modeled with an LTS in which each edge label can involve presence and absence tests of several signals, while in CSP, CCS, and SDL each label consists of a single symbol.

Signals are distinguished among inputs and outputs. Transitions, unlike dataflow networks, can be conditioned to the *absence* of an event over a signal. Hence CFSMs are not continuous in Kahn's sense [35] (arrival of two events in different orders may change the behavior).

The semantics of a CFSM network is defined based on a *global* explicit notion of time (imposing a total ordering of events). Thus CFSMs can be formally considered as *synchronous* with *relaxed timing*. I.e., while a global consistent state of the signals is required in order to perform a transition, no relationship is required between the tags of input events involved in a given transition, nor between those of its output events. There is only a partial order relationship between input and output events of the transition (inputs must have tags smaller than outputs).

Finite buffering without blocking write implies, as mentioned above, that events can be *overwritten*, if the sending end is faster than the receiving end. This sort of "deadline violation" in the CFSM context may or may not be a problem, depending both on the application and on the type of event.

The designer can make sure that "critical" events are never lost:

- either by providing an explicit handshaking mechanism, built by means of pairs or sets of signals, between the CFSMs,
- or by using scheduling techniques that ensure that no such loss can ever occur [5].

4.7. EXAMPLES OF CFSM BEHAVIORS

In this sections we provide a few examples of what it means to specify a "behavior" with a relaxed timing model such as CFSMs, and we discuss the notion of *behavior equivalence classes*.

Consider a simple case of three CFSMs as shown in figure 10. These CFSMs specify an "almost dataflow" behavior. CFSMs A and B take the same input stream i, and perform two different kinds of (unimportant) processing on it, by producing an output event for every input event. CFSM C takes an event from each input $i1$ and $i2$ and produces

- either an event on its o output if there is no error (e.g., its inputs are within a specified range)
- or an event on its *err* output, if some problem in the input stream or the CFSM state is detected.

The *err* signal causes A and B to perform some recovery action (e.g., realign their state variables).

The intuitive behavior specified by these three CFSMs is, in the designer's eyes, the same regardless of the scheduling in time of CFSM transitions, as long as:

1. no events are lost (they are all "critical" in this case), and
2. (possibly) some latency constraint is satisfied (e.g., o may be needed earlier than the next external input arrival).

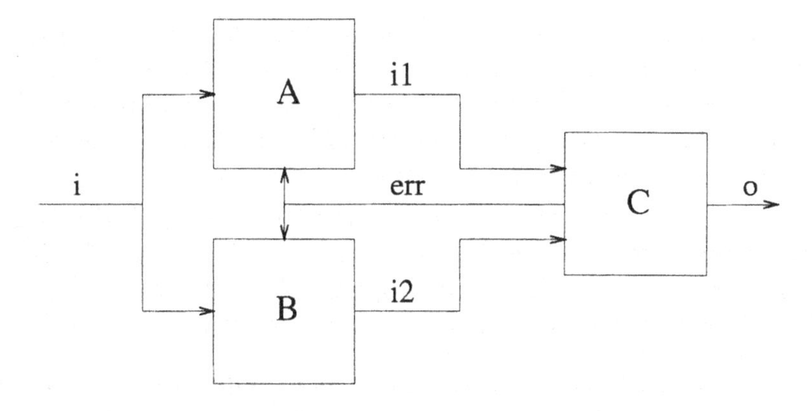

Figure 10. Example of CFSM network.

This means that, given the choice of possible timed executions of these CFSMs, they are partitioned into a set of *equivalence classes*. Let us consider, for the sake of simplicity, only reaction times and no other scheduling constraints (each CFSM is allocated its own hardware resource or processor). Let us assume that every CFSM has the same reaction time of n_r time units if there are no errors, and of $2n_r$ in case of errors (C when it emits *err*, A and B when they detect *err*). Let us also assume that input events arrive at a regular rate of n_i time units, and that there are only "no-missed-event" constraints and no other no latency constraints.

In this case, we can consider the following equivalence classes with respect to the above mentioned intuitive behavior:

1. Zero-delay executions: $n_r = 0$. These are logically inconsistent (non-causal in the Esterel terminology [11]), since if C detects an error, A and B should instantaneously react and produce different outputs (conceivably without the error conditions). This is clearly absurd.
2. Executions in which the execution delay of A, B and C is larger than the inter-arrival time of inputs: $n_i < n_r$. These executions clearly do not satisfy the intuitive behavior requirements listed above.
3. Executions in which the delay of A, B and C is smaller than the inter-arrival times of external inputs, but larger than half of that: $n_i/2 < n_r \leq n_i$. These executions handle correctly the *normal* flow of data, but "miss" an input in the case of error. This happens because the execution time of C is too long and causes it to miss the first input events after the error. Also A and B are too slow and miss an input event when recovering.
4. Executions in which the delay of A, B and C is smaller than half of the inter-arrival times of external inputs: $n_r \leq n_i/2$. In this case, no event is lost.

If errors are infrequent enough, the designer may want to consider the two last classes to be equally good, and accept the cheapest one. On the other hand, if errors are frequent, or if every single input really matters (there is zero redundancy in the input stream), only the last, conceivably most expensive equivalence class is acceptable.

Note how the notion of equivalence classes can be applied to analyze also executions in which a scheduler coordinates CFSMs by enforcing mutual exclusion constraints (this is an appropriate model, e.g., for a single-processor implementation). In that case event loss can occur due to

- Timing constraints. E.g., $n_i < 3n_r$ would imply that an execution falls into the second class above, and misses deadlines due to excessive processor occupation.
- "Incorrect" scheduling. E.g. if $3n_r \leq n_i < 4n_r$ and the scheduler activates the CFSMs in the fixed order ACBC. In that case, the first activation of C is valid, since it has an input stimulus, but redundant, since it always results in an empty execution (assumed with the same n_r delay), and causes the system to miss a deadline even during "normal operation". On the other hand, another valid schedule ABC will not miss deadlines in normal operation.

Obviously this definition of "equivalence classes" between behaviors of a CFSM network is very application-dependent, and as such difficult to formalize. Here we can suggest only a few criteria that could be used for this purpose, such as:

1. Equality between streams of values produced at some output by two different timed behaviors of the same network, given the same stream of values on the inputs ("dataflow" equivalence).
2. Compatibility with a given partial ordering between events ("Petri net" equivalence).
3. No missed critical events, possibly qualified as, e.g., "no missed events except for the first n events after abnormal event x" ("quasi-dataflow" equivalence).
4. Equality of input-output sequences, possibly modulo reordering of "concurrent events", with respect to a completely deterministic reference specification ("golden model" equivalence).
5. Equality of input-output sequences modulo *filtering* by some testbed entities that model the external, physical system constraints ("filtered" equivalence).

While we are still far from a formalization of these criteria, we believe that the richness of the CFSM model stems, among other factors, from the ability to exploit all these sorts of equivalence while, for example, dataflow networks can exploit only one (dataflow equivalence). We will further elaborate on this in the next section.

5. Conclusions

The relative advantages and disadvantages of the various MOCs have been described in the previous sections. We are still far from having a single agreed-upon standard MOC that is suitable for all types of embedded system designs. Some authors ([17]) advocate heterogeneity at the MOC level as an essential requirement of embedded systems. However, based on the discussions above, we can identify a new MOC that is expressive enough to capture most practical embedded systems, and formal to permit efficient verification and synthesis of some special cases.

This model is that of CFSMs with *initially unbounded FIFO buffers*. Bounds on buffers (essential for implementability) are imposed by *refinement*, exactly as timing information is refined in the original CFSM model. The motivations for this proposal are as follows.

1. Local synchrony: concentrating the control inside relatively large atomic synchronous entities helps the designer to better understand the overall coordination. Models such as Colored Petri Nets, which view control at a finer level of granularity, are difficult to use for large realistic designs.
2. Global asynchrony: breaking synchronicity helps resolving composition problems and mapping to a heterogeneous architecture. Synchronous models, as argued previously, cannot handle multi-rate efficiently, especially when the rates of different signals are totally uncorrelated. This is essentially due to the need to always consider all signals, including those that are not present in most clock cycles.
3. Unbounded buffers: leaving buffers unbounded at the outset offers opportunities to perform *static and quasi-static scheduling* whenever possible ([40, 18, 47]). As a result, some buffers become statically sized, as determined by the static schedule of reader and writer processes. The designer sizes the remaining buffers to ensure implementability, and to use simulation, formal verification or Real-time scheduling [4, 5] to validate the design in presence of finite FIFO buffers.

The use of *lossy buffers* (i.e., with non-blocking write) is somewhat arguable, because

- there are cases in which loss is essential (in general, any time there are tight timing constraints, that make some of the signals irrelevant under some conditions, e.g., an emergency), and
- there are cases in which loss is problematic (in general, any time one would like to model dataflow computations or any other blocking communication mechanism, such as, e.g., Remote Procedure Call).

Our choice is to *keep buffers lossy in the formal model*, and give the designer tools to *verify a priori if loss can occur*, as well as to *enforce no loss for some buffers in the implementation*. In general this can be enforced at an acceptable cost only under some specific conditions, e.g., that the lossless buffer must be local to a processor, or that the communicating processes must be statically scheduled with respect to each other.

The resulting model combines interesting properties of the main MOCs seen above, while still keeping a strong link to verifiability and implementability. In particular:

- At the initial, untimed level it describes a partial ordering between signal tags, and hence captures a whole class of possible implementations on a variety of architectural options. These options include software on multiple processors, pipelined hardware, and so on.
- It keeps computation (in the FSM), communication (in the buffers) and timing (in the architectural mapping) as separate as possible.
- After architectural mapping, it becomes essentially a Discrete Event model, and thus lends itself to performance and power consumption analysis, in order to evaluate architectural trade-offs.
- A subset, such that CFSMs have a deterministic behavior (i.e., behave as SDF actors) can be statically scheduled as SDF [40]. A larger subset can also be quasi-statically scheduled (thus performing static buffer sizing) by means of a mapping to Petri Nets [47].

While the opportunities for system-level optimization offered by this choice still need to be fully explored, we can already envision a design flow in which the designer uses multiple languages, depending on the domain of application and other requirements (e.g., tool availability, company policy or personal preference), that all have a semantics in terms of CFSMs. Then multiple scheduling, allocation, partitioning, hardware and software synthesis algorithms can be applied on the CFSM network, possibly depending on the identification of special cases that admit an especially efficient implementation. Formal verification and simulation can be used throughout the design process, thanks to the refinement-based design applied to a formal model. Refinement occurs both at the functional level, implementing CFSMs ([6]), and at the communication level, implementing communication ([56]) and scheduling ([6, 47]). This scheme has been adopted in the POLIS system and has been also followed by a commercial product of the Alta Group of Cadence Design Systems, Inc.

Acknowledgments We thank Prof. Ed Lee for the work that led to the development of the Tagged Signal Model that has been used throughout this paper, and for his many contributions to the field of system design for many years. Part of this paper has been adapted from an earlier version coauthored with Prof. Lee and Dr. Stephen Edwards. We wish to thank the Polis team and the Felix team of Cadence Design Systems, Inc., who shared the discoveries and developments which lead to this work. In particular, we wish to thank Dr. Jim Rowson for the key role he has played in putting together the design methodology and the architecture in the Cadence work, and Dr. Felice Balarin for introducing the notion of equivalence classes between CFSM behaviors. We are indebted to our colleagues of the University of California at Berkeley, of the Politecnico di Torino, of Cadence Berkeley and European Labs and of PARADES for the many discussions and their contributions. This work has been partially sponsored by grants of CNR and Cadence Design Systems.

References

1. W. B. Ackerman. Data flow languages. *Computer*, 15(2), 1982.
2. R. Alur and D. Dill. Automata for Modeling Real-Time Systems. In *Automata, Languages and Programming: 17th Annual Colloquium*, volume 443 of *Lecture Notes in Computer Science*, pages 322–335, 1990. Warwick University, July 16-20.
3. Arvind and K. P. Gostelow. The U-Interpreter. *Computer*, 15(2), 1982.
4. F. Balarin, H. Hsieh, A. Jurecska, L. Lavagno, and A. Sangiovanni-Vincentelli. Formal verification of embedded systems based on CFSM networks. In *Proceedings of the Design Automation Conference*, 1996.
5. F. Balarin and A. Sangiovanni-Vincentelli. Schedule validation for embedded reactive real-time systems. In *Proceedings of the Design Automation Conference*, 1997.
6. F. Balarin, E. Sentovich, M. Chiodo, P. Giusto, H. Hsieh, B. Tabbara, A. Jurecska, L. Lavagno, C. Passerone, K. Suzuki, and A. Sangiovanni-Vincentelli. *Hardware-Software Co-design of Embedded Systems – The POLIS approach*. Kluwer Academic Publishers, 1997.
7. A. Balluchi, M. Di Benedetto, C. Pinello, C. Rossi, and A. Sangiovanni-Vincentelli. Hybrid control for automotive engine management: The cut-off case. In *First International Workshop on Hybrid Systems: Computation and Control*. LNCS - Springer-Verlag, April 1997.
8. A. Benveniste and G. Berry. The synchronous approach to reactive and real-time systems. *Proceedings of the IEEE*, 79(9):1270–1282, 1991.
9. A. Benveniste and P. Le Guernic. Hybrid dynamical systems theory and the SIGNAL language. *IEEE Transactions on Automatic Control*, 35(5):525–546, May 1990.
10. G. Berry. A hardware implementation of pure Esterel. In *Proceedings of the International Workshop on Formal Methods in VLSI Design*, January 1991.
11. G. Berry. The foundations of esterel. In G. Plotkin, C. Stirling, and M. Tofte, editors, *Proof, Language and Interaction: Essays in Honour of Robin Milner*. MIT Press, 1998.
12. S. S. Bhattacharyya, P. K. Murthy, and E. A. Lee. *Software Synthesis from Dataflow Graphs*. Kluwer Academic Press, Norwood, Mass, 1996.
13. G. Bilsen, M. Engels, R. Lauwereins, and J. A. Peperstraete. Static scheduling of multi-rate and cyclo-static DSP applications. In *Proc. 1994 Workshop on VLSI Signal Processing*. IEEE Press, 1994.
14. F. Boussinot and R. De Simone. The ESTEREL language. *Proceedings of the IEEE*, 79(9), 1991.
15. R. Bryant. Graph-based algorithms for boolean function manipulation. *IEEE Transactions on Computers*, C-35(8):677–691, August 1986.
16. J.A. Brzozowski and C-J. Seger. Advances in Asynchronous Circuit Theory – Part

I: Gate and Unbounded Inertial Delay Models. *Bulletin of the European Association of Theoretical Computer Science*, October 1990.
17. J. Buck, S. Ha, E.A. Lee, and D.G. Messerschmitt. Ptolemy: a framework for simulating and prototyping heterogeneous systems. *Interntional Journal of Computer Simulation*, special issue on Simulation Software Development, January 1990.
18. J. T. Buck. *Scheduling Dynamic Dataflow Graphs with Bounded Memory Using the Token Flow Model.* PhD thesis, U.C. Berkeley, 1993. UCB/ERL Memo M93/69.
19. J. R. Burch. *Automatic Symbolic Verification of Real-Time Concurrent Systems.* PhD thesis, Carnegie Mellon University, August 1992.
20. N. Carriero and D. Gelernter. Linda in context. *Comm. of the ACM*, 32(4):444–458, April 1989.
21. M. Chiodo, P. Giusto, H. Hsieh, A. Jurecska, L. Lavagno, and A. Sangiovanni-Vincentelli. Hardware/software codesign of embedded systems. *IEEE Micro*, 14(4):26–36, August 1994.
22. F. Commoner and A. W. Holt. Marked directed graphs. *Journal of Computer and System Sciences*, 5:511–523, 1971.
23. J. Cortadella, M. Kishinevsky, L. Lavagno, and A. Yakovlev. Deriving Petri nets from finite transition systems. *IEEE Transactions on Computers*, 47(8):859–882, August 1998.
24. J. Desel and J. Esparza. *Free choice Petri nets.* Cambridge University Press, New York, 1995.
25. J. Desel, K.-P. Neuendorf, and M.-D. Radola. Proving non-reachability by modulo-invariants. *Theorectical Computer Science*, 153(1-2):49–64, 1996.
26. D. Drusinski and D. Harel. On the power of bounded concurrency. I. Finite automata. *Journal of the Association for Computing Machinery*, 41(3):517–539, May 1994.
27. P. Godefroid. Using partial orders to improve automatic verification methods. In E.M Clarke and R.P. Kurshan, editors, *Proceedings of the Computer Aided Verification Workshop*, 1990. DIMACS Series in Discrete Mathematica and Theoretical Computer Science, 1991, pages 321-340.
28. N. Halbwachs. *Synchronous Programming of Reactive Systems.* Kluwer Academic Publishers, 1993.
29. N. Halbwachs, P. Caspi, P. Raymond, and D. Pilaud. The synchronous data flow programming language LUSTRE. *Proceedings of the IEEE*, 79(9):1305–1319, 1991.
30. D. Harel. Statecharts: A visual formalism for complex systems. *Sci. Comput. Program.*, 8:231–274, 1987.
31. D. Harel, H. Lachover, A. Naamad, A. Pnueli, M. Politi, R. Sherman, A. Shtull-Trauring, and M. Trakhtenbrot. Statemate: A working environment for the development of complex reactive systems. *IEEE Trans. on Software Engineering*, 16(4), April 1990.
32. T.A. Henzinger. The theory of hybrid automata. Technical Report UCB/ERL M96/28, University of California, Berkeley, 1996.
33. C. A. R. Hoare. Communicating sequential processes. *Communications of the ACM*, 21(8), 1978.
34. R. Jagannathan. Parallel execution of GLU programs. In *2nd International Workshop on Dataflow Computing*, Hamilton Island, Queensland, Australia, May 1992.
35. G. Kahn. The semantics of a simple language for parallel programming. In *Proc. of the IFIP Congress 74*. North-Holland Publishing Co., 1974.
36. D. J. Kaplan et al. Processing graph method specification version 1.0. The Naval Research Laboratory, Washington D.C., December 1987.
37. R. M. Karp and R. E. Miller. Properties of a model for parallel computations: Determinacy, termination, queueing. *SIAM Journal*, 14:1390–1411, November 1966.
38. R. P. Kurshan. *Automata-Theoretic Verification of Coordinating Processes.* Princeton University Press, 1994.
39. R. Lauwereins, P. Wauters, M. Adé, and J. A. Peperstraete. Geometric parallelism

and cyclostatic dataflow in GRAPE-II. In *Proc. 5th Int. Workshop on Rapid System Prototyping*, Grenoble, France, June 1994.
40. E. A. Lee and D. G. Messerschmitt. Synchronous data flow. *IEEE Proceedings*, September 1987.
41. E. A. Lee and T. M. Parks. Dataflow process networks. *Proceedings of the IEEE*, May 1995.
42. E. A. Lee and A. Sangiovanni-Vincentelli. The tagged signal model - a preliminary version of a denotational framework for comparing models of computation. Technical report, Electronics Research Laboratory, University of California, Berkeley, CA 94720, May 1996.
43. E.A. Lee and A. Sangiovanni-Vincentelli. Comparing models of computation. In *Proceedings of the International Conference on Computer-Aided Design*, pages 234–241, 1996.
44. F. Maraninchi. The Argos language: Graphical representation of automata and description of reactive systems. In *Proc. of the IEEE Workshop on Visual Languages*, Kobe, Japan, October 1991.
45. K. McMillan. *Symbolic model checking.* Kluwer Academic, 1993.
46. R. Milner. *Communication and Concurrency.* Prentice-Hall, Englewood Cliffs, NJ, 1989.
47. M.Sgroi, L,Lavagno, and A.Sangiovanni-Vincentelli. Quasi-static scheduling of free-choice petri nets. Technical report, UC Berkeley. `http://www-cad.eecs.berkeley.edu/~sgroi/qss.ps`, 1997.
48. M. Nielsen, G. Rozenberg, and P.S. Thiagarajan. Elementary transition systems. *Theoretical Computer Science*, 96:3–33, 1992.
49. Institute of Electrical and Electronics Engineers. *IEEE standard VHDL language reference manual.* IEEE, 1994.
50. J. L. Peterson. *Petri Net Theory and the Modeling of Systems.* Prentice-Hall Inc., Englewood Cliffs, NJ, 1981.
51. C. A. Petri. *Kommunikation mit Automaten.* PhD thesis, Bonn, Institut für Instrumentelle Mathematik, 1962. (technical report Schriften des IIM Nr. 3).
52. J. Rasure and C. S. Williams. An integrated visual language and software development environment. *Journal of Visual Languages and Computing*, 2:217–246, 1991.
53. W. Reisig. *Petri Nets: An Introduction.* Springer-Verlag, 1985.
54. F. Rocheteau and N. Halbwachs. Implementing reactive programs on circuits: A hardware implementation of LUSTRE. In *Real-Time, Theory in Practice, REX Workshop Proceedings*, volume 600 of *LNCS*, pages 195–208, Mook, Netherlands, June 1992. Springer-Verlag.
55. T. G. Rokicki. *Representing and Modeling Digital Circuits.* PhD thesis, Stanford University, 1993.
56. J. Rowson and A. Sangiovanni-Vincentelli. Interface-based design. In *Proceedings of the Design Automation Conference*, pages 178–183, 1997.
57. E.M. Sentovich, K.J. Singh, C. Moon, H. Savoj, R.K. Brayton, and A. Sangiovanni-Vincentelli. Sequential Circuit Design Using Synthesis and Optimization. In *Proc of the ICCD*, pages 328–333, October 1992.
58. T. R. Shiple, G. Berry, and H. Touati. Constructive analysis of cyclic circuits. In *Proceedings of the European Design and Test Conference*, March 1996.
59. J. E. Stoy. *Denotational Semantics: The Scott-Strachey Approach to Programming Language Theory.* The MIT Press, Cambridge, MA, 1977.
60. P. A. Suhler, J. Biswas, K. M. Korner, and J. C. Browne. Tdfl: A task-level dataflow language. *J. on Parallel and Distributed Systems*, 9(2), June 1990.
61. W. Takach and A. Wolf. An automaton model for scheduling constraints in synchronous machines. *IEEE Tr. on Computers*, 44(1):1–12, January 1995.
62. D.E. Thomas and P. Moorby. *The Verilog Hardware Description Language.* Kluwer Academic Publishers, 1991.
63. A. Valmari. A stubborn attack on state explosion. *Formal Methods in System*

Design, 1(4):297–322, 1992.

64. P.H.J. van Eijk, C. A. Vissers, and M. Diaz, editors. *The Formal description technique Lotos.* North-Holland, 1989.
65. M. von der Beeck. A comparison of statecharts variants. In *Proc. of Formal Techniques in Real Time and Fault Tolerant Systems*, volume 863 of *LNCS*, pages 128–148. Springer-Verlag, 1994.

MULTILANGUAGE SPECIFICATION FOR SYSTEM DESIGN

A.A. JERRAYA, M. ROMDHANI, PH. LE MARREC, F. HESSEL, P. COSTE, C. VALDERRAMA, G.F. MARCHIORO, J.M.DAVEAU, N.-E. ZERGAINOH
TIMA Laboratory
46 avenue Félix Viallet
38000 Grenoble France

1. Introduction

This chapter discusses specification languages and intermediate models used for system-level design. Languages are used during one of the most important steps of system design: the specification of the system to be designed. A plethora of specification languages exists. Each claims superiority but excels only within a restricted application domain. Selecting a language is generally a trade off between several criteria such as the expressive power of the language, the automation capabilities provided by the model underlying the language and the availability of tools and methods supporting the language. Additionally, for some applications, several languages need to be used for the specification of different modules of the same design. Multilanguage solutions are required for the design of heterogeneous systems where different parts belong to different application classes e.g. control/data or continuous/discrete.

All system design tools use languages as input. They generally use an intermediate form to perform refinements and transformation of the initial specification. There are only few computation models. These may be data- or control-oriented. In both cases, these may be synchronous or asynchronous.

The next section details three system-level modeling approaches to introduce homogeneous and heterogeneous modeling for codesign. Each of the modeling strategies implies a different organization of the codesign environment. Section 3 deals with intermediate forms for codesign. Section 4 introduces several languages and outlines a comparative study of these languages. Finally, section 5 deals with multilanguage modeling and cosimulation.

2. System Level Modeling

The system-level specification of a mixed hardware/software application may follow one of two schemes [1]:

A. A. Jerraya and J. Mermet (eds.), System-Level Synthesis, 103–136.

1. Homogeneous specification: a single language is used for the specification of the overall system including hardware parts and software parts.
2. Heterogeneous modeling: specific languages are used for hardware parts and software parts, a typical example is the mixed C-VHDL model.

Both modeling strategies imply a different organization of the codesign environment.

2.1 HOMOGENEOUS MODELING

Homogeneous modeling implies the use of a single specification language for the modeling of the overall system. A generic codesign environment based on homogeneous modeling is shown in Figure 1. Codesign starts with a global specification given in a single language. This specification may be independent of the future implementation and the partitioning of the system into hardware and software parts. In this case codesign includes a partitioning step aimed to split this initial model into hardware and software. The outcome is an architecture made of hardware processors and software processors. This is generally called virtual prototype and may be given in a single language or different languages (e.g. C for software and VHDL for hardware).

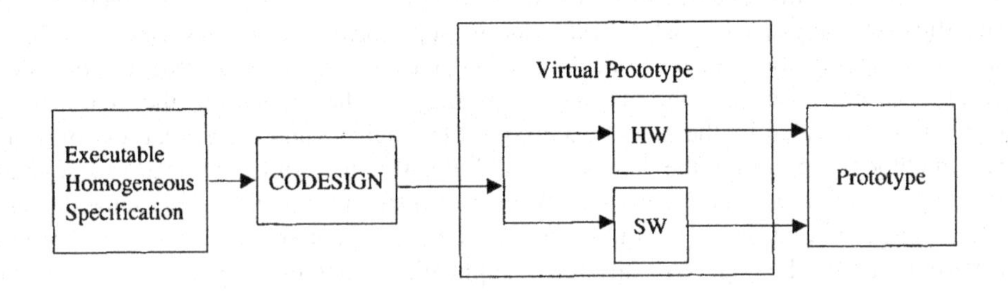

Figure 1 Homogeneous Modeling

The key issue with such codesign environments is the correspondence between the concepts used in the initial specification and the concepts provided by the target model (virtual prototype). For instance the mapping of the system specification language including high-level concepts such as distributed control and abstract communication onto low-level languages such as C and VHDL is a non trivial task [2, 3].

Several codesign environments follow this scheme. In order to reduce the gap between the specification model and the virtual prototype, these tools start with a low-level specification model. Cosyma starts with a C-like model called C^x [4, 5]. VULCAN starts with another C-like language called hardware C. Several codesign tools start with VHDL [6]. Only few tools tried to start from a high-level model. These include Polis [7] that starts with an Esterel model [8, 9], Spec-syn [10, 11] that starts from SpecCharts [12, 13, 14] and [3] that starts from LOTOS [15]. [2,66] details COSMOS, a codesign tool that starts from SDL.

2.2 HETEROGENEOUS MODELING OF HARDWARE/SOFTWARE ARCHITECTURES

Heterogeneous modeling allows the use of specific languages for the hardware and software parts. A generic codesign environment based on a heterogeneous model is given in Figure 2. Codesign starts with a virtual prototype when the hardware/software partitioning is already made. In this case, codesign is a simple mapping of the software parts and the hardware parts on dedicated processors.

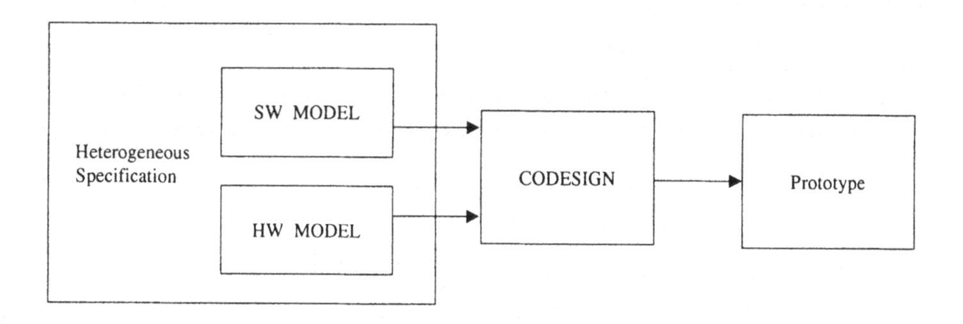

Figure 2 Heterogeneous Modeling

The key issues with such a scheme are validation and interfacing. The use of multilanguage specification requires new validation techniques able to handle a multiparadigm model. Instead of simulation we will need cosimulation and instead of verification we will need coverification. Cosimulation issues will be addressed in section 5. Additionally, multilanguage specification brings the issue of interfacing subsystems which are described in different languages. These interfaces need to be refined when the initial specification is mapped onto a prototype.

Coware[16] and Seamless[17] are typical environments supporting such a codesign scheme. They start from mixed description given in VHDL or VERILOG for hardware and C for software. All of them allow for cosimulation. However, only few of these systems allow for interface synthesis [18]. This kind of codesign model will be detailed in section 5.

2.3 MULTILANGUAGE SPECIFICATION

Most of the existing system specification languages are based on a single paradigm. Each of these languages is more efficient for a given application domain. For instance some of these languages are more adapted to the specification of state-based specification (SDL or Statechart), some others are more suited for data flow and continious computation (LUSTRE, Matlab), while many others are more suitable for algorithmic description (C, C++).

When a large system has to be designed by separate groups, they may have different cultures and expertise with different modeling styles. The specification of such large designs may lead each group to use a different language which is more suitable for the specification of the subsystem they are designing according to its application domain and to their culture.

Figure 3 shows a typical complex system, a mobile telecommunication terminal, e.g. a G.S.M. handset. This system is made of four heterogeneous subsystems that are traditionally designed by separate groups that may be geographically distributed.

a) The protocol and MMI subsystem :
This part is in charge of high-level protocols and data processing and user interface. It is generally designed by a software group using high-level languages such as SDL or C++.

b) The DSP subsystem :
This part is in charge of signal processing and error correction. It is generally designed by "DSP Group using specific tools and methods such as Matlab, Simulink [67], SPW or COSSAP [68].

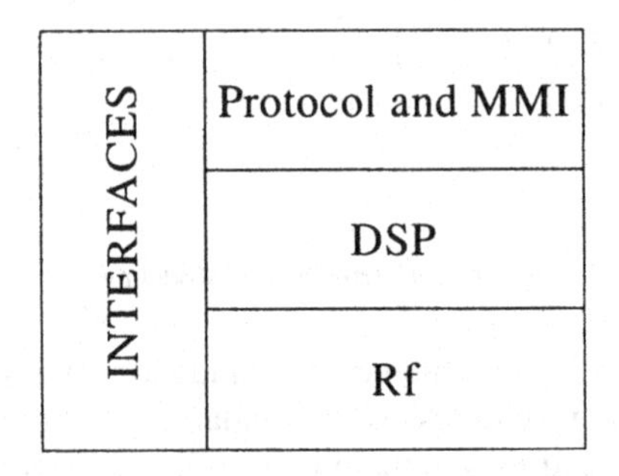

Figure 3 Heterogeneous Architecture of a Mobile Telecom Terminal e.g. G.S.M. handset

c) the DSP subsystem :
This part is in charge of the physical connection. It is generally made by an analog design group using another kind of specific tools and method such as CMS [69].

d) The interface subsystem :
This part is in charge of the communication between the three other parts. It may include complex buses and a sophisticated memory system. It is generally designed by a hardware group using classical EDA tools.

The key issue for the design of such a system is the validation of the overall design and the synthesis of the interfaces between the different subsystems. Of course, most of these subsystems may include both software and hardware.

Figure 4 shows a generic flow for codesign starting from multi-level specification. Each of the subsystems of the initial specification may need to be decomposed into hardware and software parts. Moreover, we may need to compose some of these subsystems in order to perform global hardware/software sub-systems. In other words, partitioning may be local to a given subsystem or global to several subsystems. The codesign process also needs to tackle the refinement of interfaces and communication between subsystems.

As in the case of the heterogeneous modeling for system architecture, the problems of interfacing and multilanguage validation need to be solved. In addition, this model brings another difficult issue: language composition. In fact, in the case where a global partitioning is needed, the different subsystems need to be mapped onto a homogeneous model in order to be decomposed. This operation would need a composition format able to accommodate the concepts used for the specification of the different subsystems and their interconnection.

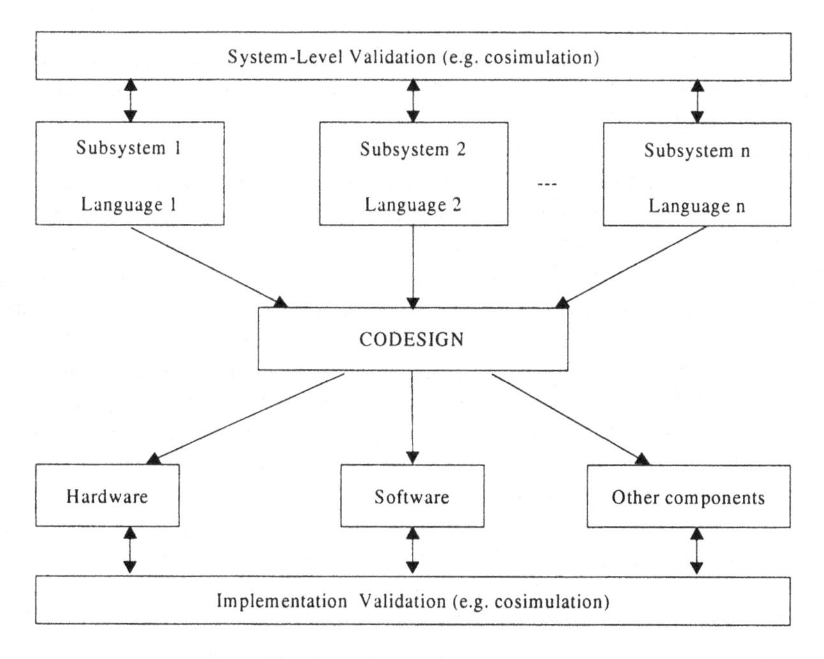

Fig. 4 Multilanguage codesign

In all cases, the key issue with multilanguage codesign is the synthesis of interfaces between subsystems. In fact, the global configuration of the system is a kind of "system-level netlist" that specifies the interconnection between different subsystems. Since different languages may be based on different concepts for data exchange, the interpretation of the link between subsystems is generally a difficult task. These issues will be discussed later in this chapter.

Only few systems in the literature allow such a codesign model. These include RAPID [64] and the work described in [54]. Both systems provide a composition format able to accommodate several specification languages. This codesign model will be detailed in section 5.

3. Design Representation For System Level Synthesis

This section deals with the design models used in codesign. The goal is to focus on the computation models underlying specification languages and architectures. The key message delivered here is that despite the proliferation of specification languages (see next section), there are only few basic concepts and models underlying all these languages.

In fact, most of codesign tools start by translating their input language into an intermediate form that corresponds to a computation model which is easier to transform and to refine. The rest of this section details the intermediate forms and the main underlying concepts and models.

3.1 BASIC CONCEPTS USED IN SYSTEM MODELS

Besides classic programming concepts, system specification is based on four basic concepts. These are concurrency, hierarchy, communication and synchronization. These are detailed in [10].

Concurrency allows for parallel computation. It may be characterized by the granularity of the computation [10] and the expression of operation ordering. The granularity of concurrent computation may be at the bit level (e.g. n-bits adder) or the operation level (e.g. datapath with multiple functional units) or the process level (multiprocesses specification) or at the processor-level (distributed multiprocessor models). The concurrency may be expressed using the execution order of computations or the data flow. In the first case, we have control-oriented models. In this case, a specification gives explicitly the execution order (sequencing) of the element of the specification. CSP like models and concurrent FSMs are typical examples of control-oriented concurrency. In the second case we have data-oriented concurrency. The execution order of operations is fixed by the data dependency. Dataflow graphs and architectures are typical data-oriented models.

Hierarchy is required to master the complexity. Large systems are decomposed into smaller pieces which are easier to handle. There are two kinds of hierarchies. Behavioral hierarchy allows constructs to hide sub-behaviors [11]. Procedure and sub-states are typical forms of behavioral hierarchies. Structural hierarchy allows to decompose a system into a set of interacting subsystems. Each component is defined with well defined boundaries. The interaction between subsystems may be specified at the signal (wire) level or using abstract channels hiding complex protocols.

Communication allows concurrent modules to exchange data and control information. There are two basic models for communication, message passing and shared memory [19]. In the case of message passing, subsystems will execute specific operations to send and to receive data from other components. In the case of shared memory, each subsystem needs to write and to read data from specific memory locations in order to exchange data with other subsystems.

Remote procedure call is a hybrid model allowing to model both message passing and shared memory [20]. In this model communication is performed through primitives that may perform memory read/write or message passing. These primitives are called like procedures and correspond to operations that will be executed by a specific remote communication unit.

Synchronization allows to coordinate the communication or information exchange between concurrent subsystems [20, 21]. There are mainly two synchronization modes; the synchronous mode and the asynchronous mode. When message passing communication is used, synchronization may be achieved using several schemes like message queues (asynchronous mode) or rendez-vous (asynchronous). In the case of shared memory more synchronization schemes are available. These include semaphores, critical regions and monitors. These may be implemented for both synchronous and asynchronous mode.

3.2 COMPUTATION MODELS

The key properties of a specification language, such as its expressiveness, derive from its underlying computation model. In fact, specification languages differ mainly on the manner they provide a view of the basic components, the link between these components and the composition of components. Basic components are described as behavior, these may be control or data oriented. The links fix the inter-module communication and the composition fixes the hierarchy.

Many classifications of specification languages have been proposed in the literature. Most of them concentrated on the specification style. The taxonomy proposed by D. Gajski [10], for example, distinguishes five specification styles: (1) state-oriented, (2) activity-oriented, (3) structure-oriented, (4) data-oriented and (5) heterogeneous. The state-oriented and activity-oriented models respectively provide a description of the behavior through state machines and transformations. The structure-oriented style concentrates on the system structural hierarchy. The data-oriented models provides a system specification based on information modeling. Chapter 2 uses another classification technique and distinguishes nine different computation models.

An objective classification of specification languages is better defined when based on the computation model rather than on the style of writing. In fact, the style of the specification reflects the syntactic aspects and cannot reflect the underlying computation model. The computation model deals with the set of theoretical choices that built the execution model of the language.

The computation model of a given specification can be seen as the combination of two orthogonal concepts: (1) the communication model and (2) the control model. The communication model of a specification language can be fit into either the synchronous or single-thread execution model, or the distributed execution model. The synchronous model is generally a one-thread-based computation, while the distributed model is a multi-thread-based computation with explicit communication between parallel processors. The synchronization arbitrates the exchange of information between processes. The control model can be classified into control-flow or data-flow. This gives the execution order of operations within one process. The synchronous execution model is suitable for mono-processor applications, while the distributed model is adequate for multi-processor applications. In a synchronous model, there is a unique global time reference and the computation is deterministic. Such an execution model is like the one found in DSP applications where data are acquired, processed, then communicated at a fixed rate and in a cyclic manner. The control-oriented model focuses on the control sequences rather than on the computation itself. The data-oriented model focus expresses the behavior as a set of data transformations.

According to this classification we obtain mainly four computation models that may be defined according to concurrency and synchronization. Figure 5 shows these classes and different languages related to these models. Most of these languages will be discussed in the next sections.

Concurrency \ Communication Model	Single-thread	Distributed
Control-driven	SCCS, StateChart Esterel, SML	CCS, CSP,VHDL OCCAM,SDL
Data-driven	SILAGE LUSTRE SIGNAL	Asynchronous Data flow

Figure 5 Computation models of specification languages

3.3 SYNTHESIS INTERMEDIATE FORMS

Most codesign tools make use of an internal representation for the refinement of the input specification into architectures. The input specification is generally given in a human readable format that may be a textual language (C, VHDL, SDL, JAVA, ...) or a graphical representation (StateCharts, SpecCharts, ...). The architecture is generally made of a set of processors. The composition scheme of these processors depends on the computation model underlying the architecture. The refinement of the input specification into architecture is generally performed into several refinement steps using an internal representation also called intermediate form or internal data structure. The different steps of the refinement process can be explained as a transformation of this internal model.

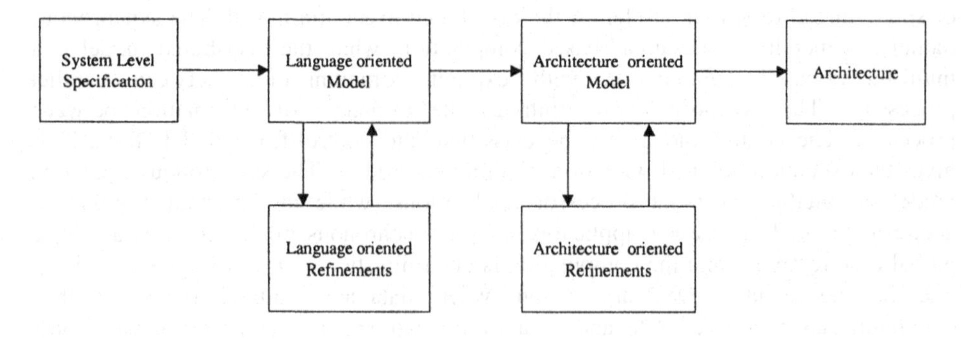

Figure 6 Language-oriented vs Architecture-oriented intermediate forms

There are mainly two kinds of intermediate forms: the first is language-oriented and the latter is architecture-oriented. Both may be used for system representation and transformations. Language-oriented intermediate forms are generally based on graph representation. They model well the concepts handled in specification languages. Architecture-oriented intermediate forms are generally based on FSM models. These are close to the concepts needed to represent architectures.

Figure 6 shows a generic codesign model that combines both models. In this case, the first codesign step translates the initial system-level specification into a language-oriented intermediate form. A set of refinement steps may be applied to this representation. These may correspond to high level transformations. A specific refinement step is used to map the resulting model into architecture-oriented intermediate form. This will be refined into an architecture.

Although general, this model is difficult to implement in practice. In fact, most existing tools make use of a unique intermediate form. The kind of selected intermediate form will restrict the efficiency of the resulting codesign tool. Language-oriented intermediate forms make easier the handling of high level concepts (e.g. abstract communication, abstract data types) and high level transformations. But, it makes difficult the architecture generation step and the specification of partial architectures and physical constraints. On the other side, an architecture-oriented intermediate form makes difficult the translation of high level concepts. However, it makes easier the specification of architecture related concepts such as specific communication and synchronization. This kind of intermediate form generally produces more efficient design.

3.4 LANGUAGE ORIENTED INTERMEDIATE FORMS

Various representations have appeared in the literature [22, 23, 24, 25], mostly based on flow graphs. The main kinds are Data, Control and Control-Data flow representations as introduced below.

3.4.1. *Data Flow Graph (DFG)*

Data flow graphs are the most popular representation of a program in high level synthesis. Nodes represent the operators of the program, edges represent values. The function of node is to generate a new value on its outputs depending on its inputs.

A data flow graph example is given in Figure 7, representing the computation $e:=(a+c)*(b-d)$. This graph is composed of three nodes v_1 representing the operation +, v_2 representing - and v_3 v representing *. Both data produced by v_1 and v_2 are consumed by v_3. At the system level a node may hide complex computation or a processor.

In the synchronous data flow model, we assume that an edge may hold at most one value. Then, we assume that all operators consume their inputs before new values are produced on their inputs edges.

In the asynchronous data flow model, we assume that each edge may hold an infinite set of values stored in an input queue. In this model, we assume that inputs, arrivals and computations are performed at different and independent throughputs. This model is powerful for the representation of computation. However it is restricted for the representation of control structures.

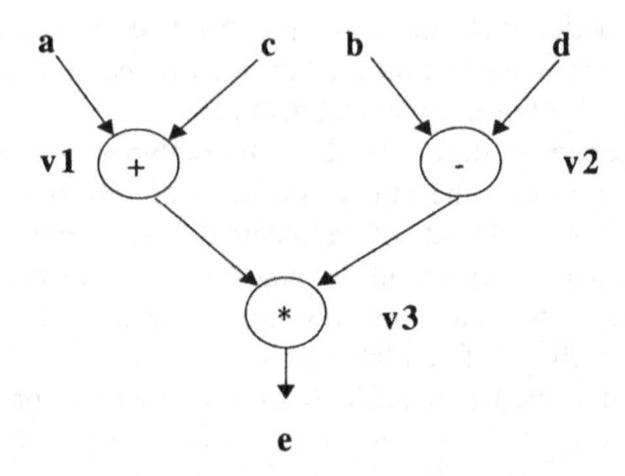

Figure 7 Example of a simple data flow graph

3.4.2. *Control Flow Graph (CFG)*

Control flow graphs are the most suited representation to model control design. These may contain many (possibly nested) loops, global exceptions, synchronization and procedure calls; in other words, features that reflect the inherent properties of controllers. In a CFG, nodes represent operations and edges represent sequencing relations.

While this kind of graph models well control structure including generalized nested loops combined with explicit synchronization statement (wait), control statements (if, case), and exceptions (EXIT), it provides restricted facilities for data flow analysis and transformations.

3.4.3. *Control Data Flow Graph (CDFG)*

This model extends DFG with control nodes (If, case, loops) [26]. This model is very suited for the representation of data flow oriented applications. Several codesign tools use CDFG as intermediate form [4] [Lycos, Vilcar].

3.5 ARCHITECTURE ORIENTED INTERMEDIATE FORMS

This kind of intermediate form is closer to the architecture produced by codesign than to initial input description. The data path and the controller may be represented explicitly when using this model. The controller is generally abstracted as an FSM. The data path is modeled as a set of assignment and expressions including operation on data. The main kind of architecture-oriented forms are FSM with data path model FSMD defined by Gajski [27] and the FSM with coprocessors(FSMC) defined in [28].

3.5.1. *The FSMD representation*

The FSMD was introduced by Gajski [27] as a universal model that represents all hardware design. An FSMD is an FSM extended with operations on data.

In this model, the classic FSM is extended with variables. An FSMD may have internal variables and a transition may include arithmetic and logic operation on these variables. The FSMD adds a datapath including variables, operators on communication to the classic FSM.

The FSMD computes new values for variables stored in the data path and produces outputs. Figure 8 shows a simplified FSMD with 2 states Si and Sj and two transitions. Each transition is defined with a condition and a set of actions that have to be executed in parallel when the transition is fixed.

3.5.2. *The FSM with coprocessors model (FSMC)*

An FSMC is an FSMD with operations executed on coprocessors. The expressions may include complex operations executed on specific calculation units called coprocessors. An FMSC is defined as an FSMD plus a set of N coprocessors C. Each coprocessor Ci is also defined as an FSMC. FSMC models hierarchical architecture made of a top controller and a set of data paths that may include FSMC components as shown in Figure 9.

Coprocessors may have their local controller, inputs and outputs. They are used by the top controllers to execute specific operations (expressions of the behavioral description). Several codesign tools produce an FSMC based architecture. These include COSYMA [5], VULCAN [29] and Lycos [70].

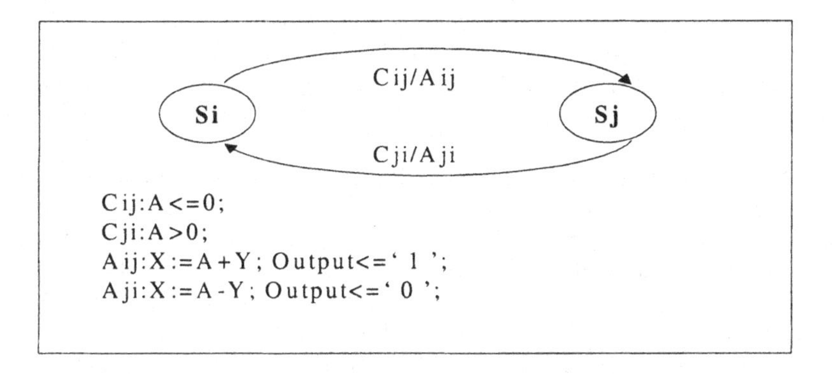

Figure 8 FSMD Model

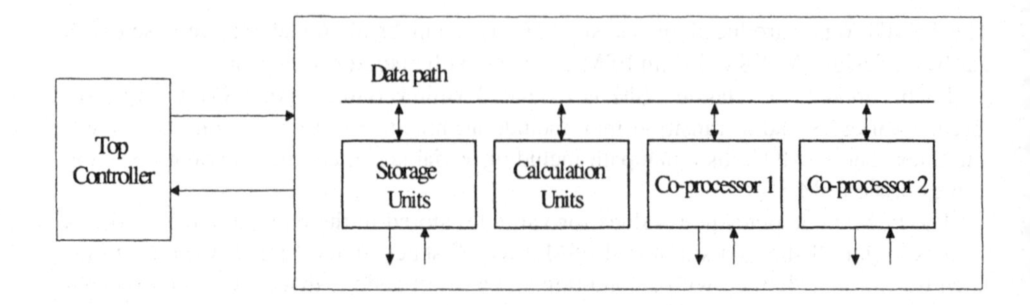

Figure 9 FSMC Architecture Model

In these codesign tools the top controller is made of a software processor and the co-processor are hardware accelerators.

3.6 DISTRIBUTED INTERMEDIATE FORMS

In order to handle concurrent processes several codesign tools make use of intermediate forms that support multithreading. At this level also intermediate forms may be language oriented or architecture oriented. A distributed language oriented intermediate form is generally made of communicating graphs [30]. The most popular format of this type is the task graph [6, 31].

In this model each node represents a simple CFG, DFG or CDFG and the edges represent execution orders that may be control oriented or data oriented. Inter task communication may follow any of the schemes listed above.

Most codesign tools that make use of task graphs, assume that all the tasks are periodic. This assumption induces that the execution time of each task is bounded and allows for all kind of rate analysis needed for task scheduling and performances estimation.

Distributed architecture-oriented intermediate forms are generally made of communicating FSMs. These generally used an extended FSM model in order to allow for data computation within the FSMs. Only few codesign tools support communicating FSMs. SpecSyn [10] and Polis [7] make use of interconnected FSMs. The model used in Polis is based on a specific kind of FSMDs, called Codesign FSMs (CFSM). In this model the communication is made at the signal level and a system is organized a DAG where each node is a CFSM. The COSMOS system is based on a communicating FSMC model called SOLAR[32]. This model allows for a generalized composition model and makes use of RPC [20] for inter-modules communication.

4 System Level Specification Languages

4.1 THE PLETHORA OF SYSTEM SPECIFICATION LANGUAGES

Specification languages were firstly introduced in software engineering in order to support the early steps of the software development [33]. The high cost of the software development and maintenance raised the need of concentrating on the specification and the requirements analysis steps. The software quality and productivity is expected to be improved due to formal verification and gradual refinement of software starting from higher level specification.

For hardware design, the first languages were aimed to the specification of computers. Von Newman used an ad hoc hardware description language for the description of its machine. DDL [34] PMS and ISP are typical examples introduced in the late 60's and the early 70's [10] for hardware specification. Since that time, a plethora of languages is presented in the literature. These are the results of several fields of research. The most productive areas are:

1. VLSI System Design: Research in this area produced what is called hardware description languages (HDLs). ISPS [35], CONLAN [36] and more recently HardwareC [37] , SpecCharts [13, 14], SpecC [12] and VHDL [38] are typical HDLs. These languages try to deal with the specific characteristics of hardware design such as abstraction levels, timing and data flow computation. The community also produced several specialized languages such as SPW [3] and COSSAP [68] for DSP systems modeling and design and several functional models.

2. Protocol specification: several languages have been created for protocol specification. In order to allow for protocol verification, these languages are based on what is called formal description technique (FDT) [39]. SDL [40], LOTOS [15] and ESTELLE [41, 42] are the main languages in this area.

   LOTOS [15] (LOgical Temporal Ordring Specification) is a formal specification language for protocols and distributed systems. It is an OSI standard. The LOTOS specification is composed of two parts:

   - A behavioral part based on the theory of process algebra.
   - A facultative part for data definition based on abstract data types.

   The formal basis of LOTOS is well defined. The specification approach consists in producing a first executable specification, then validate it, and derive an implementation.

   The SDL language [40] was designed in order to specify telecommunication systems. The language is standardized by the ITU (International Telecommunication Union) as Z.100. SDL is particularly suitable for systems where it is possible to represent the behavior by extended finite state machines.

An SDL specification can be seen as a set of abstract machines. The whole specification includes the definition of the global system structure, the dynamic behavior of each machine and the communication in between. SDL offers two forms of representations: a graphical representation named SDL-GR (SDL Graphical Representation) and a textual representation named SDL-PR (SDL Phrase Representation).

ESTELLE [42] is also an OSI standard language for the specification of protocol and their implementation. The specifications are procedural having a Pascal-like construction. ESTELLE is rather a programming language than a specification language. In fact, the ESTELLE specification includes implementation details. These details are generally not necessary during the specification phase.

ESTELLE adopts an asynchronous model for the communication based on message passing. It presents several limitations in the specification of the parallelism between concurrent processes.

3. Reactive system design: reactive systems are realtime applications with fast reaction to the environment. ESTEREL [8], LUSTRE [43] and Signal [44] are typical languages for the specification of reactive systems. Petri Nets [45] may also be included in this area.

   ESTEREL [8] is an imperative and parallel language which has a well defined formal basis and a complete implementation. The basic concept of ESTEREL is the event. An event corresponds to the sending or receiving of signals that convey data. ESTEREL is based on a synchronous model. This synchronism simplifies reasoning about time and ensures determinism.

   LUSTRE [43] particularly suits to the specification of programmable automata. Few real-time aspects have been added to the language in order to manage the temporal constraints between internal events.

   The SIGNAL language [44] differs from LUSTRE in the possibility of using different clocks in the same program. The clocks can be combined through temporal operators, allowing flexible modeling.

   StateCharts [46] is visual formalism for the specification of complex reactive systems created by D. Harel. StateCharts describes the behavior of those systems using a state-based model. StateCharts extend the classical finite state machine model with hierarchy, parallelism and communication. The behavior is described in terms of hierarchical states and transitions in-between. Transitions are triggered by events and conditions. The communication model is based on broadcasting, the execution model is synchronous.

   Petri Nets [45] are tools that enable to represent discrete event systems. They do enable describing the structure of the data used, they describe the control aspects and the behavior of the system. The specification is composed of a set of transitions and places. Transitions correspond to events, places correspond to activities and waiting states.

   Petri Nets enable the specification of the evolution of a system and its behavior. The global state of a system corresponds to a labeling that associates for each place a value. Each event is associated to a transition, firable when the entry places are labeled.

The original Petri Nets [45] were based on a well defined mathematical basis. Thus, it is possible to verify the properties of the system under-specification. However, literature reports on plethora of specialized Petri Nets which are not always formally defined.

4. Programming languages: most (if not all) programming languages have been used for hardware description. These include Fortran, C, Pascal, ADA and more recently C++ and JAVA. Although these languages provide nice facilities for the specification hardware systems, these generally lack feature such as timing or concurrency specifications. Lots of research works have tried to extend the programming languages for hardware modeling with limited results because the extended language is no more a standard. For instance, in order to use C as a specification language, one needs to extend it with constructs for modeling parallel computation, communication, structural hierarchy, interfaces and synchronization. The result of such extensions will produce a new language similar to HardwareC or SpecC (see chapter 10).

5. Parallel programming languages: parallel programs are very close to hard- ware specification because of the concurrency. However, they generally lack timing concepts and provide dynamic aspects which are difficult to implement in hardware. Most of the works in the field of parallel programming are based on CSP [47] and CCS [47]. This produced several languages such as OCCAM and Unity [48] that has been used for system specification.

6. Functional programming and algebraic notation: several attempts were made to use functional programming and algebraic notations for hardware specification. VDM, Z and B are examples of such formats. VDM [49] (Vienna Development Method) is based on the set theory and predicate logic. It has the advantage of being an ISO standard. The weaknesses of VDM are mainly the non support concurrency, its verbosity and the lack of tools. Z [50] is a predicative language similar to VDM. It is based on the set theory. Z enables to divide the specification in little modules, named "schemes". These modules describe at the same time static and dynamic aspects of a system. B [51] is composed of a method and an environment. It was developed by J.R. Abrial who participated to the definition of the Z language. B is completely formalized. Its semantics is complete. Besides, B integrates the two tasks of specification and design.

7. Structural Analysis : In order to master the development of large software applications, several design methodologies were introduced. The structural analysis was initially introduced by [Demarco]. It provides a systematic approach for structuring code and data in the case of designing large software. The key idea behind structural analysis is to decompose large systems into smaller and more manageable pieces. Several improvements of the initial structural analysis were introduced with SART. After the appearance of object oriented programming

several new analysis techniques were reported in literature. These include HOOD and OMT [71]. The latest evolution of these analysis techniques is The Unified Modeling Language (UML) [72]. The goal of UML is to gather several notations within a single language. All these techniques provide powerful tools for structuring large applications, but, they lack full executable models in order to be used by synthesis and verification tools.

8. Continuous languages : These are based on differential equations and are used for very high-level modeling of all kinds of systems. The most popular languages are Matlab [67], Matrixx [73], Mathematica [74] and SABER [75]. Chapter 4 details the characteristics of Matrixx. Most of these languages provide large libraries for modeling systems in different fields. These are very often used for DSP design, mechanical design and hydraulic design. These languages provide a large expression power and make intensive use of floating point computation which make them very difficult to use for synthesis. Additionally the intensive use of specialized libraries make them very flexible and powerful for different application domains but also make them difficult to analyze and to verify.

4.2 COMPARING SPECIFICATION LANGUAGES

There is not a unique universal specification language to support all kinds of applications (controller, heavy computation, DSP, ...). A specification language is generally selected according to the application at hand and to the designer culture. This section provides 3 criteria that may guide the selection of specification languages. These are [52]:

1. Expressive power: this is related to the computation model. The expressive power of a language fixes the difficulty or the ease when describing a given behavior

2. Analytical power: this is related to the analysis, the transformation and the verification of the format. It is mainly related to tool building.

3. Cost of use: this criterion is composed of several debatable aspects such as clarity of the model, related existing tools, standardization efforts, etc.

As explained earlier, the main components of the expressive power of a given language are: concurrency, communication, synchronization, data description and timing models.

The analytical power is strongly related to the formal definition of the language. Some languages have formal semantics. These include Z, D, SCCS and temporal logic. In this case, mathematical reasoning can be applied to transform, analyze or proof properties of system specification. The formal description techniques provide only a formal interpretation : these may be translated using a "well defined" method, to another language that has formal semantics. For instance, the language Chill [53] is used for the interpretation of SDL. Finally, most existing language have a formal syntax which is the weakest kind of formal description. The existence of formal semantics allows an easier analysis of specification. This also make easier the proof of properties such as coherence, consistency, equivalence, liveness and fairness.

The analytical power includes also facilities to build tools around the language. This aspect is more related to the underlying computation model of the language. As stated earlier, graph based models make easier the automation of language-oriented analysis and transformation tools. On the other side, architecture-oriented models (e.g. FSMs) make easier the automation of lower level transformation which are more related to architecture.

The cost of use includes aspects such as standardization, readability and tool support. The readability of a specification plays an important role in the efficiency of its exploitation. A graphical specification may be quoted more readable and reviewable than a textual specification by some designers. However, some of the designers may prefer textual specification. The graphical and textual specifications are complementary.

The availability of tools' support around a given specification language is important in order to take best benefit from the expressiveness of this language. Support tools include editors, simulation tools, proovers, debuggers, prototypers, etc.

The above mentioned criteria show clearly the difficulty of comparing different specification languages. Figure 10 shows an attempt for the classification of some languages. The first four lines of the table show the expressive power of the languages, the fifth line summarizes the analytical power and the three last lines give the usability criteria.

Each column summarizes the characteristic of a specific language:

*** : the language is excellent for the corresponding criteria

** : the language provides acceptable facilities

* : the language provides a little help

+ : coming feature

? : non proved star

Expressive power

	HDL	SDL	Statecharts Esterel	SPW COSSAP	Matlab	C, C++	JAVA, UML, ...
Abstract Communication	*	***	/	*	*	/	**
TIME	***	**	**	*	***	/	*
Computation Algorithms	***	*++	***	***	***	***	***
Specific Libraries	HW cores	Protocols+	?	DSP	DSP Math Mechanical +++	***	??
FSMs, Exceptions Control	**	***	***	/	???	**	**
Analytical Power (Formal Analysis)	*	***	**	/	/	/	/
Cost of Use (standard, learning curve, hosts)	***	***	*	*	***	***	???

Figure 10 Summary of some specification languages

VHDL provides an excellent cost of use. However, its expression power is quite low for communication and data structure.

SDL may be a realistic choice for system specification. SDL provides acceptable facilities for most criteria. The weakest point of SDL is timing specification where only the concept of timer is provided. Additionally, SDL restricts the communication to the asynchronous model. The recent version of SDL introduces a more generic communication model based on RPC. The new versions will also include more algorithmic capabilities.

StateCharts and Esterel are two synchronous languages. Esterel provides powerful concepts for expressing time and has a large analytical power. However its communication model (broadcasting) is restricted to the specification of synchronous systems. StateCharts provide a low cost of use. However, like Esterel, it has a restricted communication model. The expression power is enhanced by the existence of different kinds of hierarchies (activities, states, modules). Synchronous languages make very difficult the specification of distributed systems. For example, Esterel and StateCharts assume that the transitions of all parallel entities must take the same amount of time. This means that if one module is described at the clock-cycle level, all the other modules need to be described at the save level.

SPW and COSSAP are two DSP oriented environments provided by EDA vendors. The fact that they are based on proprietary language make their cost of use quite high because of lack of standardization and generic tools supporting these languages. These environments also make extensive use of libraries which make formal verification difficult to apply.

Matlab [67] is used by more than 400 000 engineers all over the world and in several application domains. This makes it a de facto standard which also reduces the cost of use. Matlab provides a good expression power for high level algorithm design. It also supports well the concept of time. The main restrictions of Matlab are related to the communication model (simple wire carrying floating values), the expression of state based computation (new versions included statechart like models) and formal semantics. The latter is mainly due to the extensive use of libraries.

C++, C provide a high usability power but they fail to provide general basic concepts for system specification such as concurrency and timing aspects. The use of C and C++ for system specification is very popular. In fact most designers are familiar with C and in many application domains the initial specification is made in C or C++. It is also very easy to build run-time libraries that extend the languages for supporting concurrency and timing. Unfortunately, these kinds of extensions are not standardized and make all the system design tools specific to a given environment.

JAVA and UML are generating lots of hope in the software community. They are also well studied for codesign. UML is just an emerging notation [72]. However Java seems to be adopted by a large section of the research community. If we exclude the dynamic features of JAVA such as garbage collection, we obtain a model with a very high expression power. Although JAVA includes no explicit time concepts , these can be modeled as timers using the RPC concept. The usability power of JAVA is still unknown. The same remark made about extending C applies for extensions to JAVA.

5 Heterogeneous Modeling and Multilanguage Cosimulation

Experiments with system specification languages [54] show that there is not a unique universal specification language to support the whole life cycle (specification, design, implementation) for all kinds of applications. The design of a complex system may require the cooperation of several teams belonging to different cultures and using different languages. New specification and design methods are needed to handle these cases where different languages and methods need to be used within the same design. These are multilanguage specification design and verification methods. The rest of this section provides the main concepts related to the use of multilanguage and two examples of multilanguage codesign approaches. The first starts with a heterogeneous model of the architecture given in C-VHDL. The second makes use of several system-level specification languages and is oriented towards very large system design.

5.1 BASIC CONCEPTS FOR MULTILANGUAGE DESIGN

The design of large systems, like the electronic parts of an airplane or a car, may require the participation of several groups belonging to different companies and using different design methods, languages and tools. The concept of multi-language specification aims at coordinating different modules described in different languages, formalisms, and notations. In fact, the use of more than one language corresponds to an actual need in embedded systems design.

Besides, multi-language specification is driven by the need of modular and evolutive design. This is due to the increasing complexity of designs. Modularity helps in mastering this complexity, promotes for design re-use and more generally encourages concurrent engineering development.

There are two main approaches for multi-language design: the compositional approach and cosimulation based approach.

The compositional approach (cf. Figure 11) aims at integrating the partial specification of sub-systems into a unified representation which is used for the verification and design of the global behavior. This allows to operate full coherence and consistency checking, to identify requirements for traceability links, and to facilitate the integration of new specification languages [33].

Several approaches have been proposed in order to compose partial programming and/or specification languages. Zave and Jackson 's approach [55] is based on the predicate logic semantic domain. Partial specifications are assigned semantics in this domain, and their composition is the conjunction of all partial specification. Wile's approach [56] to composition uses a common syntactic framework defined in terms of grammars and transformations. Garden project [57] provides multi-formalisms specification by means of a common operational semantics. These approaches are globally intended to facilitate the proofs of concurrent systems properties.

Polis, Javatime and SpecC detailed respectively in chapters 2, 10, 11 introduce a compositional based codesign approach. Polis uses an internal model called Codesign FSMs for composition. Both Javatime and SpecC use another specification language (Java and SpecC) for composition.

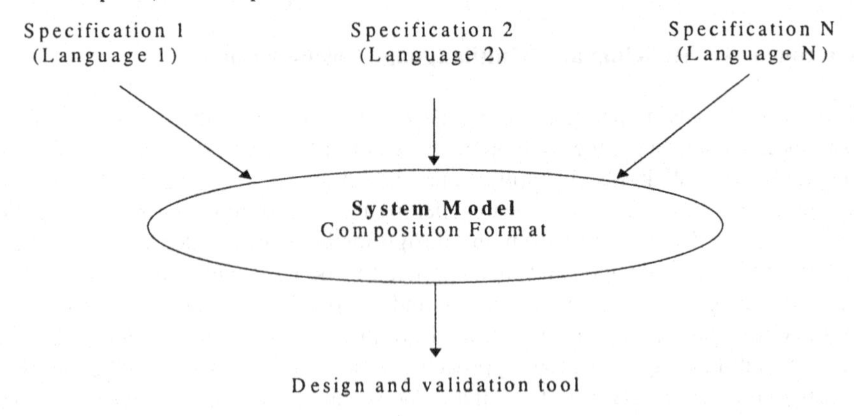

Figure 11 Composition-based multilanguage design

The cosimulation based approach (cf. Figure 12) consists in interconnecting the design environments associated to each of the partial specification. Compared with the deep specification integration accomplished by the compositional approaches, cosimulation is an engineering solution to multilanguage design that performs just a shallow integration of the partial specifications.

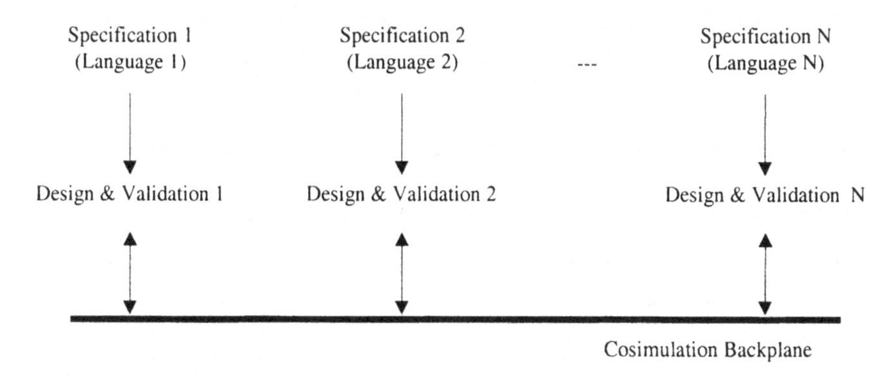

Figure 12 Cosimulation based multi language design

5.2 MULTILANGUAGE COSIMULATION

Cosimulation aims at executing several models given in different languages in a concurrent way. Cosimulation environments may be differentiated by three key factors :

a) The Engine Model
The cosimulation may be based on a single engine or on a multiple engine. The first case corresponds to the composition scheme, all the subsystems are translated into a unique notation that will be executed using a specific simulation engine. The multi engine approach corresponds to the cosimulation based on a multi-language scheme. In this case several simulators may be executed concurrently.

b) The timing model
The cosimulation may be timed or untimed. In the first case the execution time of the operation is not handled. This scheme allows only for functional validation. The time needed for computation is not taken into account during cosimulation. Inter module communication may be done only through hand shaking where the order of event is used to synchronize the data and control signal exchange. In the case of timed simulation, each block may have a local timer and the cosimulation takes into account the execution time of the computation. The time may be defined at different levels of granularity which define different levels of cosimulation, for example in the case of

HW/SW systems there are mainly three levels of cosimulation. The instruction level, the RTL level and the physical time level.

c) The synchronization scheme :
In case the cosimulation uses a multi-engine scheme, the key issue for the definition of a cosimulation environment is the communication between the different simulators. There are mainly two synchronization modes: the master slave mode and the distributed mode.

With the master-slave mode, cosimulation involves one master simulator and one or several slave simulators. In this case, the slave simulators are invoked using procedure call-like techniques. Figure 13 shows a typical master-slave cosimulation model. The implementation of such a communication model is generally made possible thanks to:

1. The possibility to call foreign procedures (e.g. C programs) from the master simulator

2. The possibility to encapsulate the slave simulator within a procedure call

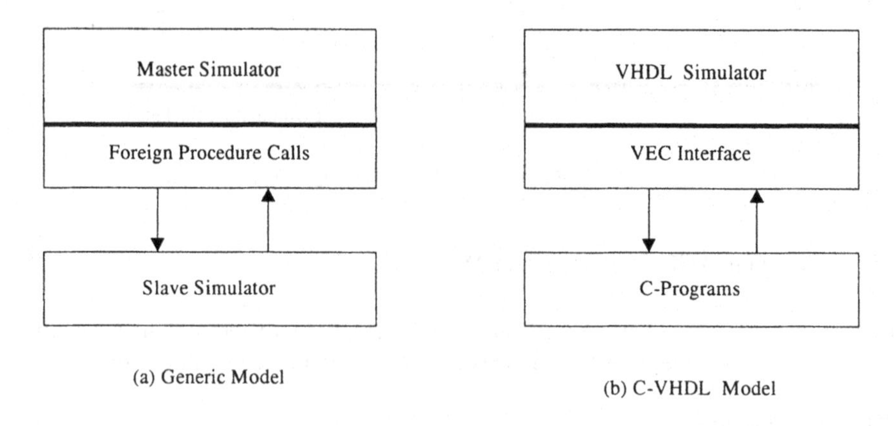

Figure 13 Master-slave Cosimulation

Most simulators provide a basic means to invoke C functions during simulation following a master-slave model. In the case of VHDL (fig.1.13b) this access is possible by using the foreign VHDL attribute within an associated VHDL architecture. The foreign attribute allows parts of the code to be written in languages other than VHDL (e.g. C procedures). Although useful, this scheme presents a fundamental constraint. The slave module cannot work in the same time concurrently to the master module. In the case of C-VHDL cosimulation, this also implies that the C module cannot hold an internal state between two calls. Then, the C-program needs to be sliced into a set of C-procedures executed in one shot and activated through a procedure call. In fact, this requires a significant style change in the C flow, specially for control-oriented applications which need multiple interaction points with the rest of the hardware parts [58]. By using this model, the software part for control-oriented applications requires a sophisticated scheme where the procedure saves the exit point of the sliced C-program for future invocations of the procedure. Moreover, the model does not allow true concurrency: when the C procedure is being executed, the simulator is idle.

The distributed cosimulation model overcomes these restrictions. This approach is based on a communication network protocol which is used as a software bus. Figure 14 shows a generic distributed cosimulation model. Each simulator communicates with the cosimulation bus through calls to foreign procedures.

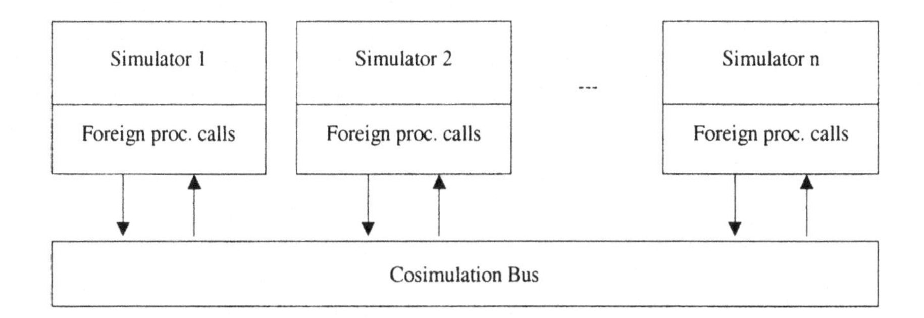

Figure 14 Distributed Cosimulation Model

The cosimulation bus is in charge of transferring data between the different simulators. It acts as a communication server accessed through procedure calls. The implementation of the cosimulation bus may be based on standard system facilities such as UNIX IPC or Sockets. It may also be implemented as an ad hoc simulation backplane [59].

In order to communicate, each simulator needs to call a procedure that will send/receive information from the software bus (e.g. IPC channel). For instance, in the case of C-VHDL cosimulation, this solution allows the designer to keep the C application code in its original form. In addition, the VHDL simulator and the C-program may run concurrently. It is important to note that the cosimulation models are independent of the communication mechanism used during cosimulation. This means that cosimulation can use other communication mechanisms than IPC (for example Berkeley Sockets or any other ad hoc protocol).

5.3 AUTOMATIC GENERATION OF INTERFACES

When dealing with multilanguage design, the most tedious and error-prone procedure is the generation of the link between different environments. In order to overcome this problem automatic interface generation tools are needed. This kind of tools take as input a user defined configuration file, which specifies the desired configuration characteristics (I/O interface and synchronization between debugging tools) and produces a ready-to-use interface [58, 60]. There are two kinds of interfaces needed : cosimulation interfaces and implementation interfaces.

Figure 15 shows a generic interface generation scheme starting from a set of modules and a configuration file, the system produces an executable model. In the case of co-simulation, the communication bus will be a software bus (such as IPC or sockets) and the module interfaces are systems calls used to link the simulators to the cosimulation bus. In the case of interface generation for implementation, the communication bus may correspond to an existing platform and the module interfaces may include drivers for software and specific hardware for other blocks [18].

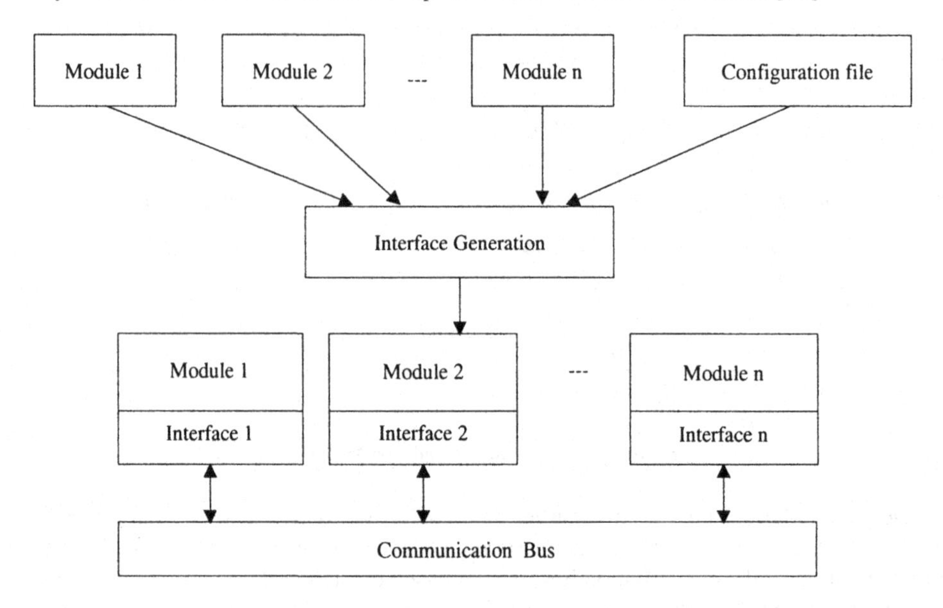

Figure 15 Intermodule interface generation for cosimulation and implementation

The configuration file specifies the inter-module interactions. This may be specified at different levels ranging from the implementation level where communication is performed through wires to the application level where communication is performed through high level primitives independent from the implementation. Figure 16 shows three levels of inter-module communication [61].

The complexity of the interface generation process depends on the level of the inter-module communication.

At the application level, communication is carried out through high level primitives such as send, receive and wait. At this level the communication protocol

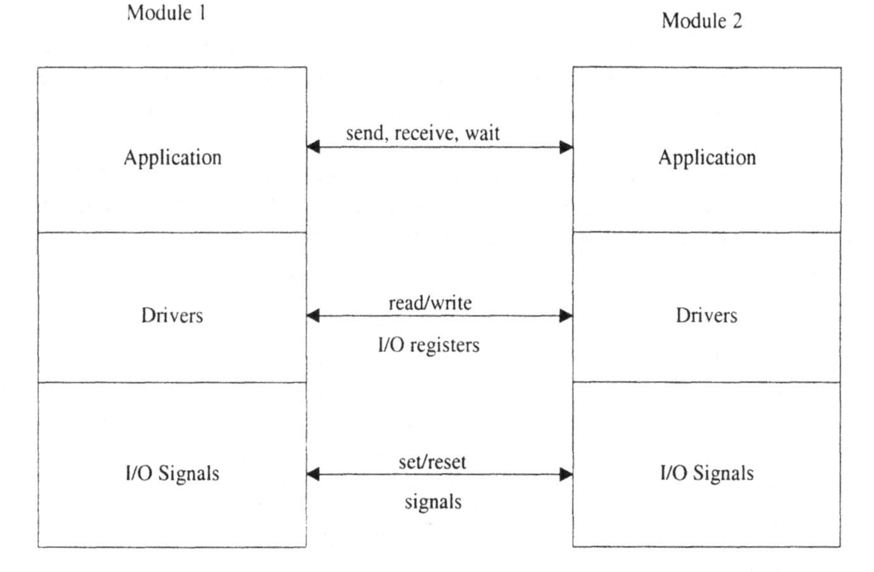

Figure 16 Abstraction Levels of Inter-module Communication

may be hidden by the communication procedures. This allows to specify a module regardless to the communication protocol that will be used later. Automatic generation of interfaces for this level of communication requires an intelligent process able to select a communication protocol. This process is generally called communication synthesis [6, 2].

The next level may be called the driver level. Communication is performed through read/write operation on I/O registers. These operations may hide physical address decoding and interrupt management. Automatic generation of interfaces selects an implementation for the I/O operations. COWARE is a typical multilanguage environment acting at this level [16].

At the lowest level, all the implementation details of the protocol are known and communication is performed using operations on simple wires (e.g. VHDL signals). The interface generation is simpler for this level.

5.4 APPLICATION 1: C-VHDL MODELING AND COSIMULATION

The design of system on chip applications generally combine programmable processors executing software and application specific hardware components. The use of C-VHDL based codesign techniques enable the cospecification, cosimulation and cosynthesis of the whole system. Experiments [63] show that these new codesign approaches are much more flexible and efficient than traditional method where the design of the software and the hardware were completely separated. More details about the design of embedded cores can be found in [76].

This section deals with a specific case of multilanguage approach based on C-VHDL. We assume that hardware/software partitioning is already made. The codesign process starts with a virtual prototype, a heterogeneous architecture composed of a set of distributed modules, represented in VHDL for hardware elements and in C for software elements. The goal is to use the virtual prototype for both cosynthesis (mapping hardware and software modules onto an architectural platform) and cosimulation (that is the joint simulation of hardware and software components) into a unified environment.

Figure 17 shows a typical C-VHDL based codesign method [58, 16]. The design starts with a mixed C-VHDL specification and handles three abstraction levels: the functional-level, the cycle accurate level and the gate-level.

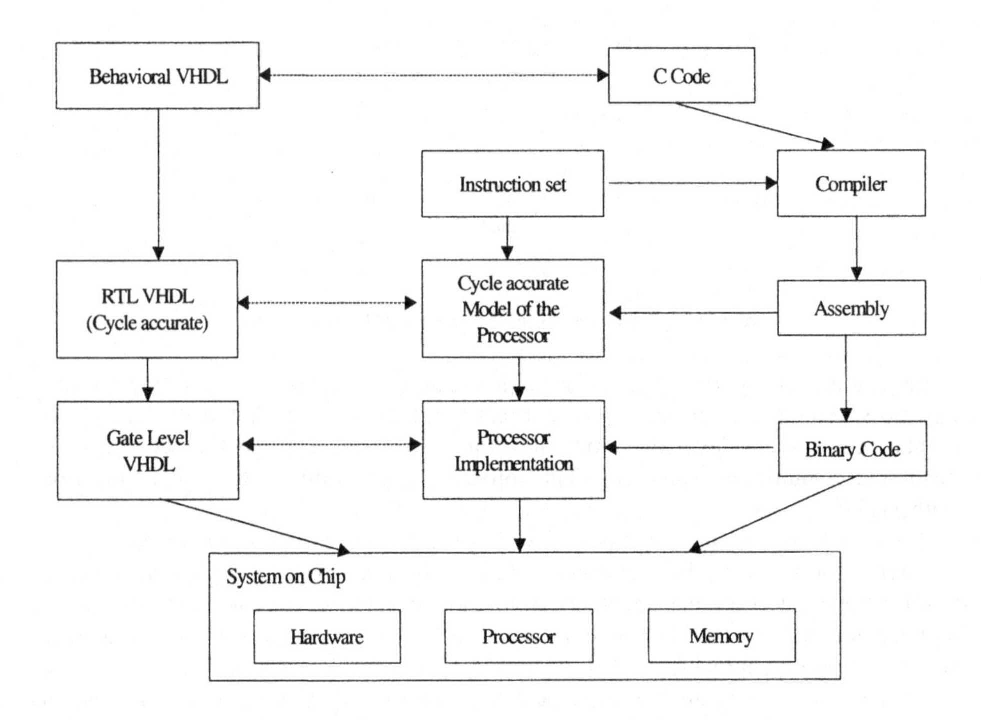

Figure 17 C-VHDL based Codesign for Embedded Processors

At the functional level, the software is described as a C program and the hardware as a behavioral VHDL model. At this level the hardware software communication may be described at the application-level. A C-VHDL cosimulation may be performed for the verification of the initial specification. The C code is executed at the workstation and the VHDL is simulated at the behavioral level. Only functional verifications can be performed at this level, accurate timing verification can be performed with the two next levels.

At the cycle accurate level, the hardware is described at the RT-level and the software at the assembly-level. A cosimulation may be performed for the verification of the behavior of the system at the cycle-level. The C code is executed on a cycle accurate model of the processor. This may be in VHDL or another software model (e.g. a C-program). At this level, the hardware/software communication can be checked at the clock-cycle level.

The cycle accurate model can be obtained automatically or manually starting from the functional level. The RTL VHDL model may be produced using behavioral synthesis. The assembler is produced by simple compilation as explained in [76]; the main difficulty is the interface refinement [16].

The gate-level is close to the implementation. The VHDL model is refined through RTL synthesis to gates. The implementation of the software processor is fixed. The processor may be a hard macro or designed at the gate-level. All communication between hardware and software is detailed at the signal level. Cosimulation may be performed to verify the correct timing of the whole system.

The C-VHDL model described in Figure 16 may handle multiprocessor design thanks to the modularity of the specification and the cosimulation model. [66] gives more details about the specification styles and interface synthesis use in a similar C-VHDL based codesign methodology.

5.5 APPLICATION 2: MULTILANGUAGE DESIGN FOR AUTOMOTIVE

This section introduces a complete design flow used for automotive design [78]. The use of electronics within cars is becoming more and more important. It is expected that the electronic parts will constitute more than 20% of the price of future cars [77]. The joint design control of various mechanical parts with electronic parts and specially micro-controllers is a very important area for automotive design. In traditional design approaches the different parts are designed by separate groups and the integration of the overall system is made at the final stage. This scheme may induce extra delays and costs because of interfacing problems. The necessity for more efficient design approaches allowing for joint design of different parts is evident. Multilanguage design constitutes an important step towards this direction. It gives the designer the ability to validate the whole system's behavior before the implementation of any of its parts.

Multilanguage system design offers many advantages including efficient design flow and shorter time to market. The key idea is to allow for early validation of the overall system through co-simulation. Figure 18 shows the methodology used to produce and to validate the initial specification for an automotive system including hardware, software and mechanical parts. The design starts with an analysis of the system requirements and a high level definition of the various functions of the system. At the top level, partitioning is done manually because we have no formal specification. The mechanical part is modeled in Matlab and the electronic part is modeled in SDL. At this stage, we obtain a multilanguage system level model given in SDL-Matlab.

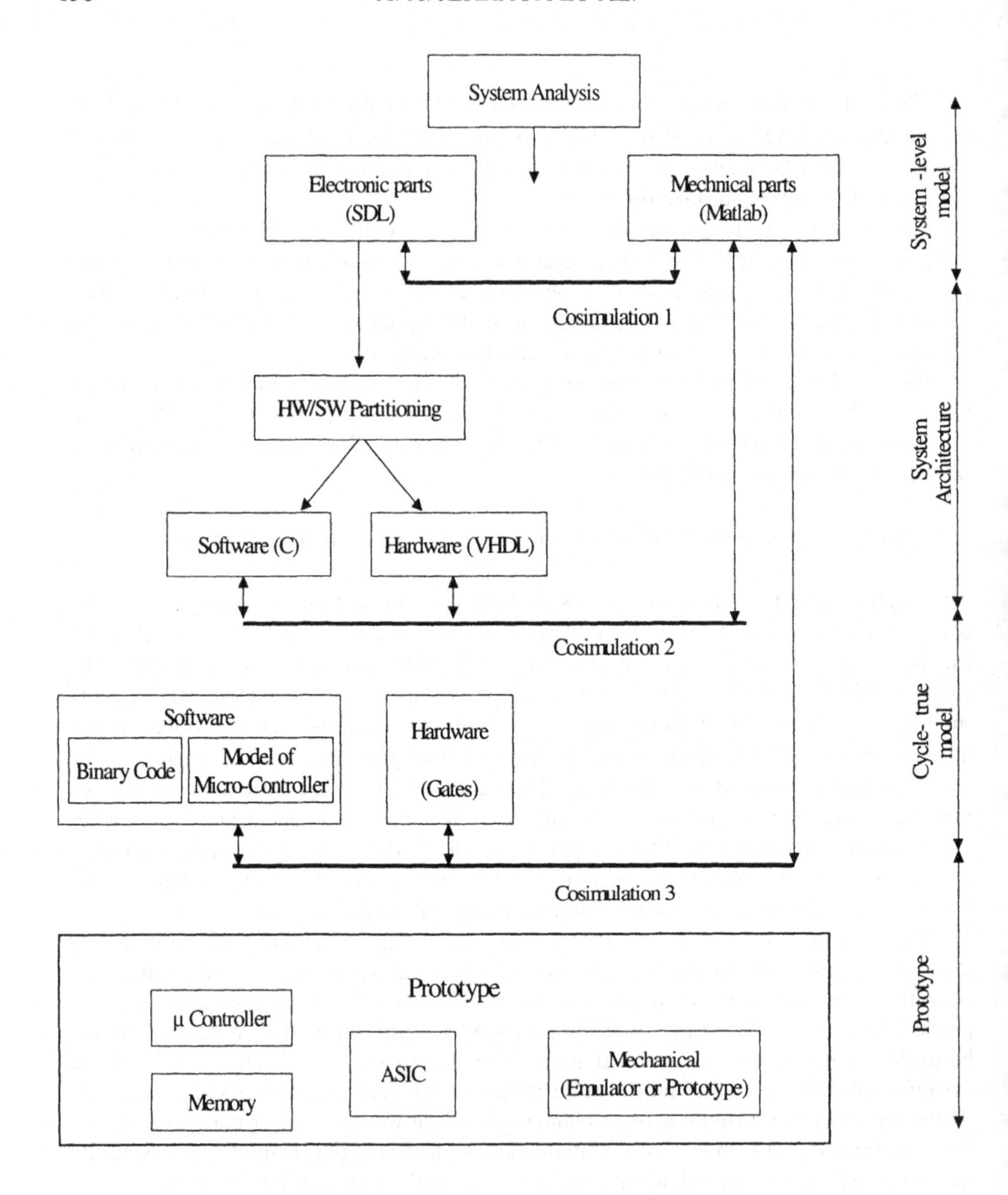

Figure 18 Multilanguage design flow for Automotive

The design flow handles three abstraction levels : the system-level, the system architecture level and the RTL or cycle accurate level.

At the system level, communication is described at the application level. The validation of the overall system may be done using system level co-simulation of SDL-Matlab models. This model may be used as a mock-up of the system in order to fix the final specification. The electronic module needs to be partitioned into hardware and software. This may be performed automatically using COSMOS-like tools [66]. This step produces a mixed hardware/software model of the electronic part.

At this stage, we obtain a system architecture Model. Hardware is modeled as behavioral VHDL, software is modeled as C-programs, Hardware/Software communication is performed through generic wires and the mechanical part remains as a Matlab model.

Cosimulation may be achieved in order to validate the partitioning and the communication protocols. Some timing verification may be achieved at this level.

At the cycle-true level, hardware is refined as an RTL model. The software is executed on a Model of the final processor and the mechanical part may be kept in Matlab. At this level all interfaces are refined to the physical level. HW/SW interfaces should include drivers in the software and some hardware adaptors to link the processor to the application. The model of the processor may be an implementation model (e.g. a gate model or a synthesizable VHDL) or a high-level model (e.g. a C program). At this level, cosimulation may be used to check timing at the clock-cycle level.

The final step in this multi-language design flow is prototyping. At this level, a prototype of the electronic system needs to be built. The mechanical part may be emulated.

Of course, this model may be extended to other languages that may cover other domains. For instance, in automotive hydraulic parts may be simulated using SABER and integrated within the design flow of Figure 18. Another extension would be to combine SDL with a Dataflow-oriented model (e.g. SPW or COSSAP) for the specification of the electronic system. Such an extension may be needed in case the design includes both control and DSP functions.

The multilanguage approach is an already proven method with the emergence codesign approaches based on C-VHDL. It is expected that for the future, multilanguage design will develop and cover multi-system specification languages allowing for the design of large heterogeneous designs. In this case, different languages will be used for the specification of the different subsystems.

ACKNOWLEDGEMENTS

This work was supported by France-Telecom/CNET, ESPRIT-OMI program under project CODAC 24139, SGS-Thomson, Aerospatiale, PSA, ESPRIT program under project COMITY 23015 and MEDEA program under Project SMT AT-403.

6. Conclusion

System-level modeling is an important aspect in the evolution of the emerging system-level design methods and tools. This includes system-level specification languages which target designers and internal format which are used by tools.

Most hardware/software codesign tools make use of an internal representation for the refinement of the input specification and architectures. An intermediate form may be language-oriented or architecture-oriented. The first kind makes easier high-level transformations and the latter makes easier handling sophisticated architectures.

The intermediate form fixes the underlying design model of the codesign process. When the target model is a multi-processor architecture, the intermediate model needs to handle multithread computation.

A plethora of specification languages exists. These come from different research areas including VLSI design, protocols, reactive systems design, programming languages, parallel programming, functional programming and algebraic notations. Each of these languages excels within a restricted application domains. For a given usage, languages may be compared according to their ability to express behaviors (expressive power), their suitability for building tools (analytical power) and their availability.

Experiments with system specification languages have shown that there is not a unique universal specification language to support the whole design process for all kinds of applications. The design of heterogeneous systems may require the combination of several specification languages for the design of different parts of the system. The key issues in this case are multilanguage validation, cosimulation and interfacing (simulator coupling).

References

1. V. Mooney T. Sakamoto and G. De Micheli. Run-time scheduler synthesis for hardware-software systems and application to robot control design. In IEEE, editor, *Proceedings of the CHDL'97*, pages 95--99, March 1997.
2. J.-M. Daveau, G. Fernandes Marchioro, and A.A. Jerraya. Vhdl generation from sdl specification. In Carlos D. Kloos and Eduard Cerny, editors, *Proceedings of CHDL*, pages 182--201. IFIP, Chapman-Hall, April 1997.
3. C. Delgado Kloos and Marin Lopez et all. From lotos to vhdl. *Current Issues in Electronic Modelling*, 3, September 1995.
4. D. HermanN., J. Henkel, and R. Ernst. An approach to the adaptation of estimated cost parameters in the cosyma system. In Proc. *Third Int'l Wshp on Hardware/Software Codesign Codes/CASHE*, pages 100--107. Grenoble, IEEE CS Press, September 1994.
5. J.Henkel and R. Ernst. A path-based technique for estimating hardware runtime in hw/sw- cosynthesis. In *8th Intl. Symposium on System Synthesis (SSS)*, pages 116--121. Cannes, France, September 13-15 1995.
6. W. Wolf. Hardware-software codesign of embedded systems. *Proceedings of IEEE*, 27(1):42--47, January 1994.
7. L. Lavagno, A. Sangiovani-Vincentelli, and H. Hsieh. *Embedded System Codesign: Synthesis and Verification*, pages 213--242. Kluwer Academic Publishers, Boston, MA, 1996.
8. G. Berry and L. Cosserat. *The esterel synchronous programming language and its mathematical semantics. language for synthesis*, Ecole Nationale Superieure de Mines de Paris, 1984.
9. G. Berry. Hardware implementation of pure esterel. In *Proceedings of the ACM Workshop on Formal Methods in VLSI Design*, January 1991.

10. D. Gajski, F. Vahid, S. Narayan, and J. Gong. *Specification and Design of Embedded Systems*. Prentice Hall, 1994.
11. D. Gajski, F. Vahid, and S. Narayan. System-design methodology: Executable-specification refinement. In *Proc. European Design and Test Conference (EDAC-ETC-EuroASIC)*, pages 458--463. Paris, France, IEEE CS Press, February 1994.
12. S. Narayan, F. Vahid, and D. Gajski. System specification and synthesis with the speccharts language. In *Proc. Int'l Conf. on Computer-Aided Design (ICCAD)*, pages 226--269. IEEE CS Press, November 1991.
13. F. Vahid and S. Narayan. Speccharts: A language for system-level synthesis. In *Proceedings of CHDL*, pages 145--154, April 1991.
14. F. Vahid, S. Narayan, and D. Gajski. Speccharts: A vhdl front-end for embedded systems. *IEEE trans. on CAD of Integrated Circuits and Systems*, 14(6):694--706, 1995.
15. ISO, IS 8807. *LOTOS a formal description technique based on the temporal ordering of observational behavior*, February 1989.
16. K.Van Rompaey, D. Verkest, I. Bolsens, and H. De Man. Coware - a design environment for heteregeneous hardware/software systems. In *Proceedings of the European Design Automation Conference*. Geneve, September 1996.
17. R. Klein. Miami: *A hardware software co-simulation environment. In Proceedings of RSP'96*, pages 173--177. IEEE CS Press, 1996.
18. I. Bolsens, B. Lin, K. Van Rompaey, S. Vercauteren, and D. Verkest. Co-design of dsp systems. In *NATO ASI Hardware/Software Co-Design*. Tremezzo, Italy, June 1995.
19. G.R. Andrews. *Concurrent Programming, Principles and Practice*. Redwood City,, California, 1991.
20. G.R. Andrews and F.B. Schneider. Concepts and notation for concurrent programming. *Computing Survey*, 15(1), March 1983.
21. G.V. Bochmann. Specification languages for communication protocols. In *Proceedings of the Conference on Hardware Description Languages CHDL'93*, pages 365--382, 1993.
22. D.C. Ku and G. DeMicheli. Relative scheduling under timing constraints. *IEEE trans. on CAD*, May 1992.
23. R. Camposano and R.M. Tablet. Design representation for the synthesis of behavioural VHDL models. In *Proceedings of CHDL*, May 1989.
24. L. Stok. *Architectural Synthesis and Optimization of Digital Systems*. PhD thesis, Eindhoven University of Technology, 1991.
25. G.G Jong. *Generalized Data Flow Graphs, Theory and Applications*. PhD thesis, Eindhoven University of Technology, 1993.
26. J.S. Lis and D.D. Gajski. Synthesis from VHDL. In *Proceedings of the International Conference on Computer-Aided Design*, pages 378--381, October 1988.
27. D. Gajski, N. Dutt, A. Wu, and Y. Lin. *High-Level Synthesis : Introduction to Chip and System Design*. Kluwer Academic Publishers, Boston, Massachusetts, 1992.
28. A.A. Jerraya, H. Ding, P. Kission, and M. Rahmouni. *Behavioral Synthesis and Component Reuse with VHDL*. Kluwer Academic Publishers, 1997.

29. R.K. Gupta, C.N. Coelho Jr., and G. DeMicheli. Synthesis and simulation of digital systems containing interacting hardware and software components. In *Proceedings of the 29th Design Automation Conference (DAC)*, pages 225--230. IEEE CS Press, 1992. References 37
30. A.H. Timmer and J.A.G. Jess. Exact Scheduling Strategies based on Bipartie Graph Matching. In *Proceedings of the European Conference on Design Automation*, Paris, France, Mars 1995.
31. A. Kalavade and E.A. Lee. A hardware/software codesign methodology for dsp applications. *IEEE Design and Test of Computers*, 10(3):16--28, September 1993.
32. A.A. Jerraya and K. O'Brien. Solar: An intermediate format for system-level design and specification. In *IFIP Inter. Workshop on Hardware/software Co-Design*, Grassau, Allemagne, May 1992.
33. Alan Davis. *Software Requirements: Analysis and Specification*. Elsevier Publisher, NY, 1995.
34. J.R. Duley and D.L. Dietmeyer. A digital system design language (ddl). *IEEE trans. on Computers*, C-24(2), 1975.
35. M.R. Barbacci. Instruction set processor specifiocation (isps): The notation and its applications. *IEEE trans. on Computer*, c30(1):24--40, january 1981.
36. P. Piloty al. Conlan report. In Springer Verlag, editor, *Lecture notes in Computer Science 151*, Berlin, 1983.
37. D.C. Ku and G. DeMicheli. Hardware C - a language for hardware design. Technical Report N! CSL-TR-88-362, *Computer Systems Lab*, Stanford University, August 1988.
38. Institute of Electrical and Electronics Engineers. *IEEE Standard VHDL Language Reference Manual*, Std 1076-1993. IEEE, 1993.
39. G.J. Holzmann. *Design and Validation of Computer Protocols*. Prentice-Hall, Englewood Cliffs, N.J, 1991.
40. *Computer Networks and ISDN Systems*. CCITT SDL, 1987.
41. S. Budowski and P. Dembinski. *An introduction to estelle: A specification language for distributed systems*. Computer Networks and ISDN Systems, 13(2):2--23, 1987.
42. *International Standard, ESTELLE* (Formal description technique based on an extended state transition model), 1987.
43. N. Halbwachs, F. Lagnier, and C. Ratel. Programming and verifying real-time systems by means of the synchronous data-flow lustre. *IEEE Tranactions on Software Enginering*, 18(9), September 1992.
44. T. Gautier and P. Le Guernic. *Signal, a declarative language for synchronous programming of real-time systems*. Computer Science, Formal Languages and Computer Architectures, 274, 1987.
45. J. L. Peterson. *Petri Net Theory and the modeling of Systems*. Prentice-Hall, Englewood Cliffs, N.J., 1981.
46. D. Harel. Statecharts : *A visual formalism for complex systems*. Science of Computer Programming, 8:231--274, 1987.
47. R. Milner. *Calculi for synchrony and asynchrony*. Theoritical Computer Science, 23:267--310, 1983.

48. E. Barros, W. Rosentiel, and X. Xiong. A method for partitioning unity language in hardware and software. In *Proc. European Design Automation Conference (EuroDAC)*, pages 220--225. IEEE CS Press, September 1994.
49. C.B. Jones. *Systematic software development using vdm*. C.A.R Hoare Series, Prentice Hall International Series in Computer Science, 1990.
50. J.M. Spivey. An introduction to z formal specifications. *Software Engineering Journal*, pages 40--50, January 1989.
51. J. R. Abrial. *The B Book. Assigning Programs to Meaning*. Cambridge University Press, 1997.
52. J. Bruijnin. *Evaluation and integration of specification languages*. Computer Networks and ISDN Systems, 13(2):75--89, 1987.
53. CCITT. *CHILL Language Definition* - Recommendation Z.200. May 1984.
54. M. Romdhani, R.P. Hautbois, A. Jeffroy, P. de Chazelles, and A.A. Jerraya. Evaluation and composition of specification languages, an industrial point of view. In *Proc. IFIP Conf. Hardware Description Languages (CHDL)*, pages 519--523, September 1995.
55. P. Zave and M. Jackson. Conjunction as composition. *ACM trans. On Software engineering and methodology*, 8(2):379--411, October 1993.
56. D.S. Wile. *Integrating syntaxes and their associated semantics*. Technical Report RR-92-297, Univ. Southern California, 1992.
57. S.P. Reis. Working in the garden environment form conceptual programming. *IEEE Software*, 4:16--27, November 1987.
58. C.A. Valderrama, A. Changuel, P.V. Raghavan, M. Abid, T. Ben Ismail, and A.A. Jerraya. A unified model for co-simulation and co-synthesis of mixed hardware/software systems. In *Proc. European Design and Test Conference (EDAC-ETC-EuroASIC)*. IEEE CS Press, March 1995.
59. W.M. Loucks, B.J. Doray, and D.G. Agnew. Experiences in real time hardware-software cosimulation. In *Proc. VHDL Int'l Users Forum (VIUF)*, pages 47--57, April 1993.
60. C.A. Valderrama, F. Nacabal, P. Paulin, and A.A. Jerraya. Automatic generation of interfaces for distributed c-vhdl cosimulation of embedded systems: an industrial experience. In *Proccedings of the 7th IEEE workshop on Rapid Systems Prototyping*. IEEE CS Press, June 1996.
61. D.E. Thomas, J.K. Adams, and H. Schmit. Model and methodology for hardware-software codesign. *IEEE Design and Test of Computers*, 10(3):6--15, September 1993.
62. D. Filo, D.C. Ku, C.N. Coelho, and G. DeMicheli. Interface optimisation for concurrent systems under timing constraints. *IEEE Trans. on VLSI Systems*, 1:268--281, September 1993.
63. P. Paulin, C. Liem, T. May, and S. Sutarwala. DSP design tool requirements for embedded systems: A telecommunications industrial perspective. *Journal of VLSI Signal Processing (Special issue on synthesis for real-time DSP)*, 1994.
64. N.L. Rethman and P.A. Wilsey. Rapid: A tool for hardware/software tradeoff analysis. In *Proc. IFIP Conf. Hardware Description Languages (CHDL)*. Elsevier Science, April 1993.

65. A.A. Jerraya al., *Languages for System-Level Specification and Design* in Hardware/Software Codesign, Ed. by J. Staunstrup & W. Wolf, Kluwer 1997.
66. C. Valderrama al., COSMOS : *A Transformational Codesign Tool for Multiprocessor Architectures in Hardware/Software Codesign*, Ed. by J. Staustrup & W. Wolf, Kluwer 1997.
67. MATLAB® 5 / SIMULINK® 2 : *Mathworks Inc.* - http://www.mathworks.com
68. Synopsys. 1997 (Jan.). *COSSAP (*Reference Manuals). Synopsys
69. *Advanced Design System.[ADS], HP 1998* http://www.tmo.hp.com/tmo/hpeesof/products/ads/adsoview.html
70. Jan Madsen et al., *Hardware/Software Partitioning using the Lycos System* in Hardware/Software Codesign, Ed. by J. Staunstrup & W. Wolf, Kluwer 1997.
71. P. Mariatos, A. N. Birbas, M. K. Birbas, I. Karathanasis, M. Jadoul, K. Verschaeve, J.L. Roux, D. Sinclair. *Integrated System Design with an Object-Oriented Methodology*, Volume 7. Object-Oriented modelling, J. M. Berge, Oz Levia, J. Rouillard editors, CIEM: Curent Issues in Electronic Modelling, Kluwer Academic Publishers, September 1996.
72. H. Karathanasis, D. Karkas, D. Metafas, DECT Source Algorithmic Components, d0.1s1 *AS PIS ESPRIT project* (EP20287) http://www.ratianal.com/uml/index.shtml
73. *MATRIXx,* Integrated Systems, Inc. 1998. http://www.Isi.com/Products/MATRIXx/.
74. WOLFRAM. 1998. *Mathematica.* http://www.wolfram.com/mathematica/.
75. Saber : *Saber Technology Ltd.* - http://www.saber.bm
76. Clifford Liem. *Retargetable Compilers for Embedded Core Processors : Methods and Experiences in Industrial Applications.* Kluwer Academic Publishers. 1997
77. A. Rault, Y. Bezard, A. Coustre, and T. Halconruy. *Systems integration in the car industry.* PSA, Peugeot-Citroen, 2 route de Gizy, 78140 Velizy Villacoublay, France, 1996.
78. Ph. Le Marrec al. HW/SW and mechanical cosimulation for Automotive Applications. *RSP'98*, Belgium, 1998.

HIGH-LEVEL SPECIFICATION LANGUAGES FOR EMBEDDED SYSTEM DESIGN

CARLOS DELGADO KLOOS[1], SIMON PICKIN[1],
LUIS SANCHEZ[1], ANGEL GROBA[2]
[1] *Área de Ingeniería Telemática, Dpto. Tecnol. Comunicaciones, Universidad Carlos III de Madrid, Av. Universidad 30, E-28913 Leganés (Madrid/Spain)*
[2] *EUIT Telecomunicación, Universidad Politécnica de Madrid, Carr. Valencia, km. 7, E-28031 Madrid (Spain)*

Abstract:

In this chapter we show how to use three complementary languages, *SDL*, *Statecharts* and the language of $MATRIX_X$, for the modelling and design of embedded systems. We base our approach on the fact that there is no perfect language for the design of embedded systems, and that one therefore has to find ways in which several formalisms, each with its own strength, can be used in conjunction. We present three ways of achieving this.

1. Introduction

Over the last few years, the level of abstraction used in the specification and the design phases of the embedded system design process has risen, the need for this increased abstraction arising from the growing complexity of such systems. There are a number of different languages for high-level description coming from different backgrounds and cultures. Each of these is better suited to a particular use and has its specific pros and cons. During the design process of a complex embedded system, one has to describe heterogeneous parts of subsystems. The heterogeneity stems from several orthogonal aspects: discrete/continuous signals, software/hardware implementation, specification/implementation, etc. Is there one optimal language suited for all purposes? Even if there were one from a technical point of view, there are other aspects to consider that make it difficult to find the perfect system specification language. Some of these are: availability of

A. A. Jerraya and J. Mermet (eds.), System-Level Synthesis, 137–174.

mature tools, adoption by a standardisation body, interoperability with other languages and tools, existence of libraries, the existence of previously designed versions of one particular system, personal preferences and tastes, etc.

In the search for a candidate general-purpose embedded system specification language, one that may at first sight seem ideal in many respects is *Java* [2], a well-designed language based cleanly on the object-oriented programming paradigm that has proved its usefulness in software development. Seldom in the history of computing has one seen a phenomenon like the "*Java* craze". In a very short period of time, *Java* has attracted an immense number of designers, many APIs have been developed for diverse purposes, tens of development environments have been produced and hundreds of books have been written. There are, nevertheless, several features needed for embedded system design which are not present in *Java*, such as analogue signals, the ability to specify in an abstract way, etc.

We therefore take a sceptical view with respect to the availability of such a general-purpose specification language at the present time. In this paper, we propose the use of not one language but several languages together when designing an embedded system. We have chosen three languages, namely, *SDL* [14] (oriented towards the description of communicating systems), *Statecharts* [4] (oriented towards the description of control-dominated systems) and the language of *MATRIX*$_X$ [7] (oriented towards the description of data-dominated, analogue systems). In this document, *MATRIX*$_X$ is often used to describe the language used in the *SystemBuild* tool of the *MATRIX*$_X$ toolset. Similarly, use of *Statecharts*, unless clear from the context, is to be understood to indicate joint use of the languages *Statecharts*, *Activity Charts* and *Module Charts* of the *Statemate* toolset.

One can envisage several levels of integration when using these specification formalisms which we will refer to as:

- Independent use
- Coordinated use
- Integrated use

In an *independent* use, the parts of the system specified in the different formalisms are only combined at the level of target code (e.g. C or VHDL).

In a *co-ordinated* use, the goal is to provide a means to interconnect subsystems specified in the different formalisms with the aim of performing specification-level cosimulation. For such a use, then, we restrict ourselves to co-ordination at the specification level. We are not concerned with the development process for the global system, so that the development of the individual specifications in the different formalisms and the generation of code from them is done independently.

In an *integrated* use, on the other hand, the aim is to define a complete multi-formalism development process. In particular, we are concerned with the development of the interconnection framework, right down to the code level, alongside that of the subsystems in the different formalisms.

In this chapter, we will describe these different ways of using the three languages and explain what will be needed to achieve each of these combined uses. The chapter is organised as follows: Section 2 sets the terminology for the rest of the chapter and introduces the problem to be handled. Section 3 presents the three main languages considered – *SDL*, *Statecharts* and $MATRIX_X$ – as well as other possible alternatives. Section 4 shows the three ways in which these languages could be used in a combined way for the design of embedded systems.

2. Overview

Let us start by having a critical look at the keywords that appear in the title of this chapter. We do this in order to clarify the terms, since some are used in different contexts with a different meaning.

2.1 WHAT IS A SPECIFICATION?

The principal purpose of a specification is to succinctly and clearly define what are considered to be the most important aspects of the behaviour and properties of a system, with the aim of checking that the initial requirements are well-defined, unambiguous and consistent. If the specification contains details superfluous to the task of checking the initial requirements, this merely makes this task more difficult. In order to fit smoothly into the development process, a specification should also lend itself well to the subsequent parameterisable derivation of an implementation, preferably in an automatic manner. It is via the parameterisation that the properties and aspects of behaviour that are considered less important are introduced. A specification should also lend itself well to the derivation of tests for checking the conformance of implementations, preferably making manifest the degree to which the generated implementations will be "testable".

A specification describes a set of acceptable implementations, either by simply giving a set of restrictions or by providing some explicit abstract model. This may take many different forms, e.g. that of a description of the output as a function of the input, that of a set of pre-conditions, post-conditions and invariants, or that of a high-level description of dynamic behaviour and data structure. The fact that many different implementations can be derived from a single specification, reflects the fact that there are many possible implementations which effect a given system functionality.

A specification language is therefore a language which is well-suited to expressing the basic system properties and basic aspects of system behaviour in a succinct and clear manner, and which lends itself well to the, preferably automatic, checking of requirements, generation of implementations and derivation of conformance test suites. Clearly, not all requirements on a system can be adequately expressed in a given specification language and, in addition, the aspects which are considered basic are context-dependent. Here we call requirements which are expressible in the semantics of the specification language, functional requirements (relative to that language), and those which are not, non-functional requirements. Concerning the choice of specification language for a given problem, the notation used for a description of the functionality of a given system must evidently contain the appropriate language constructs for expressing the required concepts.

Whenever possible, the specification language used to describe the functionality of the system should also have a formal semantics. The word "formal" here does not refer to grammar (i.e. syntax) but rather to having a mathematical or logic-based semantics. The advantage of using a language with a semantics that is completely independent of assumptions about the compiler and the run-time environment is that specifications in it are unambiguous and amenable to reasoning. If the formal semantics is also operational, then the specifications become amenable to fully automatic reasoning.

From the above definitions, it can be seen that, to some extent, the difference between a specification language and an implementation language is relative. It could be observed that, currently, the notion of specification language in the hardware world is closer to the notion of implementation language in the software world. However, the increasing complexity of modern hardware means that the need for notations with higher-levels of abstraction is increasingly being felt.

2.2 WHAT DOES HIGH-LEVEL MEAN?

High and low are relative concepts. The same level can be high or low depending on what you are comparing to. This is actually what happens when referring to a programming language. To the software community, the programming level is not the highest level. Algebraic and logic levels above, which often comprise non-executable specifications, are considered high. In the hardware community, on the contrary, an executable program, since it specifies a behaviour free of layout and other details, is high enough as a starting point.

When talking about system design, and thus comprising both software and hardware, one has to define a common standpoint. On a general setting we prefer to take a combination of both views, i.e. start the specification at a non-algorithmic level, then take design decision towards the algorithmic implementation of it, and then continue introducing further requirements. In this text, however, we will concentrate at the algorithmic level for the specification of systems and consider this to be our high level.

2.3 WHAT IS AN EMBEDDED SYSTEM?

A simple definition of an embedded system is as follows: "an electronic system that controls the operation of another system". There is a wide range of systems which can be controlled by an embedded system, some examples being a washing machine, an aeroplane or a mobile phone.

Most embedded systems have several points in common. First, the fact that an embedded system interacts with another system and not (or not much) with a user, means that a powerful PC-style user-interface is not needed. However, often there is a small user interface (e.g. the buttons of the mobile phone).

Another aspect common to all embedded systems (except those which control an electronic device) is the use of sensors and actuators, which constitute the interface between the embedded system and the controlled system. Sensors provide information about the state of the controlled system while actuators allow the behaviour of the controlled system to be modified.

As embedded systems have to react to the inputs they receive from the controlled system, they have to meet certain timing requirements. These may be hard real-time requirements, e.g. in the control of critical systems such as an aeroplane or hospital equipment, or soft real-time requirements, in the case of less critical systems, such as a TV remote control.

Many embedded systems are part of products that are sold in large series in competitive markets. In such cases, another requirement is that the costs of the embedded system be as low as possible.

When the embedded system is implemented completely or partly in a processor, it must include an Operating System (OS). OSs for embedded systems are not full-blown OSs, generally being much smaller than a PC OS. For example, they do not need to provide a command shell for interaction with a user. However, they may need specialised process-control algorithms for real-time treatment.

2.4 WHAT IS NEEDED FOR EMBEDDED SYSTEM DESIGN?

In the previous section, features common to all embedded systems were discussed. However, depending on factors such as the application domain, these systems have widely differing properties and, therefore, specification and design requirements. One consequence of this is that no single language is suitable for specifying all applications. We discuss issues influencing the decision about which specification language to use in the following section.

For a complex embedded system, separate parts of the same system may have significantly different specification requirements. In such cases, we advocate the use of a multi-formalism approach to system design, as discussed in the introduction to this chapter.

There are two main considerations concerning the use a multi-formalism design scheme. First, we need to be able to validate the system description in order to check that any implementation will have the desired functionality and to evaluate its non-functional parameters (performance and costs). An important mechanism for validation in practically all design phases is simulation. In the context of multi-formalism design, therefore, specification level cosimulation would be extremely useful.

Second, efficient code generators to obtain implementation code (for instance *C* or *VHDL* [9]) from our high-level specifications would also be of great advantage. How to interface the code obtained from the different specifications will then also have to be defined.

However, if the available code-generators are insufficiently complete or produce code of unsatisfactory quality, they may still be useful for prototyping, and the high-level specification languages are still useful in manual coding in that they provide a clean and "validatable" definition of what the system has to do.

Of course, there are also some issues related with how to specify using a combination of notations or about the level of coupling between the notations. These topics are treated in other parts of this chapter.

2.5 WHICH LANGUAGE FOR EMBEDDED SYSTEM DESIGN?

There are several aspects that influence the choice of the specification language to be used for an embedded system:

Software vs. hardware. It is usually possible to choose to implement an embedded system in software, running on a commercial or in-house processor, or in dedicated hardware. Software solutions are flexible and cheap but slower than hardware solutions. To take advantage of this hardware/software implementation trade-off, a mixed solution, including both software and dedicated hardware, can be used. Hardware-software codesign [18] attempts to obtain optimal hardware-software implementations of embedded systems. Research in this area has been particularly active in the last few years. Among the vast set of languages currently used for embedded system specification are software-oriented languages (*C*, *Java*), hardware-oriented languages (*VHDL*, *Verilog* [10]) and implementation-independent languages (*SDL*, *Statecharts*, *MATRIX*$_X$).

Continuous vs. discrete. Embedded systems are fundamentally discrete in nature though there are two kinds of situation in which we may need to model continuous components. The first is where we wish to model a continuous part of the environment of an embedded system, e.g. a control system for car brakes. The second, less-common situation, is where part of the embedded system itself is continuous in nature, e.g. a modem. There are currently many different languages available for the description of continuous systems. They can be divided into two categories depending on their origin. Some of them are part of mathematical packages, examples of these being *Matlab* and *MATRIX*$_X$, while others are extensions to digital hardware description languages, like *VHDL-AMS* and *Verilog-AMS*.

Control-oriented vs. data-oriented. Though any given application can be modelled in a more data-oriented or more control-oriented fashion, some applications lend themselves more to one than the other. Roughly-speaking, an application is control-oriented if the main aspects of its behaviour can be defined in terms of transitions between a relatively small set of states, e.g. a control system for a lift, and data-oriented, if this is not the case, e.g. image processing. An example of a language suitable for control-oriented applications is *Statecharts*, while *MATRIX*$_X$, *Java* or *SILAGE* [6] are examples of languages suitable for data oriented applications.

Centralised vs. distributed. Distribution may arise simply for performance reasons (load sharing, etc.) or due to the nature of the application. Though any common system description language should be capable of describing distributed and centralised components, some are better suited to centralised systems (e.g. *Statecharts*) and some to distributed ones (e.g. *SDL*).

Reuse. Parts of the system being designed may already be described in a given formalism (perhaps as part of a previous design). In this case, it seems reasonable to use the available description to avoid "reinventing the wheel"

Industrial use. Of course, the know-how available in industries is a major factor in deciding which specification language to use.

These different factors may often lead to contradicting recommendations for the specification language to be used. The compromise of using several of them is then often the only advisable solution to the dilemma.

3. Specification languages

In this section we present the three languages of our choice, all quite popular languages in their respective use domains, and in particular, in the industrial context, before discussing briefly the role of *Java*.

3.1 *SDL*

SDL (Specification and Description Language) [14] is a formal specification language based on communicating extended finite state machines. It includes structuring information in the form of *SDL* blocks as well as in the form of procedures, services, macros, channel substructures, signal refinement, etc. (not all of these being supported by the toolsets). It is widely used in the telecommunications field not only to describe telecommunications protocols and services, but also to specify other types of control-oriented distributed systems in which the communication aspect plays an important part. The ITU-T SDL standard is revised every four years.

The origins of *SDL* lie in informal state-transition diagrams, whose user-friendliness is preserved in the standardised graphical syntax. This syntax complements the equivalent standardised textual syntax and is to some extent responsible for the widespread (and growing) acceptance of *SDL* compared to, for example, *Estelle* [11] or *LOTOS* [12]. The importance of this graphical syntax to *SDL*'s success can be judged from the fact that a layout syntax has now also been standardised [16] with a view to ensuring that the detailed graphical presentation of *SDL* specifications is preserved between different tools.

The 1992 version of the *SDL* standard introduced object-oriented notions such as system types, block types and process types, inheritance, virtual definitions, etc. These features have only relatively recently been incorporated into the main *SDL* toolsets, *ObjectGEODE* from *Verilog* and *SDT* from *Telelogic* y the changes introduced in 1996 have still not been completely taken into account. Also introduced in 1992 was the concept of the block, process and service gates, made necessary by the introduction of block, process and service types respectively. *SDL*'96 then saw the intent to consolidate the use of these typed gates as interfaces.

Communication in *SDL* is by asynchronous signals sent via declared, FIFO[1] channels. Each process instance has a unique incoming-signal queue. Signals are either sent to a particular process instance identified by its `PId`[2] or via a particular channel (which may be implicit). In the latter case, if there are several possible destinations for a signal, one is chosen at random (since

[1] Actually, FIFO is destroyed by the use of the `save` construct or by timer signals.

[2] Note the predefined expressions of sort PId: self, parent, offspring and sender.

SDL-92). If a destination process instance does not exist, the signal is lost (also since *SDL*-92). If a signal is received in a state where it is not specified as a possible input (unspecified reception) it is simply discarded. Communication via blocking remote procedure calls (RPCs) within an *SDL* specification is also part of the *SDL*-92 language, such calls being converted into two signals and using the normal process instance queues. Since *SDL*-96, these signals can be declared in interface declarations as an alternative to declaring the importing of the remote procedure.

Communication via shared variables is also possible, though the import/export mechanism is simply a shorthand since it involves implicit signal exchange (since *SDL*-96, these implicit signals can be declared in interface declarations) and the other mechanism, the revealed variable is now discouraged[3].

The part of the *SDL* semantics concerned with the management of PIds and access to global time is referred to as the underlying *SDL* machine. This machine is assumed to perform the tasks of creating instances, assigning PIds to new instances and keeping a record of the active process instances and their addresses.

3.1.1 Data in SDL

Data in *SDL* may be specified in one of three ways: using the ADT-based *SDL* data-specification language, algorithmically, or in *ASN.1* [13], the last two being introduced in 1996.

Though the *SDL* data-description language is based on the formal ADT language *ACT ONE,* the addition of predefined sorts and generators together with variable assignments and type synonyms has meant that the tool makers have mostly bypassed formal ADTs. The main *SDL* tools also offer non-standard ways of specifying data directly in *C*, even though this may lead to certain restrictions on the simulation. Thus, most *SDL* users have little or no contact with formal ADT constructs and it appears that the intention is to remove ADTs altogether from the 2000 standard. Algorithmic data is defined on top of the ADT system using *SDL* decisions, etc., in a manner akin to a stateless procedure. The use of *ASN.1* with *SDL* is treated in a separate standard [15].

Arbitrary types can be used in signal parameters and arbitrary signals can be sent to the environment so that *SDL* allows for any predefined or user-defined datatype to figure in the interface with the environment (the latter must in this case be declared at the top level).

The *ObjectGEODE* simulator [23] allows *SDL* predefined types and *SDL* user-defined types to be exported to external *C* programs and *SDL* user-

[3] One of the aims of the modifications introduced in 1996 was to have clearly-defined interfaces and ensure that all exchange of information takes place via signals.

defined types to be imported from them. It only accepts simple kinds of user-defined ADT specifications.

3.1.2 *Time in* SDL

Time is modelled through the use of two predefined sorts: `time`, the absolute time, and `duration`, the time difference between two events. Literals of these sorts have the same syntax as those of the sort `real`. The pre-defined expression `now` of sort time is always available to any process and the time that it measures is supposed global. Apart from the use of the `now` construct, in tests for example, the main way in which time influences behaviour in *SDL* is through the use of timers. Each timer belongs to a particular process instance. When a timer expires, it sends a signal to its owning process instance. In the current standard, this signal is placed in the usual process instance queue and therefore has no priority over other signals.

The timing model used in simulation [23] is in fact different to that generally used in the generated code and restricts the timing semantics of the ITU-T standard. In simulation, time is discrete and transitions take zero time to execute; signals also spend zero time in transit, that is, between being sent and being placed in the input queue(s) of the destination process instance(s)[4]. Simulation time is advanced by the firing of an explicit time transition. This transition is only offered in a state if no other transitions are fireable and a timer is active, otherwise, time cannot be advanced.

Note that the lower priority of the time transition means that if the specification contains any zero-time loops, time can never advance. Construction of a "simulateable" model that does not have any zero-time loops can sometimes be rather difficult. In such cases, in order to simulate successfully with *ObjectGEODE* or with *Telelogic*'s *SDT*, it may be necessary to eliminate time from the specification altogether, particularly for exhaustive simulation. However, if the *SDL* specification models a reactive system but does not model its environment (that is, the system environment is not explicitly modelled as part of a closed *SDL* specification but rather implicitly modelled as the *SDL* environment of an open *SDL* specification), the case of most interest here, these difficulties should not be too frequent.

Another consequence of the timing model of the simulation is that if a delaying channel is required, it must be modelled explicitly as a process with a timeout. Care must be taken if the delay is to be random, so as not to lose the use of the *undo* facility.

[4] In the official semantics, *SDL*-92 channels may be specified as delaying or non-delaying, the former being the default, though note that this does not apply to channels between the specification and the environment which are always delaying.

3.1.3 The SDL *environment*

An *SDL* specification can communicate with its environment via signals. A specification making use of this possibility is said to be "open" (the “system environment” is modelled by the “*SDL* specification environment”), otherwise, it is said to be "closed". The standard assumes that the environment of an *SDL* specification behaves in an *SDL*-like manner and that environment process instances will have PIds distinct from any of those of the specification. The environment is generally supposed to be well-behaved in the sense that it does not send signals to non-existent process instances etc. Though it may sometimes be desirable to impose synchronisation constraints on the *SDL* environment[5], such constraints cannot be specified in an *SDL* specification other than as comments.

Process instances communicate with the *SDL* specification environment by sending signals to it by implicit addressing (along a channel leading to it) or explicit addressing (by using the predefined expression of type PId, sender), or by receiving signals from it. Treatment of SDL PId's is important in any attempt to communicate SDL specifications with other systems.

Remote procedure calls to the *SDL* specification environment were introduced in 1996, for facilitating the embedding of *SDL* specifications in other systems, notably object-based systems such as *CORBA* [19]. However, this recent introduction is not yet implemented by the two main toolsets.

3.1.4 Example

To illustrate the graphical syntax of *SDL*, in Fig. 1 we show part of an example specification: a distributed cashpoint, or ATM, system (figure constructed using postscript produced by the *ObjectGEODE* tool). The top of the figure shows the interconnection diagram of the system level, which comprises three blocks, *Netware*, *Central Manager* and *Cashpoints*. Note the communication channels between the blocks and the environment of the specification: the specification is open. The middle of the figure shows the interconnection diagram of the *Cashpoints* block, a lowest-level block containing the processes: *CardReader*, *Keyboard*, *Cashbox*, *Screen*, and *Printer*. The bottom of the figure shows the finite state machine diagram of the *Cashbox* process. The dashed lines have been drawn to make the hierarchy clear but are not part of the *SDL* graphical syntax. In the general case, there may be many levels of interconnection diagrams between the system level diagram and the lowest level block diagram.

[5] such as that it does not send signal A to the specification until it has received signal B from the specification.

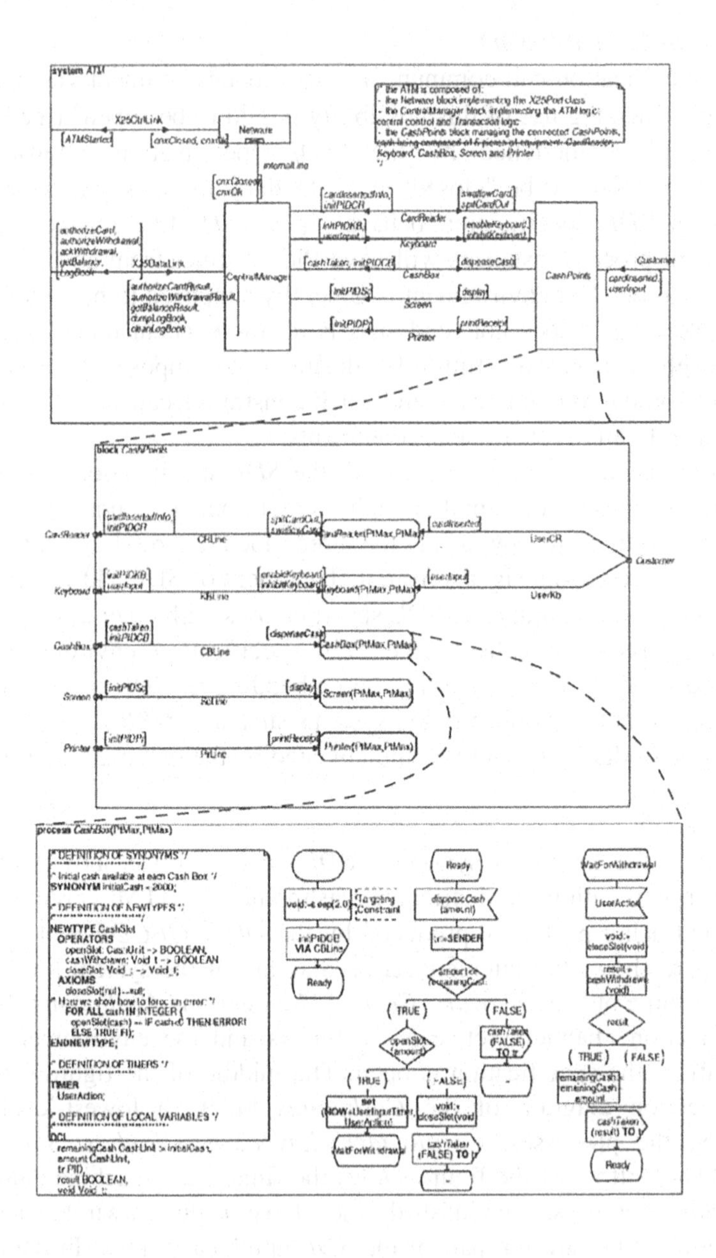

Figure 1. Illustration of *SDL* graphical syntax (courtesy of Verilog's *ObjectGEODE*). From top to bottom: system level (showing top-level blocks), lowest block level (showing processes) and process level (showing the f.s.m. implementing the process).

3.2 *STATEMATE* AND *STATECHARTS*

Statemate uses three graphical languages: *Activity-charts*, *Statecharts* and *Module-charts*. Additional non-graphical information related to the models themselves and their interconnections is provided in a Data Dictionary.

- *Activity-charts* can be viewed as multi-level data-flow diagrams. They describe functions, or activities, and data-stores, organised into hierarchies and connected via information flows.
- *Statecharts* are a generalisation of state-transition diagrams, allowing for multi-level states, decomposed in an and/or fashion, thus enabling economical specification of concurrency and encapsulation. They incorporate a broadcast communication mechanism, timeout and delay operators, and a means for specifying transitions that depend on the history of the system's behaviour.
- *Module-charts* are used to describe the implementation architecture of the system: its division into hardware and software blocks and the communication between them.

3.2.1 The Statecharts *language*

Finite state machines are a natural technique for describing dynamic behaviour and have an appealing visual representation in the form of state-transition diagrams: directed graphs, with nodes denoting states and arrows denoting transitions between them. The transitions can be labelled with the trigger events, which are dependent on a condition, and consequent actions in the form *trigger[condition]/action*. *Statecharts* ([3],[4],[5]) are enhanced state-transition diagrams with additional features such as state hierarchy (for structuring state diagrams, especially useful in large systems), orthogonality (the *and-state* notation for specifying concurrency), compound expressions and special connectors between states.

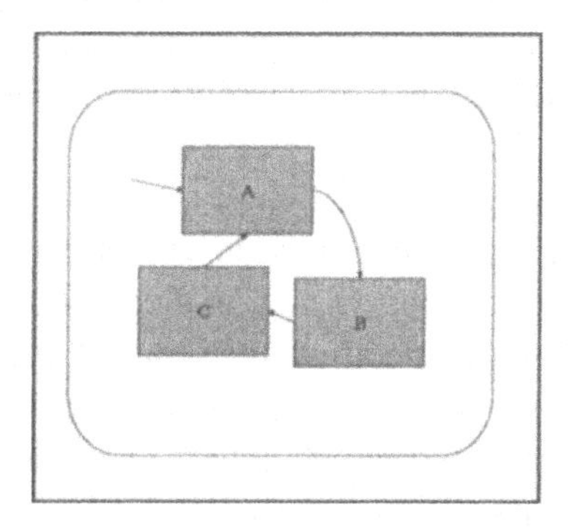

Figure 2. Example of an or-state

Being in an *or-state* means being in one, and only one, of its substates. If the system is in the *or-state* shown in Fig. 2, it is in one of the states A, B or C. Being in an *and-state* means begin in each and every one of its substates concurrently. After the transition from state INI to the *and-state* shown in Fig. 3, the system is in both state ABC and state XY. Of course, states ABC and XY may also be *and-states* or *or-states* (hierarchy).

The main information elements in *Statecharts* are the following:

- *Events* are communication signals indicating occurrences; they are very often used for synchronisation purposes. They do not convey any content or value, only the fact that they have occurred. If they are not sensed immediately they are lost.
- *Conditions* like events, are used for control purposes. However, they are persistent signals that hold for continuous time spans. They can be either true or false.
- *Data-items* are units of information that may assume values from one of the available data types. They maintain their values until these are explicitly changed and new values assigned.

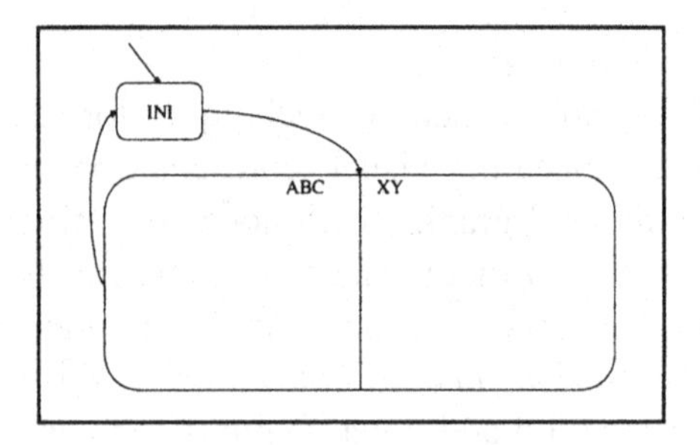

Figure 3. Example of an and-state

Transitions between states, as well as *static reactions* associated to states, such as arrival in, presence in or departure from a state, may cause *actions* - manipulations of the three types of elements named above - to occur.

3.2.2 Data in Statecharts

Statemate predefined data types include numeric types (Integer, Real, Bit, Bit-Array), alphanumeric types (String) and compound types and structures.

User-defined types are specified in terms of predefined types or other user-defined-types; this type mechanism can be used to define complex types with multiple-level structure. They may be used to define a new *subtype* for a set of characteristics shared by several data-items in a model.

3.2.3 Time in Statecharts

The input of a reactive system is a sequence of stimuli generated by its environment. The system senses these external changes and responds to them by moving from one state to another and/or performing some actions.

The system executes a step when it performs all relevant reactions whose triggers are enabled. As a result of a reaction, the system may change its states, generate internal events and modify values of internal data elements. These changes can cause derived events to occur and conditions to change their values, which may, in turn, enable other triggers that can cause other reactions to be executed in the next step.

In reactive systems, the notion of sequentiality and its relationship with time is very important. In *Statecharts* not only is temporal ordering important but time quantification must also be respected since timeout events and scheduled actions may be used. Time is measured in terms of some abstract time unit, common to a whole statechart. Two different time schemes are possible when dealing with model execution: synchronous and asynchronous. In both, time is assumed not to advance during the execution of a step, i.e. step duration is zero. The actual meaning of this assumption is that no external changes occur during execution of a step, and the time information needed for any timeout events and scheduled actions in a step is computed using a common clock value.

The synchronous, or step-dependent, time scheme assumes the system executes a single step every time unit. This time scheme is well suited to the modelling of digital electronic systems, where execution is synchronised with clock signals and external events occur between steps.

The asynchronous, or step-independent, time scheme, allows several steps to take place at a single point in time. There is no strict rule to be followed in order to implement this scheme, though an execution pattern that fits many real systems responds to external changes by executing the sequence of all steps these changes imply, in a chain-reaction manner, until it cannot advance without further external input and/or without advancing the clock. Once this situation is reached the system is ready to react to further external changes (c.f. the notion of "reasonable" environment).

3.2.4 Example

Fig. 4 presents an example of a model developed with *Statemate*: a simple clock that incorporates a chronometer.

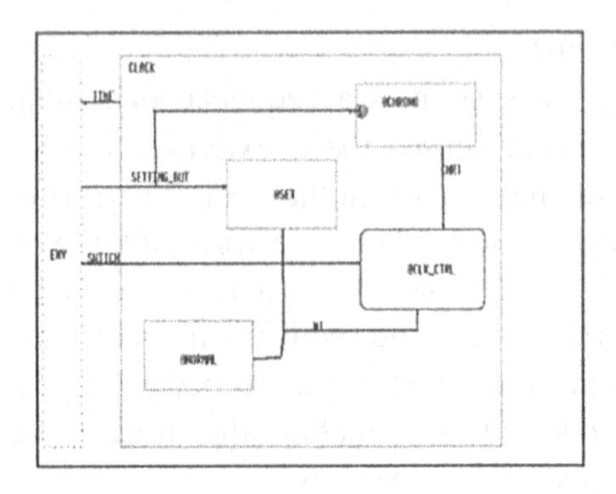

Figure 4. Clock

The model shown in Fig. 4 is composed of 3 *Activity Charts* plus a Statechart (CLK_CTRL, see Fig. 5) that controls which of the *Activity Charts* are active in each simulation step and a module representing the environment. The *Activity Charts* control different tasks that can be performed by the Clock:

- NORMAL: computes the current time,
- SET: changes the current time,
- CHRONO: the Clock works as a chronometer.

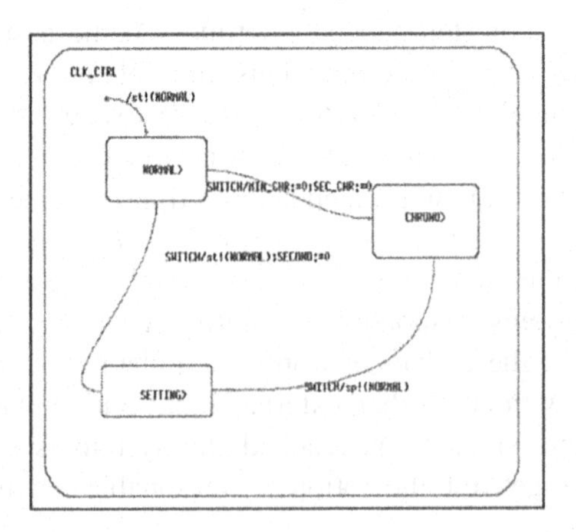

Figure 5: CLK_CTRL

The state in which CLK_CTRL rests in a given instant (NORMAL, CHRONO, SETTING) indicates which of the Activity Charts are active. The initial state is NORMAL. Whenever the SWITCH input is active the state changes, as shown in the figure. When the statechart enters the CHRONO state, the value of the minutes and seconds are initialised to 0 (in this state, no hour value is available). In this state, the NORMAL Activity Chart is also active to update the time.

3.3 *MATRIX$_X$/SYSTEMBUILD*

MATRIX$_X$ is a system analysis and design toolset which includes the *SystemBuild* tool [7], for the design of continuous and discrete systems.

SystemBuild is a graphical programming environment in which systems are described as hierarchies of blocks, the leaf blocks being taken from a predefined library. *SystemBuild* offers capabilities for editing, analysis and simulation of system descriptions. The blocks in a *SystemBuild* model include the SuperBlock (the basic hierarchy unit), Data Stores, algebraic blocks, piecewise linear blocks, dynamic blocks, including integrators, transfer functions and differential equations, trigonometric blocks, signal generator blocks, like pulse trains, ramps, etc. and logical blocks.

SuperBlocks can be of four different kinds. *Continuous SuperBlocks* are used to model continuous systems. *Discrete SuperBlocks* are used to model systems that sample and hold their inputs at a specified sample rate. *Trigger SuperBlocks* are executed each time the leading edge of the associated trigger signal is detected. *Procedure SuperBlocks* are special constructs designed to represent generated software procedures.

The datatypes supported in *SystemBuild* are: *floating point*, *integer*, *logical* and *fixed-point*. In each block there are four classes of elements that should have an associated datatype: inputs, outputs, states and parameters.

3.3.1 Time in SystemBuild

To perform a simulation with *SystemBuild*, the inputs to the model and the time vector, in which the initial and final simulation time and the time step (Δt) are indicated, need to be defined. The external input values must be defined for the entire simulation time at the beginning of the simulation. However, the tool provides several ways of defining input values during the simulation (by ensuring that they are not external). The first is through the use of Interactive Animation Icons. Another is through the use of *User Code Blocks* (UCB) whose outputs are defined by external C routines, and can serve as inputs for the rest of the model.

SystemBuild SuperBlocks can be classified into two basic categories according to whether they are discrete or continuous. The discrete parts of a specification are simulated by iterating the discrete state equations from the initial conditions until the specified final time.

The SuperBlocks of a discrete nature are classified according to how the time at which they are executed is calculated:

- *Discrete (free running)*: executed periodically at assigned Sample Rate.

- *Discrete (enabled)*: executed periodically at assigned Sample Rate if their parent SuperBlocks have enabled them (using their enable signal).
- *Trigger*: discrete in nature, executed once each time the leading edge of their trigger signal is detected rather than periodically. They are divided into 3 types in function of when the new values appear in their outputs:
  - *At Next Trigger*: The new outputs appear when the next trigger is given to the subsystem.
  - *At Timing Requirement*: The new outputs appear after a user-specified amount of time.
  - *As Soon As Finished*: The new outputs appear in the next scheduled simulation time after the SuperBlock finishes its computations.

The order in which the discrete SuperBlocks are scheduled is based on the principle of rate-monotonic scheduling; SuperBlocks with lower sample rate (or time requirement for *Trigger SuperBlocks*) have higher priority.

For the *Continuous SuperBlocks*, a numerical algorithm must be used to obtain a solution to the differential equations involved in the model. There are two parameters that can be defined by the user for the simulation of *Continuous SuperBlocks*. The first is the numerical algorithm itself. The default is the Variable-Step Kutta-Merson algorithm, though many other algorithms are also available. The second is the integration step length. This parameter is usually equal to the least common multiple of the sample times of the discrete SuperBlocks and the Δt in the input time vector. However, it is also possible to define an upper limit for the step length.

3.3.2 Example

As an example of a system that can be described with *MATRIX*$_X$ we use part of a Window lift developed by BMW AG and shown in Fig. 6. The inputs to the subsystem are two control signals that indicate if the window should be moved up or down. Note that these signals are not directly related to the state of the buttons. For instance, if the button that closes the window is pressed then the window starts to rise (the control line for move up is activated) but if it is pressed again, the window stops (the control line for move up is deactivated).

The main components of the model are the following: the component that computes the percentage during normal operation of the window lift (NormalMode), the component that computes the percentage before the initialisation condition (the first time the window is completely closed) (InitialMode) and the model of the position of the window (Position).

The inputs to NormalMode are the two control signals from the environment (mu for move up and md for move down) and the position of the window (pos). It outputs the percentage of the voltage that should be supplied to the motor. The InitialMode is like NormalMode, but it has also an input from Position that specifies if the window is blocked (it is either in the top or the bottom position). This is used to detect the first time that the window is completely closed, which starts NormalMode (by means of signal start). Position computes the position of the window from the percentage of the maximum voltage (Vin) supplied to the motor (perc).

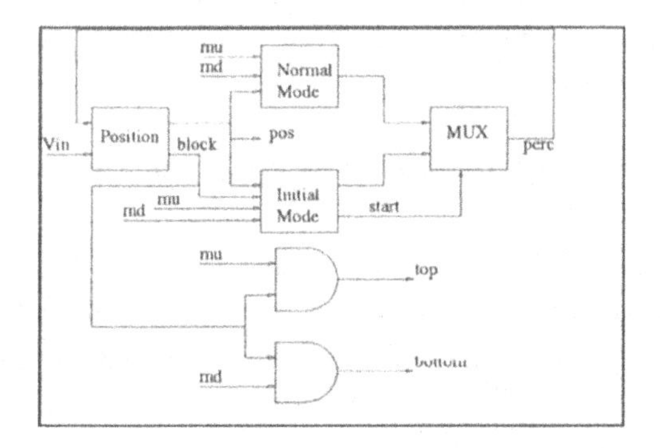

Figure 6. Window lift

The outputs from the $MATRIX_X$ model are the position (continuous signal) plus the top and bottom events, obtained from the signals block, mu and md.

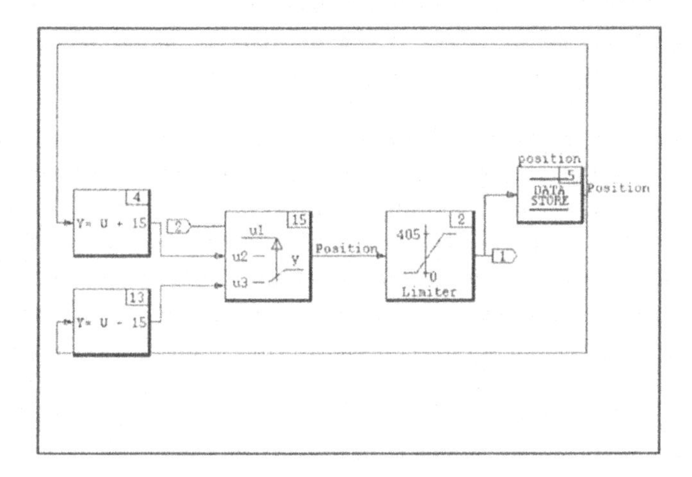

Figure 7. Example of description with $MATRIX_X$

Fig. 7 shows the detail of the part of the model of Fig. 6 responsible for updating the value of the Position signal. This position value ranges from 0 and 405 millimetres and is updated in steps of 15 millimetres by a sensor attached to the motor. Whenever the sensor detects a 15-millimetre movement of the motor, it activates the *SuperBlock* shown in Fig. 7. Each block of which this *SuperBlock* is composed is identified by a number on its upper right-hand corner, these blocks are:

- a data store (5) that stores the current value of the position,
- two algebraic blocks (4 and 13) that increment and decrement the value of the position,
- a selector (15) that selects one of the values produced by 4 and 13 depending on the sense of movement of the window,
- and a limiter (2) that constrains the value of the position to between 0 and 405.

3.4 JAVA AN EMBEDDED SYSTEM SPECIFICATION LANGUAGE?

The role of *Java* [2] in the embedded world has recently attracted a lot of attention, particularly since the release by *Javasoft* of the *Personal Java* [21] and *Embedded Java* [22] specifications. The advantages cited for its use in embedded systems are mostly those which are not specific to *E-Java* or *P-Java*: its inherent portability, its built-in security and its lack of some of the main problem-causing programming-language features, as well as its support for threading and its ability to interface easily to other languages. *E-Java* is a subset of *Java* which is tailored to the needs of embedded systems in that it is more modular, scalable and configurable and has reduced system-resource needs, especially memory. *P-Java* is a larger subset of Java which includes support for networked devices, in particular, for the downloading of applets and (optionally) for the use of RMI, as well as more sophisticated display capabilities.

However, some serious difficulties will need to be overcome if *Java* is to be successful in the embedded-system world. The main problem concerns the difficulty of achieving determinism (in the real-time sense). In particular, system operations such as context-switching are non-deterministic and the duration and occurrence of garbage-collection is not predictable. General performance is also a problem: apart from the well-known decrease arising from the use of byte-code interpretation instead of compilation and machine-code execution, garbage collection is not optimal and the consequent excessive memory fragmentation also contributes to reducing performance, as does the reliance on "bloated" software libraries. The use of *a Java Virtual Machine* also means that minimum memory requirements are relatively high. The performance problems are to some extent addressed by *E-Java* and *P-Java* but still not sufficiently for many applications. *Just In Time* compilers are no panacea, increasing memory requirements while not increasing performance sufficiently due to the lack of a global optimisation capability.

These difficulties mean that the use of *Java* in hard real-time systems, even in its embedded or personal incarnations, is currently problematic. However, the use of *Java* in devices such as webphones (in this case *P-Java*) seems to be assured. The worst drawbacks of *Java* can be eliminated by compiling, when this is feasible and practical. When it is not, they be mediated by running the JVM on top of a proven real-time operating system (which *E-Java* and *P-Java* have been designed to do). In the future the language and/or the JVM may be more adapted to the real-time setting.

Nevertheless, the above discussion concerns the use of *Java* as a suitable *implementation* language (albeit relatively high-level) for embedded systems whereas this chapter centres on the use of *specification* languages for these systems. In our view, *Java* lacks the required level of abstraction to be considered a high-level specification language. Furthermore, it does not contain all the language constructs required, and, as far as hardware specification is concerned, it does not lend itself well to the generation of implementations. For our purposes then, the arguments of the above discussion about the suitability of *Java*, *P-Java* or *E-Java* for embedded systems are largely irrelevant. *Java* may well provide a very suitable language for describing the software implementations which are to be derived from the specifications of interest to us here. However, this should only concern us in that we may wish for it to be taken into account in the code-generation part of the specification language toolsets.

4. Scenarios for a combined use of specification languages

4.1 INDEPENDENT USE

The simplest way of developing embedded systems using several specification formalisms is through, what we term, independent use of these formalisms. In this approach, the system is initially split into several subsystems, which may then be designed independently as far as the implementation phase, each using a different formalism. The source code obtained from each subsystem is then integrated in the target architecture at which point, the interfaces between the different subsystems must be built.

The main advantage of this approach is its simplicity. To be able to design an embedded system we only need the sum of all the tools that are needed for each notation. On the other hand, the lack of tools, semantic models and notations that support the overall design cycle makes the work of the designer more difficult and more costly. Local errors, in a particular subsystem, can be detected in the initial stages of the design but global errors, affecting models in several notations, are difficult to detect until the design is in its final stages.

Embedded system development using the independent use approach requires a methodology in which different design flows can be followed for each notation and later combined. This methodology should be suitable for the development of complex systems, be customisable for different industrial uses, and be suitable for hardware-software and continuous-discrete systems. In the rest of this section, we present the development phases treated in this methodology and the individual design flows used for each notation.

4.1.1 Development phases

Requirements capture and specification. The objective of this phase is to obtain as much information as possible about the system being designed, in terms of what the system has to do rather than how it has to accomplish it. Both functional and non-functional aspects, such as performance and cost, should be considered. The requirements should be as clearly defined as possible in order to validate with respect to them in later design phases. For this reason it is useful to use formal notations already in the requirements phase.

System architecture design. In this phase we define the specification architecture (not the target architecture: processors and communication resources etc.) of the system. In this phase, the following tasks can be distinguished:

1. **Identification of the system components**. The division into subsystems can be influenced by many factors, such as the degree of internal/external coupling, common timing requirements, target architecture restrictions, re-use considerations, grouping of computationally-costly functions, development team expertise, etc.
2. **Selection of the notation to be used for each subsystem**. The considerations to be taken into account in this choice are treated in Section 2.5. Note that the system division may already have been influenced by the desire to use certain notations, in which case the choice may already be made for some subsystems.
3. **Definition of the information to be exchanged between subsystems**. This information should be defined at both the conceptual level and the implementation level. The former treats the data exchanged between the different subsystems and the synchronisation points between them, while the latter treats the concrete flows and events that form the interfaces, and the data types of the data carried in these flows.

4. **Validation of the architectural decisions taken so far**. The objective of this task is to check that the architectural decisions taken are reasonable (judged against the requirements) since, at this point, it is not possible to validate all the functional and non-functional requirements that have been defined.

After this phase the design flow divides and for the next two phases, the design of each subsystem is carried out in isolation of the others.

System modelling or detailed design. Several refined versions of the specifications to be developed in each notation are produced. In each refinement cycle, the system is validated via simulation of each specification in isolation. In this phase, a set of complementary tests between the different models should be planned, so that the inputs to each model are the values that we expect the other models to produce.

Hardware/software partitioning. In this phase, the decision of which parts are to be implemented in hardware and which in software is made. Prior to partitioning, estimates are obtained for the non-functional parameters (area, CPU consumption, etc.) of the components of the system in two main ways: static analysis (also called non-standard interpretation) - analysis of the code - and dynamic analysis - here understood to mean simulation.

Partitioning may be done manually or with machine-assistance using partitioning algorithms. Among the factors which influence the partitioning, apart from performance and costs, are the suitability for implementation in hardware or software, the availability of a previous implementation and the target architecture.

Integration and implementation. In this phase the interfaces between the different pieces of code that have been obtained from each component specification are built, the software is compiled to obtain target processor code, the hardware components are synthesised, and the different components are integrated in the target architecture.

4.1.2 Independent design flows

In the ESPRIT project COMITY, a toolset [1] and a design methodology [20] to support multi-formalism specification and design of complex embedded systems have been developed. COMITY supports the following specification languages: *SDL*, *Statecharts*, *MATRIX*$_X$ and *Lustre*.

Based on the tools used in the COMITY project, two independent design flows (see Fig. 8), one for *SDL*, and another combining *Statecharts* and *MATRIX*$_X$, have been defined.

The tools considered in the *SDL* design flow are:

- *ObjectGEODE*: specification, design, simulation, code generation and performance analysis using *SDL*
- *COSMOS*: a hardware/software partitioning tool that accepts several languages as input, apart from *SDL*. It supports the partitioning phase and the interface generation.

In the *SDL* design flow, first, the *SDL* subsystem is specified. This specification is then refined in the high-level design phase, each version being validated through simulation. When an efficient enough version has been obtained, an interactive hardware-software design phase, including system partitioning and communication synthesis, is performed using the *COSMOS* environment [8]. For the parts of the system to be mapped to hardware, a high-level synthesis step follows to obtain an RTL description.

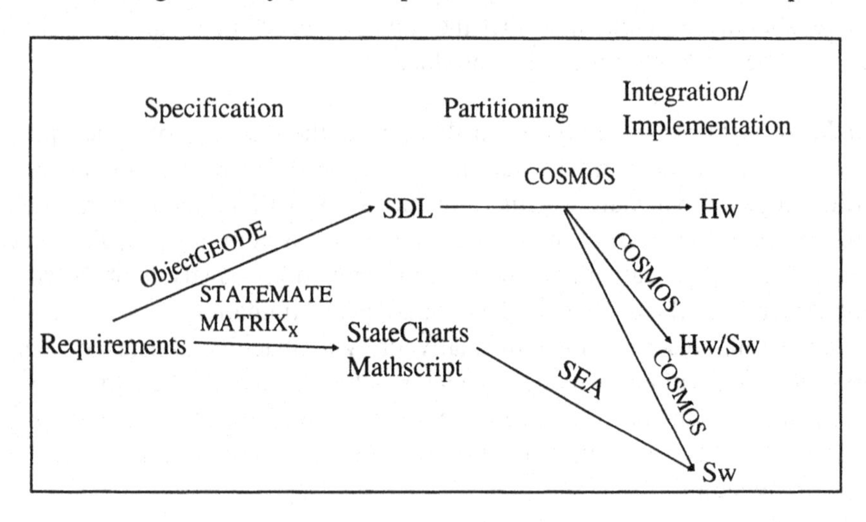

Figure 8: Independent design flows

The tools considered in the *Statecharts/MATRIX*$_X$ design flow are:

- *MATRIX*$_X$: specification, design, simulation and code generation of discrete and continuous systems, using a proprietary notation.
- *SEA*: a software/software partitioning tool (allocating components to processors) that accepts as input *Statecharts* and *MATRIX*$_X$.
- *STATEMATE*: specification, design, simulation and code generation using Statecharts.

In the *Statecharts/MATRIX*$_X$ design flow, first, each component is specified in the corresponding notation (*Statecharts* or *MATRIX*$_X$), then these specifications are refined and validated through simulation using the *Statemate* and *MATRIX*$_X$ simulation tools. Next, the SEA tool [17] is used to derive a description of each component in an intermediate format - extended Predicate/Transition-Nets (Pr/T-Nets) - from its specification. The resulting

Pr/T-Nets are then combined into one net, using a library of interface models. In a subsequent analysis phase, certain system properties are formally verified. Finally, C code is generated from the Pr/T-Net.

4.2 COORDINATED USE

4.2.1 Interconnection framework

Interconnecting subsystems specified using different formalisms inevitably involves the definition of a framework in which to embed them and give unambiguous meaning to this interconnection. The main aspects of such a semantic framework (both in the context of a coordinated use and an integrated use) are:

1. **a global timing model** and a definition of the relation of the timing models of each formalism to this global timing model,
2. **a common data model**, needed to make sense in one formalism of the data from another, covering at least the datatypes to be interchanged,
3. **the global connectivity**, that is, the specification of the communication channels between the different subsystems: which subsystem or (part of a subsystem) is to communicate with which subsystem (or part of a subsystem), this being the major part of the global architecture,
4. **global communication mechanisms**, defining the structure and semantics of both the interfaces offered by the subsystems and the information exchange between them.

Point 4 may be considered to include point 3, in particular, if we do not wish to statically specify explicit communication channels between subsystems.

We now examine each of the main aspects of the interconnection framework defined above, with emphasis on a coordinated use.

4.2.2 Global timing model

Zero time delay communications and transitions. One of the first issues to address is whether or not inter-subsystem communications have zero delay (that is, take zero time). We will assume this to be the case. To simplify, we will further assume here that actions in each of the discrete-time formalisms used take zero time. In the synchronous case, this assumption is part of what is often referred to as the "perfect synchrony hypothesis". In the asynchronous case, it is more polemical, but it is the semantics that is generally used in simulation and is adequate for the applications with which we are dealing.

Nature of global time. Concerning the nature of the global timing model, for simplicity, we assume here that the global time is discrete so that, in the context of interconnection, we may need to consider the embedding of continuous subsystems in a discrete system but not vice-versa. As a consequence, we assume that if two subsystems specified in a continuous formalism are to communicate, they do so via the discrete global model. If it is required they communicate in a continuous-time manner, they should be merged.

An important issue, once it has been decided that the global time is discrete, concerns this time scale and its relation to the time scale of the different subsystems. If the time scale in the (discrete-time) component subsystems is absolute, it seems reasonable to choose the global time scale to be as fine as the finest of the these local time scales, though other arrangements are also possible. Otherwise, the relation between the different local time units must be explicitly defined, reflecting assumptions about relative speeds of the different parts of the system.

Synchronous or asynchronous global temporal behaviour. Perhaps the main issue, once it has been decided that the global time is discrete, is to decide whether the global time behaviour is to be synchronous or asynchronous. Ideally, as in the *Statecharts* language, the interconnection framework should be defined to cope with both synchronous and asynchronous behaviour, allowing the most appropriate to be chosen in function of that which is being modelled. A cosimulation toolset would then also offer a choice of asynchronous or synchronous global temporal behaviour. However, if one of the formalisms to be used is asynchronous, an asynchronous time scheme is the most logical choice for the global time behaviour since embedding an asynchronous specification in a synchronous one could be considered counter-intuitive.

Synchronous case. In the synchronous case, at each global instant, each of the discrete subsystems performs the (zero-time) actions it has scheduled in that instant, the external inputs of each subsystem for that instant being the external outputs sent by other subsystems in the previous instant.

A subsystem with asynchronous temporal behaviour, and with zero-time transitions and zero-delay internal communications, can be embedded in a synchronous global time scheme by executing it under the assumption that its environment is "reasonable" (though as stated above, this may be rather counter-intuitive). A "reasonable" environment is one which only stimulates the system when it is in a stable state, that is, when it has no more actions to perform; a "cooperative" environment is one which may stimulate the system in unstable states.

A continuous time subsystem can be embedded in a synchronous discrete time scheme with zero-delay communications and where all actions of the discrete-time formalisms take zero-time by considering that:

- throughout each global instant, the external inputs to the continuous subsystem from the discrete global system are the external outputs sent by other subsystems in the previous instant
- external inputs and outputs of the continuous system to the discrete global system remain constant between global instants

Asynchronous case. In the asynchronous case, the global system only executes a transition at moments of time when one of its components has a scheduled action and at such moments, all actions that can be scheduled in the global system are executed.

Again for simplification, we assume here that inter-subsystem parallelism is implemented through interleaving so that if, at any given moment, more than one subsystem has scheduled actions, these actions are interleaved. There are two main choices with respect to the granularity of this inter-subsystem interleaving:

- **fine-grain interleaving**: interleaving of individual transitions from each subsystem
- **coarse-grain interleaving**: interleaving of maximal sequences of (zero-time) transitions from each subsystem[6].

That is, whether, during its turn, a subsystem is only allowed to execute a single transition or it is allowed to execute as many transitions as it can execute in that time instant (the fact that such a maximal sequence may be infinite should be taken into consideration[7]). The choice of which policy to adopt can either be made for all the subsystems together or for each subsystem independently, in the latter case perhaps reflecting relative speed assumptions.

Under the assumption that inter-subsystem communications are delivered immediately and are available to be consumed immediately (zero delay), the choice between using fine-grain or coarse-grain interleaving of a subsystem's actions with those of its environment corresponds to the choice of executing it in a "cooperative" or "reasonable" environment, respectively. Note that, unlike the synchronous time scheme case, here the output from one subsystem can be used as input in another subsystem *in the same time instant*, so that the problem of zero-time loops involving more than one subsystem may arise.

[6] c.f. the *Statemate* notion of "superstep".

[7] The presence of zero-time loops in *SDL* or *Statecharts* cannot be ruled out; in *Statemate* an explicit limit is placed on the length of such sequences.

A subsystem executing according to a synchronous time scheme can be embedded in an asynchronous global time scheme by ensuring that it is only scheduled once at each instant for which it has executable actions, using the outputs of the previous instant as its inputs in the current instant.

A continuous time subsystem can be embedded in an asynchronous discrete time scheme with zero-delay communications and where all actions of the discrete-time formalisms take zero-time by considering that:

- the values of its inputs from the rest of the system are constant between time instants where actions are scheduled in some part of the global system, and change "after" the execution of all the zero-time actions of the discrete time formalisms scheduled for such an instant,
- the values of its outputs to the rest of the system are constant between time instants where actions are scheduled in some part of the global system, and change "before" the execution of all the zero-time actions of the discrete time formalisms scheduled for such an instant,
- the continuous subsystem can schedule outputs to the rest of the system at future discrete global instants thus influencing the global time advance

The last point needs a notion of communication event to be defined for the continuous subsystem. This point is dealt with in Section 4.2.5.

4.2.3 Common data model

For data produced in a subsystem specified in one formalism to be communicated to a subsystem specified in another formalism, some type of equivalence between the data types of the two formalisms must be established.

Defining an equivalence between the pre-defined types of each formalism is a relatively simple matter, though care must be taken with respect to the actual implementation (e.g. max size of integers etc.). A common data model for the data to be communicated between the different subsystems is therefore easily obtained if there are no user-defined types, as is the case in *MATRIX*$_X$, or if user-defined types cannot be communicated between a specification and its environment in any of the formalisms. Thus the simplest possibility is to impose this restriction on all specifications in each of the formalisms.

If we do not wish to impose such a severe restriction on our multi-formalism specifications, a way of translating user-defined datatypes between formalisms must be found. However, defining inter-formalism type equivalences for each pair of formalisms is exponential in the number of formalisms. A better solution, therefore, is to define a common data model and map the data types of each formalism to this model. In view of the widespread acceptance which it has now achieved, and with an eye to facilitating the generation of (possibly distributed) implementations, the data part of *CORBA-IDL* [19] would constitute a good choice for a common data language. However, other choices may be more suitable depending on the specific circumstances. Another possible candidate is the ITU-T standard data language *ASN.1*, which, in our case, has in its favour that it can be used to describe the data part of *SDL*.

4.2.4 Global connectivity

As mentioned previously, it is not always useful to separate the connectivity questions from those concerning the communication mechanisms, particularly if we do not wish to statically define the inter-subsystem communication channels. Here we assume that we do wish to statically specify these channels, in order to obtain a more complete specification of the global system.

If the communication at the global level is by broadcast, the interconnectivity can be thought of as being already specified; here we assume that this is not the case. The connectivity of the global system can be defined in a specification which is completely external to the subsystems specified in the different formalisms, or it can be fully or partially defined within them.

An important consideration is whether elements specified in one formalism can contain elements specified in another, either by embedding whole subsystems (complete specifications) in other subsystems or by embedding parts of subsystems (specification blocks) in other subsystems.

In both cases, the encapsulating subsystem is effectively imposing some structure on its environment, and its specification partially defines its connectivity with respect to other subsystems. In both cases also, the formalism used to specify the encapsulating subsystem must evidently contain block constructs suitable for this purpose (the *Statecharts* language, for example, provides for such a possibility by allowing activities with no specified content). In order to cosimulate, moreover, the corresponding simulator must be able to deal with "behaviourless" blocks.

In the second case, if more than one subsystem makes use of this facility, the coherence between the views each subsystem has of the global system must be checked. If only one of the subsystems makes use of this facility, coherence problems are avoided but symmetry is lost.

For these reasons, (partially) specifying global connectivity by embedding parts of subsystems in other subsystems (via either mutual, symmetric embedding or one-way, asymmetric embedding) is not a generic approach, and is only really appropriate to a master-slave relationship and, even then, not in an integrated use context.

In the case where an entire subsystem is embedded in another, leading to proper inter-formalism hierarchies, only the formalism of the encapsulating subsystem needs the appropriate constructs and simulation possibilities, and there are no coherence problems. However, care must be taken concerning the rather counter-intuitive situation of a subsystem using asynchronous temporal behaviour, such as one specified in *SDL*, being embedded in one using synchronous temporal behaviour, such as one specified in *Statecharts*.

Defining the connectivity in an external specification is best carried out using a purpose-designed composition language, particularly in the case of an integrated use of the different formalisms; such a language need not be very large. In an integrated use, it will play a role in all the development phases from analysis to coding and testing.

4.2.5 Global communication mechanisms

As was the case for data exchange between subsystems specified in different formalisms, communication mechanisms could be defined for each pair of formalisms. But such a procedure is exponential in the number of formalisms. A more generic solution is to define a common communication semantics and then map the communication mechanisms of each formalism to this common communication semantics.

At the system level, we define a generic communication mechanism and then examine how it is mapped to the communication mechanisms of immediate interest to us: those of *MATRIX*$_X$, *SDL* and *Statecharts*. We stipulate that all inter-subsystem communications are events, possibly with associated data exchange. As we see below, this enables the time scales of the continuous and discrete parts to be decoupled, reflecting the fact that the discrete part of the global specification is only interested in the values of the continuous part outputs when a certain threshold is passed, rather than at each global instant.

The mapping of *MATRIX*$_X$ communication mechanisms to the global communication mechanism requires a notion of event in *MATRIX*$_X$. We choose to do this by defining Boolean input and output "event channels". An event can then be defined either as a change in the signal on an event channel from false to true, or as any change in the signal on an event channel. When an event is generated by a subsystem specified in *MATRIX*$_X$, data values on each of the output channels are made available to the rest of the system. Thus it is the *MATRIX*$_X$ system which explicitly models the firing condition for each communication with the rest of the system, that is, the continuous-discrete interface is modelled in *MATRIX*$_X$. This notion of event in *MATRIX*$_X$ enables us to decouple the frequency of a *MATRIX*$_X$ subsystem, or rather of its interface, from the frequency of communication.

The mapping of *SDL* communication mechanisms to the global communication mechanism is done simply by making the correspondence between signals and events, and between signal parameters and the data associated to the event. However, due to the addressing used in SDL, events are actually chosen to correspond to signal name + originating process name + an instance number for the originating process.

The mapping of *Statecharts* communication mechanism to the global communication mechanism is done simply by ensuring that, for communications between a subsystem specified in *Statecharts* and its environment, no data exchange takes place without an associated event. Events are communicated between a subsystem specified in *Statecharts* and its environment via activity chart event flow lines and data is communicated via activity chart data flow lines.

4.2.6 Cosimulation

The term cosimulation is used to describe the simulation of a system composed of subsystems specified in one of several different formalisms. This is achieved using an instance of the appropriate simulation tool for each subsystem together with mechanisms to manage the interaction between these simulator instances[8]. The interconnection framework described in previous sections can provide the basis for the interconnection of simulators via a generic coupling mechanism.

Here, we discuss cosimulation using a strong-coupling methodology, that is, where the different simulators are tightly integrated via a central scheduler, which controls the progress of each simulator and mediates all inter-simulator communications. A more optimistic possibility would be to allow each simulator to evolve independently, use time-stamped messages for inter-simulator communication, and make systematic use of a rollback facility in the case where, on reception of a time-stamped message, a simulator realises that it has advanced its local time too far. A more efficient approach would be a semi-optimistic one in which each simulator evolves independently within certain limits. However, we concentrate here on the simplest approach, the centralised implementation, in which the only interface that each simulator offers is to the scheduler or backplane.

We define a simulator-backplane interface such that each simulator sees the backplane as its environment and has no awareness of other simulators. In defining this interface, care must be taken not to make over-restrictive assumptions, since the view the simulators have of their environment is not uniform. Moreover, an interface which is as generic as possible increases the extensibility of the system.

In the centralised approach, one of the first concepts to define is that of a cosimulation step for the discrete simulators, based on the interconnection

[8]In the general case, a system may contain more than one subsystem specified in the same formalism. A cosimulation, therefore, may include more than one instance of the same simulator and we therefore refer to simulator instances in the following.

framework used. Cosimulation runs are then built from this cosimulation step and different cosimulation modes can be defined in terms of it.

For a cosimulation with *synchronous* global temporal behaviour, the scheduling of each of the participant simulators is simpler. The cosimulation backplane control loop is as follows:

1. Advance all simulator instances in lock-step to the instant after the earliest of the following:
   - the earliest global time at which one of the asynchronous instances has a scheduled action,
   - the earliest global time at which one of the synchronous or continuous instances has an external communication to make.
2. deliver the external outputs from the previous instant to the destination simulator instances (note that the external outputs are only available for consumption at the destination instance in the instant after the one in which they were generated).

Notes:

- case of continuous simulator instance: during a global advance, there is no change in the external inputs,
- case of asynchronous simulator instance: in a time instant where it has scheduled actions, it is allowed to execute a "superstep",
- case of synchronous or continuous simulator instance: implementing a global advance may involve rollback.

For a cosimulation with *asynchronous* temporal behaviour, the scheduling of each of the participant simulators is more complex. One reason for this is that if more than one simulator proposes an action at any one instant, the backplane must decide on the how to schedule these actions (implementing parallelism as interleaving), assuming the granularity of interleaving has already been defined. The cosimulation backplane control loop is as follows:

1. advance time in each of the simulator instances (executing the necessary actions in the continuous and synchronous simulator instances) to the earliest of the following:
   - the earliest global time at which one of the asynchronous instances has a scheduled action,
   - the earliest global time at which one of the synchronous or continuous instances has an external communication to make.
2. schedule a step in *one of* the simulator instances (recall the notion of granularity of interleaving), in the case where more than one simulator instance proposes an action at this time; if this is a synchronous instance, mark it as already having evolved in this instant so that it will not be evolved again without evolving time

3. if this step involves sending to an asynchronous simulator instance, deliver the communication and interrogate this receiver instance to see if it then has scheduled actions in the current instant; if this step involves sending to a continuous or synchronous simulator instance ensure that the communication is not available to be consumed by the receiver instance until time advances,
4. continue scheduling steps until none of the simulator instances can advance without evolving time (recall that this may be never!).

Notes:

- case of continuous simulator instance: during a global advance, there is no change in the external inputs,
- case of synchronous or continuous simulator instance: implementing a global advance may involve rollback.

4.2.7 A cosimulation style of specification?

The specification may be affected by the fact that it is to be used in cosimulation. For example, in the situation where there is a choice of whether to define a data type in the environment and import it into the specification, or define it in the specification and export it to the environment.

There are other aspects of the externally-implemented environment which could feed back into the specification via a "cosimulation style" of specification, for example, the use of a *CORBA*-based simulation backplane. If the cosimulation context in which the specification is to be used can affect the specification itself, this has some bearing on the choice of a cosimulation mechanism. This is since any requirements on the specification style that a cosimulation mechanism may have should be consistent with any other uses to be made of the specification[9]. For example, if the final implementation is to be used as a *CORBA* component, a *CORBA*-influenced specification style would be useful, and the choice of *CORBA* for the simulation bus would then be complementary.

The chosen simulation methodology may therefore inevitably have some influence on the type of specification developed. However, such influence should be minimised and should certainly not obstruct any other uses to which the specification is to be put. If such a style were to be defined, it should evidently have as little dependence as possible (preferably none at all) on the other simulation tools and languages taking part in the cosimulation.

[9] Of course, the cosimulation mechanism should not make any specification style obligatory and any style requirements it may have should not be too specific.

4.2.8 Implementation of cosimulation

$MATRIX_X$ is of most interest here as a continuous time simulator. Cosimulation using this simulator is implemented through the use of a User Code Block (UCB) which encapsulates the *C* code implementing the interface to the rest of the system. All blocks which are to communicate with the rest of the system must be connected to this block. Through this *C* code, the simulator can be instructed by the backplane to advance to a certain time, that at which the next global event is scheduled, or to rollback to a certain time, in the case where an event was generated by another continuous simulator instance or by a synchronous simulator instance.

The *SDL* language has asynchronous temporal behaviour. Cosimulation using the *ObjectGEODE* simulator is carried out with the cosimulation backplane playing the role of the *SDL* environment, by making use of the interface between the simulator proper and the simulator GUI. By means of this API, a real *SDL* environment can not only exchange signals with the simulator, but can also send simulation advance commands to it and receive time progression and other trace information from it.

The *ObjectGEODE* simulator provides access to this API to simulator X-clients so the interface between the backplane and the *SDL* simulator must be implemented through such an X-client. As stated previously, *SDL* simulation is performed under the assumption that transitions take zero time and signals spend zero time in transit. In conformity with this viewpoint, the *ObjectGEODE* simulator only offers the time transition when no other transition is fireable and does not really need to provide any other way to advance time. However, for cosimulation, a command to force time to advance by a fixed amount is needed; fortunately, *ObjectGEODE* provides such a command, the *wait* command, designed in order to cover the case of a modelling error or that of an incomplete specification.

The *Statemate* simulator allows either synchronous or asynchronous temporal behaviour. It provides a means to communicate with the outside world via interface code (though it has the tendency to view this communication as master-slave type). However, for a reasonably efficient and complete cosimulation implementation using this simulator, use should be made of the simulator API, and in particular, of primitives to send commands to control the behaviour of the simulator. Such primitives use the interface between the simulator proper and the simulator GUI[10] in a manner

[10] Unfortunately, the owner of the software, *i-Logix*, does not have an open policy regarding external control of the simulator: apparently, such primitives are only made available in the context of a co-operation with the company. Directly implementing the communication with the X-interface, that is, implementing such primitives from scratch, would be considerably more involved and any attempt to do so suffers from the almost complete lack of documentation, apparently also a reflection of *i-Logix'* closed policy.

similar to that of the *ObjectGEODE SDL* simulator. Implementing the control of the simulator via an SCP (Simulator Control Program) suffers certain drawbacks but may be adequate for feasibility checking. Time can be explicitly advanced by use of the `GoAdvance` command.

4.2.9 Distributed cosimulation

If it is required that the simulator instances do not all execute on the same machine, the overhead involved in using a centralised architecture, in which a cosimulation backplane controls the progress of each simulator and mediates all inter-simulator communications, may prove excessive. Performance demands may mean that the use of a looser integration philosophy is inescapable. We do not explore alternative architectures here but simply make a few remarks about distributed cosimulation which are valid for all the different cosimulation architctures.

One way of distributing a cosimulation would be to use the *CORBA* communication bus. The choice of *CORBA* then also provides a common data model (which the data in the different formalisms must then be mapped to) as described in Section 4.2.3. However, it implies a particular communication semantics. *CORBA* communications (using static invocation) are either of the datagram type, with *best-effort* semantics, or of the RPC type, with *exactly-once*, and in case of exception *at-most-once*, semantics. Using best-effort semantics may not be good enough for reliable cosimulation so that it may be required that all communications be of the RPC type. For the centralised architecture, the blocking aspect of such communications may be useful for the control of synchronous or continuous simulators. In the case of a looser integration of simulators, this blocking may restrict the looseness of the integration in that it may limit the independent advance of the simulators.

4.3 INTEGRATED USE

In this section, we will sketch what could be expected for an integrated use of specification languages. With the co-ordinated use we have cosimulated at the specification level, but there we only had a very simple glue language to combine the specification components in the different formalisms. Now we want to have a sophisticated glue language in which to express the interrelationship of the different elements in a convenient and flexible way. We just give a few hints that have to be further developed and just stay at the requirements level for such a glue language.

4.3.1 Definition of structure

In the co-ordinated use, we envisaged a simple mechanism for composition by joining channels. When joining two channels coming from different formalisms, some adaptation module was needed in order to take into account the different computation models. Abstracting away from this fact, it would be useful to have a number of predefined higher order functions or functionals (we use here the terminology of functional programming languages) with which to combine modules with one another in a simple way. The naming mechanism for channels is simple and useful when there is no regularity.

4.3.2 Definition of hierarchy

In the co-ordinated use, we have assumed that the components in the different formalisms are used side by side. But, what about using one in another? This would allow maximum flexibility in the use of the languages.

4.3.3 Definition of parameterised systems

For an overall system specification language, the definition of parameterised systems is of prime importance. The more parameterisable a description is, the more useful it will be in different contexts. Often, when designing a system one of the most important issues is finding the appropriate generalisations of the problem to be solved, such that the system description is most useful in the future. At any rate, the formalism we are looking at should contain parameterisation and instantiation mechanisms for different purposes.

4.3.4 Other features

There are many other features that could be of interest. Let us mention a few. It would be useful to have some support for the handling of libraries of predefined modules at a global level. Furthermore, some methodological support would be also quite useful. By this we mean constructs to aid in the development of the system, allowing e.g. the specification of requirements at an early stage or some pre-/post-condition predicates that could help the formal verification.

There are many features one could think of, but as a rule one should keep in mind that if a feature already exists in one of the three embedded languages, it should not be replicated at the glue level language.

5. Conclusion

We started with the premise that there is no perfect language for the definition of embedded systems, rather that there are a number of complementary formalisms suited for different purposes under different perspectives. We have studied how three of these formalisms can be used jointly in an effective manner. We finished with an integrated use of these formalisms, that involving the use of another language for describing the communication between them. It is clear that this is not an ideal situation, but a pragmatic one. Ideally one would like to have one clean wide-spectrum language, or what in the end is the same, a set of neatly adapted application specific languages, that would cover the whole spectrum of needs. Java, with all its commercial interests, could be one potential candidate, but for this to be true in the field of embedded system design still a lot needs to be done, starting with a Java Virtual Machine adequate to hard real time modelling.

Acknowledgements

This research has been partially carried out in ESPRIT project No.23015 COMITY (*Co-design Method and Integrated Tools for Advanced Embedded Systems*), and in the "Accion Integrada hispano-alemana" HA97-0019 *"Metodología de diseño de sistemas reactivos"*, a collaboration with Tec. Univ. of Munich, Univ. of Augsburg and Med. Univ. of Lübeck, funded by the Spanish Government, The authors wish to acknowledge the collaboration of the COMITY consortium partners: Aerospatiale, ISI, TIMA, Verilog (France), BMW, C-Lab (Germany) and Intracom (Greece), and thank our local colleagues A. Alonso, N. Martínez and J.A. de la Puente for their collaboration in the context of the COMITY project.

References

1 Alonso, A., Martínez, N., Pickin, S. and Sánchez, L. (1998) *COMITY Method Guide. Advanced Version*, Deliverable D21.2.1 of the Esprit project 23015 COMITY.

2 Arnold, K. and Gosling, J. (1997) *The Java Programming Language*, Addison-Wesley.

3 von der Beeck M. (1994) A Comparison of Statecharts Variants, in *Formal Techniques in Real-Time and Fault-Tolerant Systems*, LNCS 863, Springer-Verlag.

4 Harel, D. (1987) Statecharts A visual formalism for complex systems, *Science of Computer Programming*, vol. 8, pp. 231-274,.

5 Harel, D. and Naamad (1996) A. The Statemate semantics of Statecharts. *ACM Trans. Soft. Eng. Method.* 5:4, Oct. 1996.

6 Hilfinger, P. (1985) SILAGE, A High-Level Language and Silicon Compiler for Digital Signal Processing, in *Proc. of Custom Integrated Circuits Conference* CICC, May 1985.

7 Integrated Systems Incorporated (ISI). (1996) *System Build Reference Guide.*

8 Ismail, T. and Jerraya, A. (1995) Synthesis Steps and Design Models for Codesign, *IEEE Computer*, 28(2):44-52, February 1995.

9 IEEE 1076-1987 (1987) *IEEE Standard VHDL Language Reference Manual*, IEEE, New York.

10 IEEE 1364-1995 (1995) *Verilog Hardware Description Language Reference Manual*, IEEE, New York.

11 ISO/IEC. (1990) *ISO 9074. Estelle: a formal description technique based on an extended state transition model*, ISO/IEC.

12 ISO/IEC. (1989) *ISO 8807. Lotos: a formal description technique based on the temporal ordering of observational behaviour*, ISO/IEC.

13 ITU-T. (1994) *Recommendation X.680. Abstract syntax notation one (ASN.1)*, ITU-T.

14 ITU-T. (1993) *Recommendation Z.100. CCITT Specification and description language SDL*, ITU-T.

15 ITU-T. (1995) *Recommendation Z.105. SDL combined with ASN.1 (SDL/ASN.1)*, ITU-T.

16 ITU-T. (1996) *Recommendation Z.106. Common interchange format for SDL*, ITU-T.

17 Kleinjohan, B., Tacken, J. and Tahedl, C.. (1997) Towards a Complete Design Method for Embedded Systems Using Predicate/Transition-Nets in *Proc. CHDL '97*, Toledo, Spain (eds.Delgado Kloos, C. and Cerny, E.).

18 de Micheli, G. and Sami, M. (eds.) (1996) *Hardware/Software Co-Design*. NATO ASI Series, E: Applied Sciences, Kluwer Academic Publishers.

19 Object Management Group (OMG). (1998) *The Common Object Request Broker: Architecture and Specification, v2.2*, OMG.

20 Roux, J.L. (ed.) (1998) *User Manual for the COMITY Advanced Toolset* v1, Deliverable D21.3.1 of the Esprit project 23015 COMITY.

21 Sunsoft. (1998) *Personal Java Specification API, version 1.1*, July 1998.

22 Sunsoft. (1998) *Embedded Java Specification, draft 0.1.*

23 Verilog. (1997) *ObjectGeode, SDL Simulator - Reference Manual.*

TOWARDS A NEW SYSTEM LEVEL DESIGN LANGUAGE -- SLDL

STEVEN E. SCHULZ
Texas Instruments
Dallas, TX USA

It seems that system level design is on everyone's mind these days. So, it is not surprising that talk of a system level design language has been appreaing in the news. Yet, major fundamental questions quickly arise whenever a "standard" design language is proposed. What is it? Why do we need it? What will it do? How is it different from other existing languages? How will it change my design methodology and design tool environment?

In fact, no such language is available yet, and won't be for several years. However, its capabilities and priorities are being determined now, and prototyping could start as early as next year. The purpose of this article is to explain why an SLDL is needed, who's driving it, what it promises to offer designers, and when you can expect it integrated into EDA tool flows.

1. A Changing Environment

In simpler times, designers assumed that, if they could write RTL code, they could handle the mundane challenges of getting function and timing right. Today however, a rapidly increasing proportion of designs now include embedded software, making even the description and analysis of system functionality and timing performance far more complex. In this consumerization era, there are increasing demands for integration of analog and RF technologies on the same silicon. Add to this the integration of on-chip memories with multiple processors and interface protocols, and "system level design" starts to take on new meaning. Because of greater time-to-market pressures, risks to deliver all this within highly constrained time windows are growing. With leading technology wafer fabs costing $2 billion (and rising), higher design productivity is essential to ensure silicon profitability.

Of course, this change provides similar motivation to the Virtual Socket Interface Alliance (VSIA). VSIA is working hard to agree on common standards that will facilitate large-scale design reuse on silicon. SLDL and VSIA are highly

A. A. Jerraya and J. Mermet (eds.), System-Level Synthesis, 175–187.

complementary; whereas VSIA is attempting to provide better infrastructure for hardware implementation by leveraging existing technologies, SLDL is focusing on delivering an entirely new technology at the front end of the process. Arguably, SLDL hopes to facilitate reuse by enabling true hardware-software co-design and system constraint budgeting. While many approaches are more narrow and application-specific, SLDL will attempt to permit a broader, interoperable industry solution.

2. Existing Solutions are Inadequate

HDL's were originally designed for implementation of hardware. Yet today, over 50 percent of new VLSI design starts include some form of embedded software -- and this trend is accelerating. While several EDA tools have in recent times offered specific hardware-software co-verification capabilities, hardware-software co-design remains an elusive Holy Grail for the industry. Currently, the top-down design process is unable to support a full specification of tool-processable requirements and constraints, prior to entering the detailed hardware design phase at an RTL level (see Figure 1).

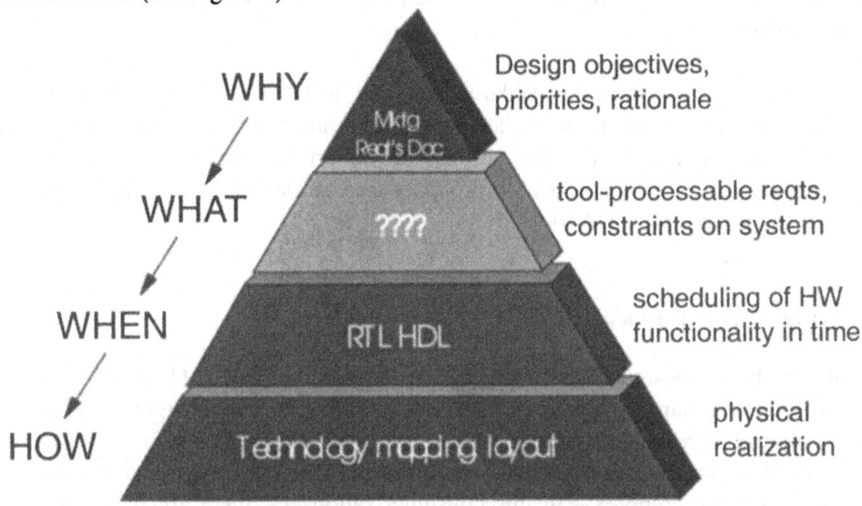

Figure 1: Top-Down Design Pyramid

Second, the problem of managing design constraints is growing exponentially. Larger designs, the integral use of software and analog/RF hardware, a wider variety of consumer applications, deep submicron effects that impact system tradeoffs, and packaging/thermal effects all impose new constraints. Until recently, designers have managed to juggle basic constraints in their heads, comparing them against waveform diagrams and textual reports from tools. Now, constraint management has more in common with trying to solve hundreds of simultaneous nonlinear equations! Battery life, heat dissipation, density, and cost are typical careabouts in addition to timing constraints.

System design requires a system view: if you can't describe it, you can't design it. Presently, no interoperable format exists to capture that system view. Industry leaders believe this is holding back an enormous potential market, both for highly integrated electronic products, as well as new EDA tools.

3. Industry-Driven Origins

In October 1996, over 40 industry leaders in the field of system design convened in Dallas, Texas for a 2-day "needs assessment" workshop. Their challenge was to determine what role, if any, EDA standards must play in response to the 1996 EDA Standards Roadmap document. This roadmap cited serious and critical gaps in EDA flows for system level design. The conclusion of this worldwide expert group was that current standards were insufficient to support "system-on-a-chip" capabilities, and that some new notation would be necessary to achieve those business objectives.

A new SLDL committee was established, with David Barton (Intermetrics) elected as chair. From that point, subcommittees were formed to focus on four interrelated areas: requirements and constraints, hardware-software co-design, system to component interfaces, and formal semantics. Since that initial workshop, numerous SLDL requirements workshops have been held (in the USA, Italy, and Switzerland). The Air Force Research Labs recently awarded a 3-year government contract to help accelerate the development and industrial adoption of SLDL. In addition, several hundred participants representing semiconductor and electronics systems companies, EDA vendors, standards development bodies, and government agencies exchange views and make recommendations using the SLDL email reflector.

Primary support is provided by the EDA Industry Council, with the Industry Council PTAB (project technical advisory board) serving as official sponsor. However, other consortia are also providing strong support. VHDL International is embracing SLDL as a natural complement to VHDL, and in fact is planning on strong synergy with development of VHDL 200x. VHDL 200x requirements are derived from multiple sources, but include SLDL requirements for integration purposes.

Strong support is also provided by key industry standards development organizations. The Design Automation Steering Committee (DASC) of the IEEE tracks SLDL progress regularly and discusses synergistic opportunities. The European CAD Standardization Initiative (ECSI) hosted a three-day SLDL workshop in Italy last summer, and SLDL was included as part of the NATO Advanced Study Institute on System Level Synthesis in August. In Japan, EIA-J members have selected SLDL as the initial system specification for system-on-a-chip designs in their EDA Industry Roadmap for 2002. In addition, many of the top university researchers in this field have participated in the discussions surrounding SLDL and how to incorporate the best of their research concepts.

4. Concepts Behind SLDL

Fundamentally, SLDL is a language notation which embodies semantics for describing system requirements prior to partitioning. The scope of SLDL is motivated by today's "system-on-a-chip" business drivers, however nothing prevents its use in large-scale electronics systems. The emphasis is on micro-electronics designs with embedded software, and tightly coupled interactions between digital, analog, and RF hardware. In addition, the scope supports mechanical and optical interfaces. SLDL will include some new language development, but will primarily leverage existing notations and semantics through an integrated approach.

But, just what are "system requirements"? One way of viewing them is according to three categories: *WHAT* the system should do, *HOW* the system is architected, and *ASSUMPTIONS* that constrain the design space. These categories may be labeled Behavior, Architecture, and Constraints (see Figure 2). Behavior describes the functional goals and objectives of the design. Architecture, which we definehere as strutural organization plus configuration, defines relationships between various components that will be used to comprise the system. Architectural decisions are often fixed early in the design phase, because exploring multiple choices is presently too difficult. Constraints describe assumptions and assertions that must be satisfied by the implemented system.

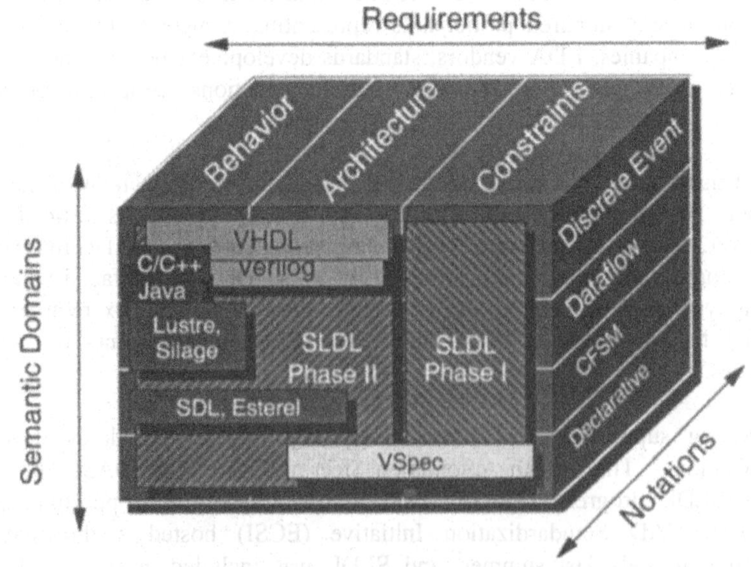

Figure 2: The SLDL Semantics Requirements Cube

It is important to understand how constraint specification and behavior specification are fundamentally different. Most engineers are used to describing functionality in terms of sequences of actions and reactions. This is known as an *operational* approach, and it lends itself to simulation-based techniques. On the other hand, constraints are described by stating them as assertions. This is a *declarative* approach, which lends itself to formal proof techniques. One of the problems with operational style specification is that it requires more complete specification for meaningful simulation. Declarative constraints cannot be simulated per se, however their consistency can be verified at any time. Thus, declarative statements permit partial descriptions to be entered and analyzed, using faster static, numerical, and formal analysis techniques. Since both VHDL and Verilog are operational notations (as are C++ and Java), declarative constraints are a highly complementary enhancement even within today's HDL environment.

Specifying behavior at the system level is more complex. The semantic meaning of statements can be fuzzier, and more inter-dependencies occur between them. For example, the semantics to describe dataflow versus control flow behavior are typically quite different. High-level dataflow algorithms written in C are commonplace, but C code algorithms cannot adequately describe the design objectives of a RISC microprocessor. So it is not surprising that existing system-level notations all assume a specific "model of computation" in their underlying semantics. These computational models may use different semantic domains, however all could very easily occur simultaneously within a single piece of silicon. Thus, there is a need for SLDL to support those models of computation in a unified environment that preserves flexibility for re-partitioning. Exactly how many diferent models of computation are needed by industry is still an open question, but current thoughts are that only two or three well-chosen ones can meet 95% of anticipated needs.

To support the set of semantic domains needed in industry, it is expected that multiple language notations will be required. While multiple notations can share the same semantic domain, and may even share the common portion of several domains, it is not possible for any notation (and it's underlying semantics for computation) to support the exclusive regions of multiple semantic domains (as shown in Figure 3).

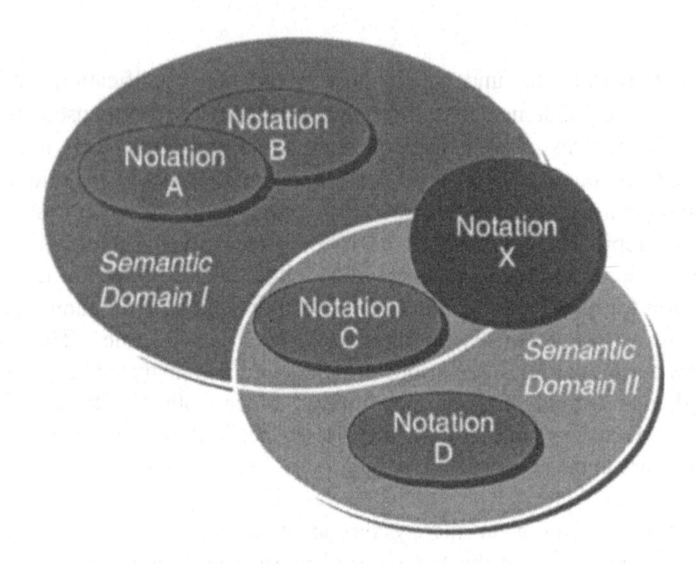

Figure 3: Semantic Domain Notational Space

Since hardware is inherently concurrent, modeling the proper granularity of interacting behavior in hardware can be critical. For example, trading off high level features of an on-chip communication bus may become too awkward if every detailed handshake must be modeled explicitly; yet refinement for timing budgeting purposes may require that detail. Time should also be represented at multiple levels of abstraction. Designers should not be forced into defining clock edges in absolute time before they know the required clock structures and clock rates. In those cases, it can be more natural to specify events using relative time references or abstract synchronization points.

In the past, most designs largely ignored the effects of interconnect between components, while today these are recognized as the key challenge in timing convergence. Similarly, our emerging focus on component-based design (e.g. "IP reuse") may need to re-focus on the increasingly complex communication channels between them. Overall trends are towards asynchronous interfaces between largely synchronous component modules.

All requirements are not created equal. Some are more important than others, and some are fixed while others are subject to change. Dynamic requirements are often derived from analysis of fixed requirements and designer experience, and evolve as more detail is learned about the system. One could even envision that tradeoffs among competing requirements might imply arbitrary functions that could vary the relative priority weighting of a function based on how well other requirements are satisfied.

Though it would be convenient to assume all constraints are driven from the top down, in fact bottom-up constraints are just as common. This has historically been due to lessons learned from similar designs, but in our coming era of design reuse will also be driven by existing virtual components integrated in the solution. Since all implementation constraints must eventually be satisfied, it is most desirable to capture and work with these known constraints as early in the design process as possible. Thus, SLDL intends to support both *specification* and *implementation* constraints.

The SLDL committee is implementing these features in a multi-phase approach. The first phase will focus on declarative features, primarily requirements and constraints. Since our existing HDL's lack broad system constraint information, this initial deliverable can be integrated into EDA tools along with VHDL and Verilog to immediately enhance our design capability. The second phase will deal with the more complex "models of computation" problem, supporting broader semantic richness for describing system-wide behavior.

System level semantics can become very complex. The requirements for SLDL have derived semantic implications, which will be resolved as a part of Phase II. For example, it is important to support the controlled use of non-determinism as a means to model an unknown external environment, and to reduce modeling/verification complexity. However, it is also critical to support compositional reasoning methods using formal techniques. This places restrictions on the semantic set for SLDL. At present, it is predicted that both static and dynamic dataflow scheduling can be supported within the SLDL language environment. Both synchronous and asynchronous control flow need to be supported as well; however, it may be necessary to restrict the models of communication supported to keep the asynchronous issues in check (see Figure 4 below).

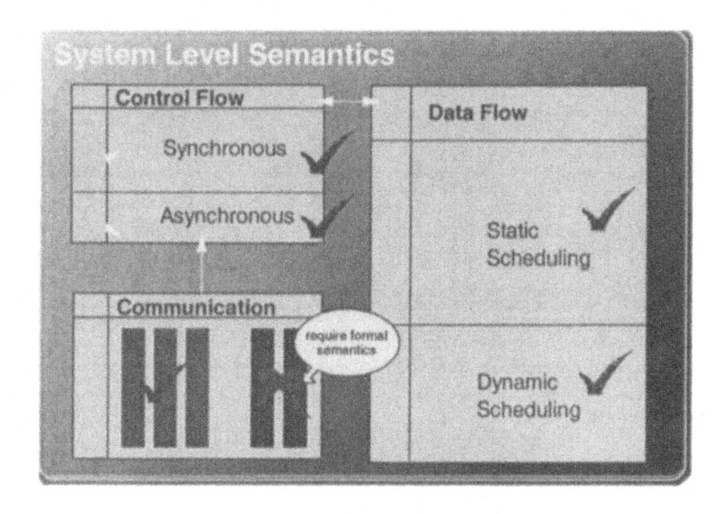

Figure 4: System Level Semantic Perspective for SLDL

Phase 1 support plans emphasize cross-domain constraint specification, including timing (latency), operational parameters such as power consumption, noise, gain, and other measurable engineering parameters. It will support information flow between components, including constraints on that exchange. Temporal abstraction will include fixed time, relative time, and statistical timing relationships. Structural descriptions of component interaction will be supported. It is likely that two classes of data types will emerge: those directly supporting tool analysis, and data types captured for designer reference. The committee plans to implement SLDL as an open ASCII format, suitable for either direct user input or generation by EDA tools.

Because research has shown it may not be possible, let alone practical, to design a self-contained semantic supporting all computational models, we instead plan to connect these "semantic islands" with semantic bridges. The overall architecture for both phases of SLDL is shown in Figure 5. It will attempt to integrate several behavioral models of computation using this concept of bridging semantics. In the figure, note that each language supporting a given model of computation implements a consistent set of semantics that are fully contained within a semantic domain. Fundamentally, we anticipate that only two definitions must be mapped using these "bridges": a) consistent abstractions of time, and b) consistent semantics for information flow (either hardware signals or software variables). It is hoped that this simpler approach will be both feasible for language design, and more easily implemented into EDA tools. Rather than an enormous language with ambiguous ways to describe the same thing, this approach should help in partitioning the language design into semantics where each computational model is easier to write, and easier for tools to process.

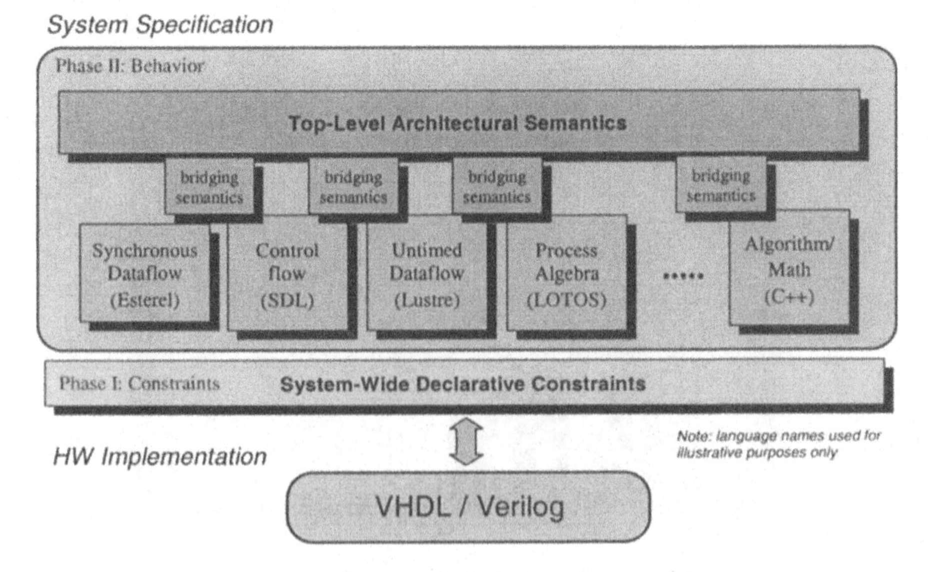

Figure 5: SLDL "Plug N' Play" Architecture Concept

5. Requirements and Features Planned

Most of the key features of SLDL have emerged and solidified over the past 18 months; they are summarized below. These should be considered high-level design objectives for SLDL, not detailed language requirements. Such requirements for the details of the language environment will follow at a later stage.

REQUIREMENTS AND CONSTRAINTS CAPTURE:

- Specify partial descriptions.
- Specify the value and the meaning of parametric values.
- Track the satisfaction of a requirement or constraint.
- Divide "responsibility" for the requirement between components.
- State things in a domain independent fashion.
- Support the use of templates to fill in system requirements.
- Cross-domain constraint specification.

HARDWARE-SOFTWARE CO-DESIGN:

- Support early abstraction performance models of microprocessors, real-time operating systems, and application algorithms.
- Provide semantic consistency for interfaces of performance models, regardless of abstraction level.
- Abstract specification of operating system models.
- Standard modeling template for device drivers.

SYSTEM LEVEL TO COMPONENT INTERFACES:

- Reflect system level constraints in the implementable component-level domain.
- Pass bottoms-up information into the SLDL environment.
- Map system level interface protocols into known or new ways of communication at the interface level.
- Describe a component without making assumptions about the implementation domain.
- Enable smooth transition from system level into implementation domains.
- Establish a structured mapping between an event at the systems level and an event, or a set of events, at the component level.
- Support specification from the point of view of the interface alone.
- Collect data gained by experience on other, similar systems.

FORMAL SEMANTICS:

- Domain theories: within a semantic domain (or application domain), permit a set of default properties and relationships to be established that apply within that domain. These properties can then be reused rather than rewritten, and encourage consistency in formal analysis.
- Partial specifications: enable incomplete descriptions that can be analyzed using formal consistency techniques.
- Composable descriptions: ensure that all semantics in SLDL encourage compositional and hierarchical techniques by defining clean and orthogonal semantic elements that do not assume flat construction.
- Extensibility: all language definitions should be made under the assumption that extensions will be required at a future date, and permit both standard and user-defined extensions wherever possible.
- Views: SLDL should support the concept of "views" into a description, to present only the information that is relevant to the current analysis; other unrelated information can be hidden for simplicity in both user understanding and tool processing.

6. Related System level specification capabilities

Since SLDL intends to take advantage of existing work where feasible, it is relevant to consider the state of the art in the field. A few well-known examples are briefly described in this secton.

System-level specification is hardly a new idea. SDL, for example, is a flow chart system intended to describe real-time systems, originally defined as a graphical notation in the 1970's for documentation purposes. However, in the 1990's several commercial tools have supported simulation and automatic code generation. SDL is based on finite state machine principles, supports abstract data types, and is well suited for describing asynchronous control intensive behavior. In particular, SDL supports dynamic process creation, which is important in communication systems. Processes may be synchronized via the use of blocking states awaiting synchronization messages, or through non-blocking remote procedure calls. SDL can monitor performance constraints using special observer processes, and all messages between processes are visible on both the sending and receiving side.

Esterel is a fully synchronous language well suited for control intensive, reactive systems. Models are written in a formal imperative style that encourages the user to "write things only once". In Esterel, the user writes activities that wait on signals, emit signals, or preempt signals. In fact, controlling the life and death of activities through preemption is a fundamental concept in Esterel. The priority of activities is implicitly defined by the nesting of preemption. Exceptions to a regular activity flow are easily handled, and arbitrary data types are supported.

Finite state machines can be described using StateCharts, originally proposed by Harel in the mid-eighties. StateCharts defined a representation of states and state transitions, however most would agree it lacked expressiveness and thorough semantic for communication. As a result, over 20 variations now exist. As one variation that built on this work, SpecCharts provides enhancements that support hierarchical behaviors, behavioral completion, and termination through exceptions. These extensions also allow VHDL programs to be attached to both states and transitions.

The SPW system (from Cadence/Alta) has become a de facto standard in the communications, multimedia, and networking fields. One of SPW's strengths is its ability to mix fixed-rate processing, multi-rate processing, and control flow systems. This is achieved by linking synchronous dataflow, dynamic dataflow, and cycle-based control flow semantic models through simulation. SPW relies heavily on pre-existing libraries of functional blocks that are graphically interconnected; the user clusters these blocks together according to their semantic domain to simulate the system. COSSAP (from Synopsys) is a stream-driven simulation environment that uses a dynamic dataflow semantic, best suited for multi-rate processing systems. COSSAP's block diagram editor is fully hierarchical, and allows parameters to be passed up from lower levels of the hierarchy.

Other proprietary system languages often use custom extensions to software programming languages to fill in the missing semantics for concurrent hardware. None of these descriptions are interoperable, and all tend to emphasize a single semantic domain.

7. Status, Schedules, and Plans

Over the last six months, a handful of industrial representative examples have been collected as reference points for testing the validity of SLDL requirements. The examples are expected to represent at least 80-85% of industry need, but additional examples for any remaining requirements are still welcomed by the committee.

The requirements for phase 1 were completed and approved in June '98. During the 2nd half of 1998, language design will begin in earnest to start implementing those requirements. Early prototyping between several of the EDA suppliers and user companies is anticipated to occur throughout 1999 and year 2000. After several prototyping iterations, first industrial use is being targeted by end of year 2000. Recently, several companies have developed a "pseudo code" of SLDL features to aid in early analysis of the many language design options. A number of important semantic and structural issues were identified from this useful exercise, and the resulting decisions should help SLDL's formal language design phase.

While phase 2 has yet to be scheduled, SLDL leadership expects it to start in mid-'99, likely requiring 3 or 4 years to complete. This is an aggressive target, however the expectation of leveraging much existing good work in behavioral semantics should help significantly.

8. Applications for SLDL Use in Design Flows

Many readers may be wondering how SLDL will be used in real design flows. There are actually three primary uses anticipated, all leveraging integration into both existing and new design automation tools:

1. Early system-wide estimation supporting rapid architectural exploration:
   - performance
   - power
   - silicon area, I/O, code size, bus size,...
2. Early analysis of system-wide behavior:
   - exploration of features vs. cost
   - formal verification
   - executable functional validation
3. Tracking of requirements, constraints, and priorities to assist large, possibly dispersed design teams.

SLDL phase 1 will assist with items (1) and (3), while item (2) will largely require the capabilities planned for phase 2.

It is hoped that the availability of such early design information, organized into architectural catagories and divisible into "what-if" partitions, will increase the opportunities for high-level estimation tools that support architectural tradeoffs. Much of this new emphasis on estimation technology should take advantage of predictive and statistical correlation heuristics, to provide hints on how the architecture can be improved. In addition, constructive methods that perform rapid behind-the-scenes synthesis will also be welcomed to facilitate system design needs in EDA tool flows.

9. Getting Involved

There are numerous ways to get involved. First, the reader is invited to visit the SLDL websites, to review all work done to date for additional background, and to evaluate the relevance of SLDL objectives against individual company needs. Readers are also encouraged to communicate this information on SLDL within one's company, and join the email reflector to participate in open discussions as SLDL evolves. Companies not already represented in the initiative may request official participation on the SLDL committee. Contact information is available on the SLDL website at: <http://www.inmet.com/SLD>.

10. Conclusions

The beginning of a large paradigm shift in design is coming, and with it new capabilities for design expressiveness. While there is significant risk in this bold initiative, there appears to be even stronger business urgency and worldwide support. New supporting design methodologies will eventually emerge to take full advantage of SLDL's potential; however, significant benefits for system-on-a-chip designs should be possible within just a few years. Because SLDL is on a fast track for year 2000 deployment, leading electronics designers and EDA suppliers alike will want to follow its progress closely.

HARDWARE/SOFTWARE CO-SYNTHESIS ALGORITHMS

WAYNE WOLF
Department of Electrical Engineering
Princeton University

This is an updated version of Chapter 2 of J. Staunstrup and W. Wolf (eds.), Hardware/Software Co-Design: Principles and Practice, Kluwer Academic Publishers, 1997.

1. Introduction

This chapter surveys methodologies and algorithms for hardware-software co-synthesis. While much remains to be learned about co-synthesis, researchers in the field have made a great deal of progress in a short period of time. Although it is still premature to declare an authoritative taxonomy of co-synthesis models and methods, we can now see commonalities and contrasts in formal models and algorithms.

Hardware-software co-synthesis *simultaneously* designs the software architecture of an application and the hardware on which that software is executed [42]. The problem specification includes some description of the required **functionality**; it may also include **non-functional requirements** such as performance goals. Co-synthesis creates a **hardware engine** consisting of one or more **processing elements (PEs)** on which the functionality is executed; it also creates an **application architecture**, which consists in part of an **allocation** of the functionality onto PEs in the hardware engine. Some of the functions may be allocated to programmable CPUs, while others may be implemented in ASICs which are created as a part of the co-synthesis process. Communication channels to implement the communication implied by the functionality must also be included in the system architecture—in general, this entails both hardware elements and software primitives. In order to complete the implementation, a schedule for execution is required, along with a component mapping and other details.

If either the hardware engine design or the application architecture were given, the problem would not be a co-design problem. Co-synthesis allows

A. A. Jerraya and J. Mermet (eds.), System-Level Synthesis, 189–217.

trade-offs between the design of the application and the hardware engine on which it executes. While such trade-offs may not be as important in data processing applications, where application programs are regularly changed and the hardware engine is large and immovable, trade-offs are particularly important in embedded computing. Embedded computing applications almost always have cost constraints; they may also have power, weight, and physical size limitations. Being able to, for example, modify the hardware engine to simplify some aspect of the application software is often critical in meeting the requirements of sophisticated embedded systems.

There are several different styles of co-design, depending on what assumptions are made about the specification and the components and what elements of the system are synthesized. The next section sets the background for co-synthesis. Section 2 reviews some background material: models for the behavior of the system to be synthesized, models of architectures, and real-time systems. Section 4 introduces **hardware/software partitioning**, which maps a behavioral specification onto a hardware architecture consisting of a CPU and multiple ASICs. Section 5 describes algorithms for **distributed system co-synthesis**, which synthesizes arbitrary hardware topologies.

There are strong reasons to use embedded CPUs to help implement complex systems. Implementing a system in software on an existing CPU is much simpler than designing logic by hand, so long as other constraints such as speed, power consumption, etc. can be met. Software can also be changed quickly and in many cases much more cheaply than is possible when changing an ASIC.

In this naive view, the implementation choices are between one CPU and an all-custom hardware solution. If we can find a CPU that can execute the code at the required speed and meet the other non-functional constraints, we use the CPU, otherwise we bite the bullet and design custom hardware. But in fact, the choices are more complex and subtle than that—we can design systems which include combinations of CPUs and special-purpose function units which together perform the required function. When we design sophisticated, heterogeneous systems, we need hardware/software co-synthesis to help us evaluate the design alternatives.

Why should we ever use more than one CPU or function unit? Why not use the fastest CPU available to implement a function? There are two reasons. The first is that CPU cost is a non-linear function of performance. Figure 1 shows the June, 1998 retail prices of two types of microprocessors at various clock rates. The higher-performance CPUs come at a cost premium. This fact should be no surprise since the maximum operating frequencies produced by IC manufacturing lines are normally distributed.

The non-linear cost/performance curve of CPUs implies that paying

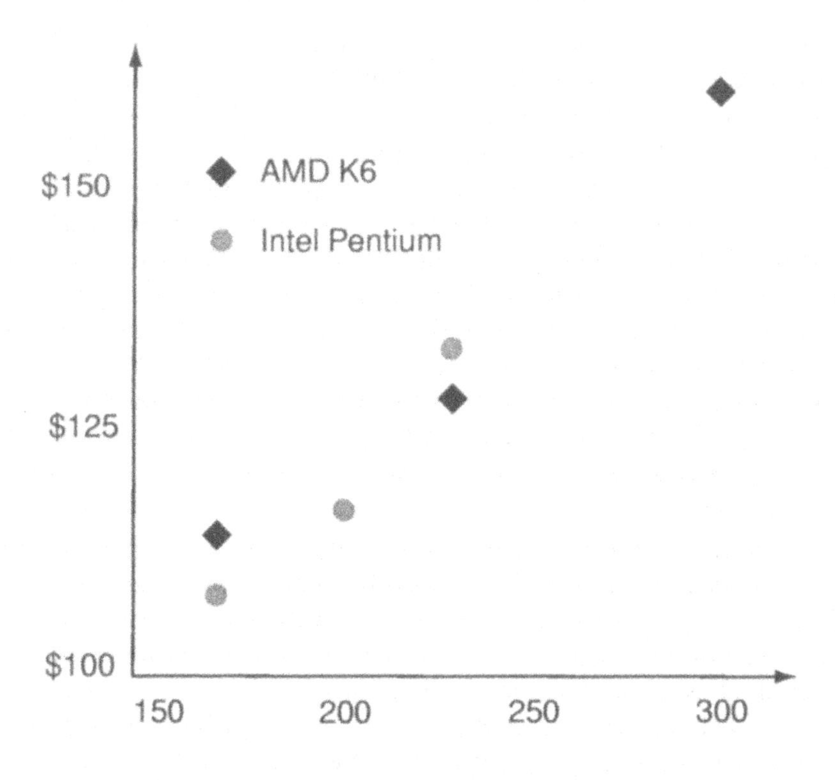

Figure 1. CPU performance as a function of price.

for performance in a uniprocessor can be very expensive. If a little extra processing power is needed to add an extra function to the system, the performance required to execute that function will be increasingly costly. It is often cheaper to use several small processors to implement a function, even when communication overhead is included. And the processors in the multiprocessor need not be CPUs—they may be special-purpose function units as well. Using a special-purpose unit to accelerate a function can result in a much cheaper implementation (as measured in manufacturing cost, not necessarily in design cost) than using even a dedicated CPU for that function. By co-designing a heterogeneous multiprocessor which uses special-purpose function units for some operations and consolidating the remaining operations on one or a few CPUs, we can dramatically reduce manufacturing cost.

The second reason not to use one large CPU is that scheduling overhead increases the cost of adding functionality even more. We will see in Section 3 that, under reasonable assumptions, a processor executing several processes cannot be utilized 100% of the time—some time (31% of the time, under the rate-monotonic model) must be reserved because of the uncertainty in the times at which the processes will need to execute. In all but the simplest systems, external and internal implementation factors conspire to make it impossible to know exactly when data will be ready for a process and how long the process will take to execute. Almost any data dependencies will, for example, change the speed at which a piece of code executes for different data values. Even if external data is produced at exact intervals (which is not always the case), when that data is fed into a process with data-dependent execution times, the process will produce its outputs at varying times. When that data is fed into the next process in a data-dependency chain, that process will see jitter in the times at which its data is available. Because we do not know exactly when computations will be ready to run, we must keep extra CPU horsepower in reserve to make sure that all demands are met at peak load times. Since that extra CPU performance comes at non-linearly-increasing cost, it often makes sense to divide a set of tasks across multiple CPUs.

The job of hardware/software co-synthesis is to create both a **hardware architecture** and a **software architecture** which implements the required function and meets other **non-functional** goals such as speed, power consumption, etc. The notion of a hardware architecture is familiar—it is described as some sort of block diagram or equivalent. The reader may be less comfortable with a software architecture, but such things do exist. While the division of a program into modules, functions, etc. can be considered its architecture, a more useful view for embedded systems is that the software architecture is defined by the **process structure** of the code. Each process executes sequentially, so the division of functions into processes determines the amount of parallelism which we can exploit in the design of the system. The process structure also influences other costs, such as data transfer requirements. However, dividing a system into too many processes may have negative effects both during and after design: during design, having too many processes may obscure design decisions; as the system executes, context-switching and scheduling overhead may absorb any parallelism gains. As a result, proper divison of a function into a set of software processes is critical for obtaining a cost-effective implementation.

2. Preliminaries

Some co-synthesis systems which deal with **single-rate** behavior while others can synthesize **multi-rate** behavior. A single-rate system would generally perform a single complex function. A good example of a multi-rate system is an audio/video decoder: the audio and video streams are encoded at different sampling rates, requiring the specification to be written as multiple components, each running at its own rate.

No matter what model is used for behavior, we are often interested in

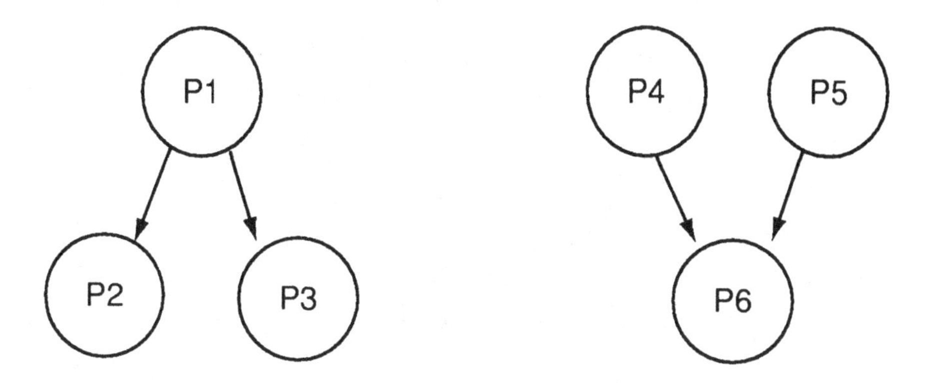

Figure 2. An example task graph.

specifying the required performance, this being the most common form of non-functional specification. We generally assume that the behavior model describes one iteration of a behavior which is repeatedly executed. Two performance parameters are commonly used:

- The **rate** at which a behavior is executed is the reciprocal of the maximum interval between two successive initiations of the behavior.
- The **latency** of a behavior is the required maximum time between starting to execute the behavior and finishing it.

Rate and latency are both required in general because of the way in which data may arrive. In simple systems, data will arrive at a fixed time in each sampling period and go through a fixed set of processing. In more complex systems, data may arrive at different points in the processing period; this may be due either to exernal causes or because the data is generated by another part of the system which has data-dependent execution times. Rate specifies the intervals in which data samples may occur, while latency determines how long is allowed to process that data.

The standard model for single-rate systems is the **control/data flow graph (CDFG)**. A CDFG's semantics imply a program counter or system state, which is a natural model for either software or a harwire-controlled datapath. The unified system state, however, makes it difficult to describe multi-rate tasks.

A common multi-rate model is the **task graph**.

The most common form of the task graph is shown in Figure 2. The task graph has a structure very similar to a data flow graph, but in general the nodes in a task graph represent larger units of functionality than is common in a data flow graph. Edges in the task graph represent data communication. This model describes multiple flows of control because one datum output from a node can be sent to two successor nodes, both of which are activated and executed in some unspecified order. Unfortunately,

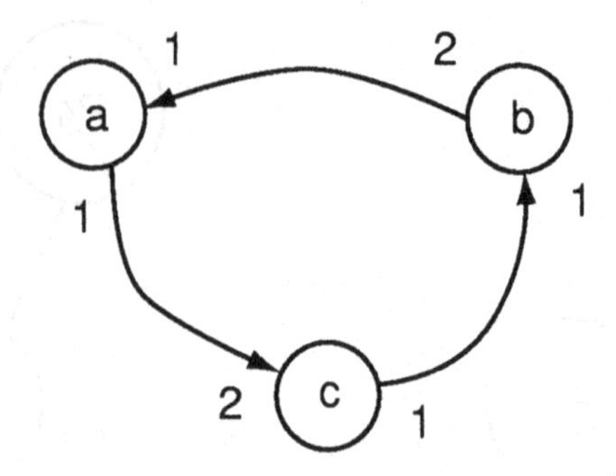

Figure 3. A synchronous data flow graph.

the terminology used for the task graph varies from author to author. We prefer to refer to the nodes as **processes**, since that is what the nodes most often represent. Some authors, however, refer to each node as a task. The task graph may contain several unconnected components; we prefer to call each component a **subtask**. Subtasks allow the description of multi-rate behavior, since we can assign a separate rate to each subtask.

A multi-rate behavior model particularly well-suited to signal processing systems is the **synchronous data flow graph (SDFG)** [25]. An example is shown in Figure 3. As in a data flow graph, nodes represent functions and arcs communication. But unlike most data flow graphs, a valid SDFG may be cyclic. The labels at each end of the arcs denote the number of samples which are produced or consumed per invocation—for example, process b emits two samples per invocation while process a consumes one per invocation. This implies that a runs at twice the rate of b. Lee and Messerschmitt developed algorithms to determine whether a synchronous data flow graph is feasible and to construct schedules for the SDFG on either uniprocessors or multiprocessors.

The **co-design finite-state machine** (CFSM) is used as a model of behavior by the POLIS system [4]. A CFSM is an event-driven state machine—transitions between states are triggered by the arrival of events rather than by a periodic clock signal. As a result, the component CFSMs in a network need not all change state at the same time, providing a more abstract model for time in the system. A simple CFSM is illustrated in Figure 4. When the machine receives an event, its action depends on its current state and the defined transitions out of those states. Upon taking a transition to a new state, the machine may emit events, which can go either to the outside world or other CFSMs in the system. (Transitions in the figure are annotated with input / output.) A network of CFSMs can be interpreted as a network of non-deterministic FSMs, though that description will generally be less compact than the CFSM network.

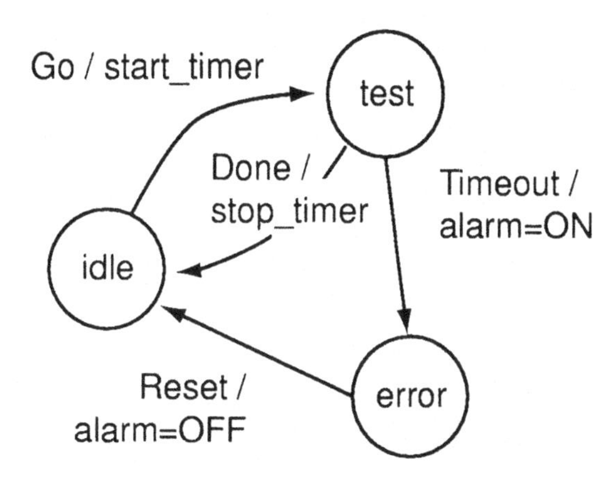

Figure 4. A co-design finite-state machine.

3. Architectural Models

We also need architectural models to describe the implementation. The allocation of elements of the functional description to components in the hardware architecture is generally described as a map, so the greatest need is for a description of the hardware engine. Chapter [1] describes CPU architectures in detail; here we concentrate on basic models sufficient for cost estimation during co-synthesis.

The engine itself is generally modeled as a graph. In the simplest form, processing elements are represented by nodes and **communication channels** by edges. However, since an edge can connect only two nodes, busses are hard to accurately model in this way. Therefore, it is often necessary to introduce nodes for the channels and to use edges to represent nets which connect the PEs and channels. Each node is labeled with its type.

When pre-designed PEs or communication channels are used to construct the hardware engine, a **component technology library** describes their characteristics. The technology library includes general parameters such as manufacturing cost, average power consumption, *etc.* The technology description must also include information which relates the PEs to the functional elements—a table is required which gives the execution time required for each function that can be executed on that type of PE.

When a custom ASIC is part of the hardware engine, its clock rate is an important part of the technology model. Detailed cost metrics, such as the number of clock cycles required to implement a particular function, must be determined by synthesizing the function using high-level synthesis or otherwise estimating the properties of that implementation. The synthesis system usually makes assumptions about the communication mechanism used to talk to the rest of the system—shared memory, for example.

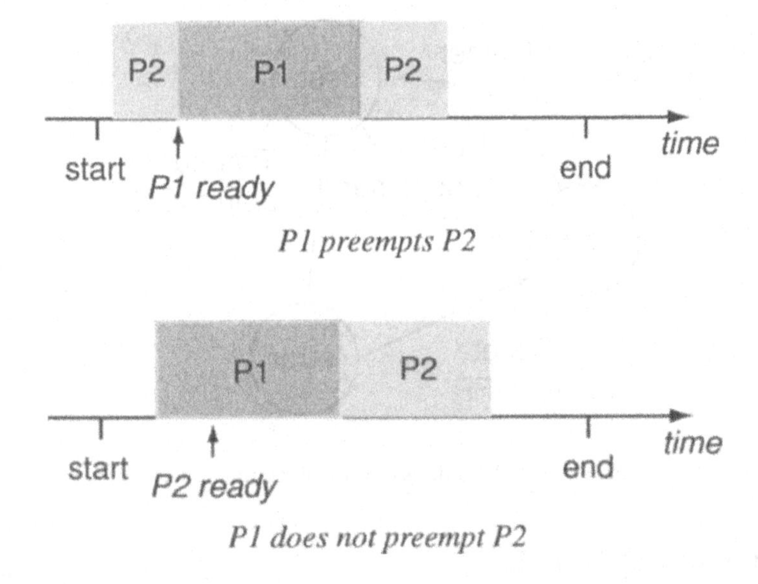

Figure 5. An example of prioritized process execution.

Some background in CPU scheduling is useful for co-synthesis since these systems must run several components of the functionality on a CPU. The units of execution on a CPU are variously called **processes**, **threads**, or **lightweight processes**. Processes on a workstation or mainframe generally have separate address spaces to protect users from each other; in contrast, a lightweight process or thread generally executes in a shared memory space to reduce overhead. We will generally use these terms interchangeably since the details of operating system implementation are not important for basic performance analysis.

The processes on a CPU must be scheduled for execution, since only one may run at any given time. In all but the simplest implementations, some decision-making must be done at run time to take into account variabilities in data arrival times, execution times, *etc.* The simplest scheduling mechanism is to maintain a queue of processes which are ready to run and to execute them in order. A more complex mechanism is to assign **priorities** to each process (either statically or dynamically) and to always execute the highest-priority process which is ready to run. Figure 5 illustrates the prioritized execution of two processes *P1* and *P2*, where *P1* has higher priority (presumably because it has a tighter deadline to meet). In the first case, *P2* has become ready to execute and obtains control of the CPU because no higher-priority process is running. When *P1* becomes ready, however, *P2*'s execution is suspended and *P1* runs to completion. When *P1* is running and *P2* becomes ready, however, *P2* has to wait for *P1* to finish. Unlike timesharing systems, embedded and real-time systems generally do not try to time-slice access to the CPU, but allow the highest-priority process to run to completion. Prioritized execution gives significantly higher utilization of the CPU resources—in general, prioritization allows a slower CPU to meet the performance requirements of a set of processes than would be required for non-prioritized execution.

The real-time systems community has studied the execution of processes on CPUs to meet **deadlines**. The survey by Stankovic *et al.* [39] gives a good overview of results in real-time scheduling. Two common real-time scheduling algorithms are **rate-monotonic analysis (RMA)** [31], which is a static-priority scheme, and **earliest deadline first (EDF)**, which can be modeled as a dynamic priority scheme. RMA assumes that the deadline for each process is equal to the end of its period. Liu and Layland analyzed the worst-case behavior of a set of processes running on a CPU at different rates. They showed that, under their assumptions, a static prioritization was optimal, with the highest priority going to the process with the shortest period. They also showed that the highest possible utilization of the CPU in this case is 69%—some excess CPU capacity must always be left idle to await possibly late-arriving process activations. EDF does not assume that deadline equals period. This scheduling policy gives highest priority to the active process whose deadline is closest to arriving.

Lehoczky *et al.* [26] formulated the response time (the time from a process becoming ready to the time it completes execution for that period) for a set of prioritized processes running on a single CPU. For ease of notation, assume that the highest-priority process is numbered 1 and that the process under consideration is numbered i. Let p_j be the period of process j and c_j be its execution time. Then the worst-case response time for process i is the smallest non-negative root of this equation:

$$x = g(x) = c_i + \sum_{j=1}^{i-1} c_j \lceil x/p_j \rceil \tag{1}$$

The first term c_i merely captures the computation time required for process i itself. The summation captures the computation time expended by higher-priority processes; the ceiling function takes into account the maximum number of requests possible from each higher-priority process.

4. Hardware/Software Partitioning

Hardware/software partitioning algorithms were the first variety of co-synthesis applications to appear. In general, a *hardware/software partitioning* algorithm implements a specification on some sort of architectural template for the multiprocessor, usually a single CPU with one or more ASICs connected to the CPU bus. We use the term *allocation* for synthesis methods which design the multiprocessor topology along with the processing elements and software architecture.

In most hardware/software partitioning algorithms, the type of CPU is normally given, but the ASICs must be synthesized. Not only must the functions to be implemented on the ASICs be chosen, but the characteristics of those implementations must be determined during synthesis, making this a challenging problem. Hardware/software partitioning problems are usually single-rate synthesis problems, starting from a CDFG specification. The ASICs in the architecture are used to accelerate core functions, while the CPU performs the less computationally-intensive tasks. This style of architecture is important both in high-performance applications, where there may not be any CPU fast enough to execute the kernel functions at the required rate, and in low-cost functions, where the ASIC accelerator allows a much smaller, cheaper CPU to be used.

One useful classification of hardware/software partitioning algorithms is by the optimization strategy they use. Vulcan and Cosyma, two of the first co-synthesis systems, used opposite strategies: Vulcan started with all functionality in the ASICs and progressively moved more to the CPU to reduce implementation cost; while Cosyma started with all functions in the CPU and moved operations to the ASIC to meet the performance goal. We call these methodologies **primal** and **dual**, respectively, by analogy to mathematical programming.

4.1. ARCHITECTURAL MODELS

Srivastava and Brodersen [38] developed an early co-synthesis system based on a hierarchy of templates. They developed a multi-level set of templates, ranging from the workstation-level to the board architecture. Each template included both hardware and software components. The relationships between system elements were embodied in busses, ranging from the workstation system bus to the microprocessor bus on a single board. At each stage of abstraction, a mapping tool allocated components of the specification onto the elements of the template. Once the mapping at a level was complete, the hardware and software components were generated separately, with no closed-loop optimization. The components at one level of abstraction mapped onto templates at the next lower level of abstraction. They used their system to design and construct a multiprocessor-based robot arm controller.

Lin *et al.* [30, 40] developed an architectural model for heterogeneous embedded systems. They developed a set of parameterizable libraries and CAD tools to support design in this architectural paradigm. Their specifications are constructed from communicating machines which use a CSP-style rendezvous communication mechanism. Depending on the synchronization regime used within each component, adapters may be required to mediate communications between components. Modules in the specification may be mapped into programmable CPUs or onto ASICs; the mapping is supported in both cases by libraries. The libraries for CPUs include descriptions of various I/O mechanisms (memory mapped, programmed, interrupt, DMA). The communication primitives are mapped onto software libraries. For ASIC components, VHDL libraries are used to describe the communication primitives. Parameterizable components can be constructed to provide higher-level primitives in the architecture.

4.2. PERFORMANCE ESTIMATION

Hardt and Rosenstiel [14] developed an estimation tool for hardware/software systems. Their architectural model for performance modeling is a co-processor linked to the CPU through a specialized bus. They model communication costs using four types of transfer—global data, parameter transfer, pointer access, and local data—and distinguish between reads and writes. They use a combination of static analysis and run-time tracing to extract performance data. They estimate speed-up based on the ratio of hardware and software execution times and the incremental performance change caused by communication requirements.

Henkel and Ernst [15] developed a clustering algorithm to try to optimize the tradeoff between accuracy and execution time when estimating the performance of a hardware implementation of a data flow graph. They used a heuristic to cluster nodes in the CDFG given as input to Cosyma. Clustering optimized the size of the clusters to minimize the effects of data transfer times and to create clusters which are of moderate size. Experiments showed that clustering improved the accuracy of the estimation of hardware performance obtained by list scheduling the clusters.

4.3. VULCAN

Gupta and De Micheli's co-synthesis system uses a primal methodology, by starting with a performance-feasible solution and moving functionality to software to reduce cost. As a result, Vulcan places a great deal of emphasis on the analysis of concurrency—eliminating unnecessary concurrency is the way in which functions are moved to the CPU side of the partition.

The user writes the system function as a HardwareC program [23], which was originally developed for high-level synthesis. HardwareC provides some data types for hardware, adds constructs for the description of timing constraints, and provides serialization and parallelization constructs to aid in

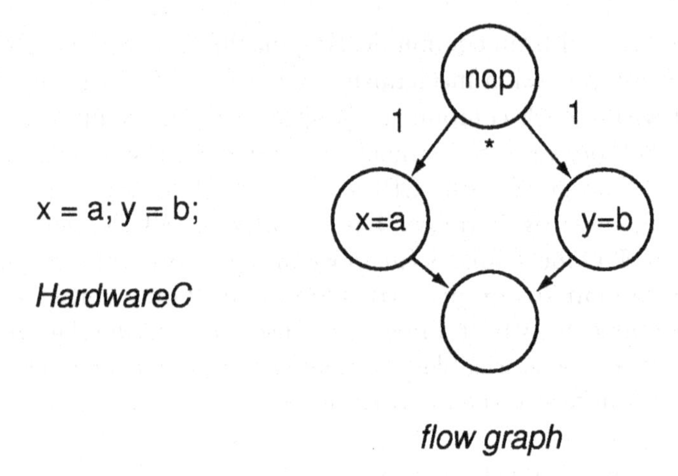

Figure 6. Conjunctive execution in HardwareC.

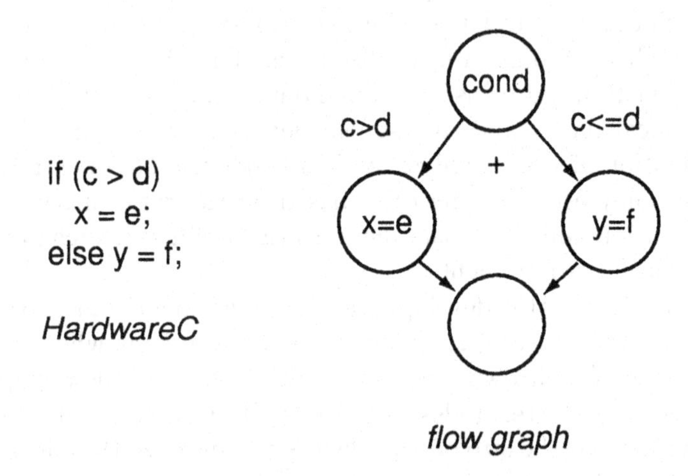

Figure 7. Disjunctive execution in HardwareC.

the determination of data dependencies. However, the fundamental representation of the system function is as a **flow graph**, which is a variation of a (single-rate) task graph. A flow graph includes a set of nodes representing functions and a set of edges specifying data dependencies. The operations represented by the nodes are typically low-level operations, such as multiplications, in contrast to the process-level functionality often represented by task graphs. Each edge has a Boolean condition under which the edge is traversed in order to control the amount of parallelism. Figure 6 illustrates a flow graph in which two assignments can be executed in parallel; a *nop* node is inserted to allow the flow of control to follow two edges to the two assignments. In this case, each edge's enabling condition is tautologically true. Figure 7 illustrates disjunctive execution—the conditional node controls which one of two assignments is created and the edge enabling conditions reflect the condition. The flow of control during the execution of the flow graph can split and merge, depending on the allowed parallelism.

The flow graph is executed repeatedly at some rate. The designer can specify constraints the relative timing of operators and on the rate of execution of an operator. Minimum and maximum timing constraints [13] between two nodes i and j specify minimum time l or on maximum time u between the execution of the operators:

$$\begin{aligned} t(v_j) &\geq t(v_i) + l_{ij} \\ t(v_j) &\leq t(v_i) + u_{ij} \end{aligned}$$

The designer can specify a minimum rate constraint $\rho_i <= R_i$, which specifies that the jth and $j+1^{th}$ executions of operator i should occur at the rate R_i. Maximum rate constraints can be similarly specified. Each non-deterministic delay operator must have a maximum rate specified for it so that the total run time of the flow graph can be bounded.

Vulcan divides the flow graph into **threads** and allocates those threads during co-synthesis. A thread boundary is always created by a non-deterministic delay operation in the flow graph, such as a wait for an external variable. Other points may also be chosen to break the graph into threads. Part of Vulcan's architectural target model is a scheduler on the CPU which controls the scheduling of all threads, including threads implemented either on the CPU or on the co-processor. Since some threads may be initiated non-deterministically, it is not possible to create a static dispatch procedure to control system execution.

The size of a software implementation of a thread can be relatively straightforwardly estimated. The biggest challenge is estimating the amount of static storage required; Vulcan puts some effort into bounding the size of the register spill set to determine a thread's data memory requirements. Performance of both hardware and software implementations of threads are estimated from the flow graph and basic execution times for the operators.

Partitioning's goal is to allocate threads to one of two partitions—the CPU (the set Φ_H) or the co-processor (the set Φ_S)—such that the required execution rate is met and total implementation cost is minimized. Vulcan uses this cost function to drive co-synthesis:

$$f(\omega) = c_1 S_h(\Phi_H) - c_2 S_s(\Phi_S) + c_3 \mathcal{B} - c_4 \mathcal{P} + c_5 |m| \quad (2)$$

where the c_i's are constants used to weight the components of the cost function. The functions S_h and S_s measure hardware and software size, respectively. $\mathcal{B}$ is the bus utilization, $\mathcal{P}$ is the processor utilization, which must be less than 1 (= 100% utilization), and m is the total number of variables which must be transferred between the CPU and the co-processor.

The first step in co-synthesis is to create an initial partition; all the threads are initially placed in the hardware partition Φ_H. The co-synthesis algorithm then iteratively performs two steps:

- A group of operations is selected to be moved across the partition boundary. The new partition is checked for performance feasibility by computing the worst-case delay through the flow graph given the new thread times. If feasible, the move is selected.
- The new cost function is incrementally updated to reflect the new partition.

Once a thread is moved to the software partition, its immediate successors are placed in the list for evaluation in the next iteration; this is a heuristic to minimize communication between the CPU and co-processor. The algorithm does not backtrack—once a thread is assigned to Φ_S, it stays there.

Experimental results from using Vulcan to co-synthesize several systems show that co-synthesis can produce mixed hardware-software implementations which are considerably faster than all-software implementations but much cheaper than all-hardware designs.

4.4. COSYMA

The Cosyma system [16] uses a dual methodology—it starts out with the complete system running on the CPU and moves basic blocks to the ASIC accelerator to meet performance objectives. The system is specified in C^x, a superset of the C language with extensions for specifying time constraints, communication, processes, and co-synthesis directives. The C^x compiler translates the source code into an extended syntax graph (ESG), which uses data flow graphs to represent basic blocks and control flow graphs to represent the program structure. The ES graph is partitioned into components to be implemented on the CPU or on the ASIC.

Cosyma allocates functionality to the CPU or ASIC on a basic block level—a basic block is not split between software and custom hardware implementations. Therefore, the major cost analysis is to compare the speed of a basic block as implemented in software running on the CPU or as an ASIC function. When a function is evaluated for reallocation from the CPU to the ASIC, the change in the basic block b's performance (not the total system speedup) can be written as:

$$\Delta c(b) = w(t_{HW}(b) - t_{SW}(b) + t_{com}(Z) - t_{com}(Z \cup b)) \times It(b) \qquad (3)$$

where $\Delta c(b)$ is the estimated decrease in execution time, the $t(b)$s are the execution times of hardware and software implementations, $t_{com}(Z)$ is the estimated communication time between CPU and co-processor, given a set Z of basic blocks implemented on the ASIC, and $It(b)$ is the total number of times that b is executed.

The weight w is chosen to drive the estimated system execution time T_S toward the required execution time T_c using a minimum number of basic blocks in the co-processor (which should give a minimal cost implementation of the co-processor). The time components of δc are estimated:

- $t_{SW}(b)$ is estimated by examining the object code for the basic block generated by a compiler.
- $t_{HW}(b)$ is estimated by creating a list schedule for the data flow. Assuming one operator per clock cycle, the depth of the schedule gives the number of clock cycles required.
- $t_{com}(Z \cup b)$ is estimated by data flow analysis of the adjacent basic blocks. In a shared memory implementation, the communication cost is proportional to the number of variables which need to be communicated.

Henkel et al. report that limiting the partitioning process to basic blocksbasic block only limited the performance gains that could be achieved—by moving only basic blocks into the co-processor, the speedup gained was typically only 2×. This observation coincides with the observations of compiler designers, who observe a limited amount of intra-basic-block parallelism. As a result of that initial experience, they implemented several control-flow optimizations to increase the number of parallel operations that can be implemented in the co-processor: loop pipelining, speculative execution of branches with multiple branch prediction, and operator pipelining. Cosyma uses the BSS synthesis system to both estimate and implement the co-processor.

Given these optimizations, Cosyma produced CPU-co-processor implementations with speedups ranging from 2.7 to 9.7 times over the all-software implementation. CPU times ranged from 35 to 304 seconds on a typical workstation.

5. Distributed System Co-Synthesis

Distributed system co-synthesis does not use an architectural template to drive co-synthesis. Instead, it creates a multiprocessor architecture for the hardware engine as part of co-synthesis. The multiprocessor is usually heterogeneous in both its processing elements, communication channels, and topologies. This style of co-synthesis often puts less emphasis on the design of custom ASICs and more on the design of the multiprocessor topology.

(The multiprocessor may consist of characterized ASICs as well as CPUs.) Heterogeneous distributed systems are surprisingly common in practice, particularly when one or two large CPUs are used in conjunction with small microcontrollers and ASICs to deliver performance and complex functionality at low cost.

5.1. AN INTEGER LINEAR PROGRAMMING MODEL

Prakash and Parker developed the first co-synthesis method for distributed computing systems [34]. They developed an integer linear programming (ILP) formulation for the single-rate co-synthesis problem and used general ILP solvers to solve the resulting system of equations.

Their methodology started with a single-rate task graph and a technology model for the processing elements, the communication channels, and the processes' execution characteristics on them. Given these inputs, they can create a set of variables and an associated set of constraints.

One set of variables describes the task graph:

- The variables $S_1, \ldots$ denote the processes.
- An input variable $i_{1,x}$ represents the xth input to process S_1.
- An output variable $o_{1,y}$ represents the yth output from process S_1.
- Input and output variables may also have parameters which specify what fraction of the process can start/finish before the input/output action.
- The volume of data transferred between processes S_1 and S_2 is written as $V_{1,2}$.

Another set of variables represents the technology model:

- The parameter $D(T, S_1)$ gives the execution time required for process S_1 on a processor of type T.
- The transfer time for a unit of data within a PE is given as D_{CL}. The time required to transfer a unit of data across a communication channel is D_{CR}.

The model allows for both local data transfers and multi-hop data transfers.

A set of auxiliary variables and help define the implementation:

- Process execution timing variables: start time, end time.
- Data transfer timing variables: transfer start time, transfer end time.
- Process-to-PE mapping variable: $\delta_{d,1} = 1$ denotes that process S_1 will be allocated to PE d.
- Data transfer type variables: local or multi-hop (non-local).

And finally, a set of constraints define the structure of the system:

- Processor-selection constraints: exactly one of the $\delta_{d,1}$ must be equal to 1.

- Data-transfer type constraints: makes sure that a data transfer is either local or multi-hop, but not both and not neither.
- Input-availability constraints: make sure that data is not used by the sink process until after it is produced by the source process.
- Output-availability constraints: make sure that the data obeys the fractional output generation parameters.
- Process execution start/end constraints: uses the processor mapping to determine the amount of time from the start of process execution to process completion.
- Data-transfer start/end constraints: similar to process execution constraints, but using data transfer times $D()$.
- Processor-usage-exclusion: makes sure that processes allocated to the same PE do not execute simultaneously.
- Communication-usage-exclusion: similarly, makes sure that multiple communications are not scheduled on the same link simultaneously.

Given that they used general ILP solvers, they could solve only relatively small problems—their largest task graph included nine processes. Solving that problem took over 6000 CPU minutes on a processor of unspecified type. However, they did achieve optimal solutions on those examples which they could solve. Their examples provide useful benchmarks against which we can compare heuristic co-synthesis algorithms.

5.2. PERFORMANCE ANALYSIS

During co-synthesis, a proposed allocation must be tested for feasibility by determining if the processes can be feasibly scheduled to meet their deadlines. One way to do this is by simulation—running a set of input vectors through the process set and measuring the longest time required for any input vector. Simulation has been used by D'Ambrosio and Hu [10, 19] and Adams and Thomas [1], among others, to determine the feasibility of a system configuration. However, if not all input combinations are tested, the longest running time measured may not be the actual worst case; exhaustive simulation is not feasible for practical examples. In fact, simulation fails to give worst-case running times on even examples of only moderate size. The only way to be sure that a given system architecture is guaranteed to meet its deadlines is to use an algorithm which bounds the worst-case execution time. Multiprocessor scheduling is NP-complete, so obtaining the tightest bound is impractical. However, it is possible to get tight bounds in a reasonable amount of CPU time.

Yen and Wolf [46] developed an algorithm which computes tight bounds on the worst-case schedule for a multi-rate task graph on a distributed system. The algorithm is given the task graph, the allocation of processes to processors, the processor graph, and the worst-case execution times of the processes. The algorithm uses a combination of linear constraint solving and a variant of the McMillan-Dill max constraint algorithm to bound the execution time. Linear constraints are induced by data dependencies in the task graph—the start time of a sink node is bounded below by the completion time of the source node. Max constraints are introduced when processes share a PE—the start time of a node depends on the max of the completion times of its predecessors.

The goal of the algorithm is to find **initiation** and **completion** times for each process in the task graph. The times computed as $[min, max]$ pairs to bound the initiation and completion times. The algorithm iteratively applies linear-constraint and max-constraint solution phases to the task graph. At each step, the initiation and completion time bounds are tightened. The algorithm terminates when no more improvement occurs in a step.

The linear-constraint procedure is a modified longest path algorithm. It computes the latest request time and latest finish time for each process by tracing through the task graph from the source nodes. A simple longest path algorithm would add worst-case completion times for each process along the path. To obtain more accurate estimates, this procedure also estimates the **phases** of the processes. If one process is known to be delayed with respect to another, we can reduce the maximum amount of interference that one process can create for another based on the phase difference between their initiation times.

The max constraint procedure computes a request time and a finish time for each process, as measured relative to the start time for the task graph. Two runs of the algorithm are used to produce both upper and lower bounds on request and finish times. The max constraints induced by co-allocation of processes model the preemption of one process by another. McMillan and Dill showed that, although solving constraint systems including both min and max constraints is NP-complete, max-only constraint systems can be solved efficiently. Because multiprocessor scheduling is NP-complete, this procedure does not guarantee the minimum-length schedule, but it does produce tight bounds in practice in a very small number of iterations (typically < 5).

Balarin and Sangiovanni-Vincentelli [6] developed an efficient algorithm for the validation of schedules for reactive real-time systems. They model a system as a directed acyclic graph of tasks, each with a known priority and execution time; all tasks run on a single processor. Events are initiated by events (either external or internal) and the validation problem is cast in terms of the times of events in the system—namely, for every critical event transmitted from task i to task j, the minimum time between two executions of i must be larger than the maximum time between executions of i and j. They compute *partial loads* at various points in the schedule to determine whether any deadlines are not met.

Memory is an important cost in any system, but is particularly important for single-chip systems. Li and Wolf [28] developed a performance model for hierarchical memory systems in embedded multiprocessors. Caches must often be used with high-performance CPUs to keep the CPU execution unit busy. However, caches introduce very large variations between possible behaviors. Naive worst-case assumptions about performance—namely, assuming that all memory accesses are cache misses—are unnecessarily pessimistic. Li and Wolf model instruction memory behavior at the process level, which allows context switching to be efficiently modeled during the early phases of co-synthesis. Process-level modeling generates much tighter bounds on execution time which take into account the actual behavior of the cache. These estimates are also conservative—they do not overestimate performance. Agrawal and Gupta [2] developed a system-level partitioning algorithm to minimize required buffer memory. They use compiler techniques to analyze variable lifetimes in a multi-rate system and use the results to find a partitioning which minimizes buffering requirements. More recently, they developed a co-synthesis methodology based on this memory hierarchy model [29].

5.3. HEURISTIC ALGORITHMS

Wolf [44] developed a heuristic algorithm to solve Prakash and Parker's single-rate co-synthesis problem. Given a single-rate task graph, this algorithm: allocates a set of heterogeneous processing elements and the communication channels between them; allocates processes in the task graph to processing elements; and schedules the processes. The input to the algorithm includes the task graph and a technology description of the PEs and communication links. The major objective of co-synthesis is to meet the specified rate for execution, and the minor objective is to minimize the total cost of PEs, I/O devices, and communication channels. The algorithm is primal since it constructs a performance-feasible solution which it then cost-minimizes.

Co-synthesis starts by allocating PEs and ignoring communication costs; it later fills in the communication and devices required. Co-synthesis proceeds through five steps:

1. Create an initial feasible solution by assigning each process to a separate PE. Perform an initial scheduling on this initial feasible solution.
2. Reallocate processes to PEs to minimize total PE cost, possibly eliminating PEs from the initial feasible solution.

3. Reallocate processes again to minimize the amount of communication required between PEs.
4. Allocate communication channels.
5. Allocate I/O devices, either by using devices internally implemented on the PEs (counters on microcontrollers, for example) or by adding external devices.

The initial reallocation to minimize PE cost is the most important step in the problem, since processing elements form the dominant hardware cost component. This step consists of three phases: **PE cost reduction**, **pairwise merge**, and **load balancing**. The PE cost reduction step tries to move several processes at a time to achieve a good solution in a small number of moves. The procedure scans the PEs, starting with the least-utilized PE. It first tries to reallocate that PE's processes to other existing PEs. If all the processes can be reallocated, it eliminates the PE. If some processes cannot be moved to existing PEs, it replaces the PE with another, lower-cost PE which is sufficient to execute the remaining processes. Once all PEs have been cost-reduced, the algorithm tries to merge a pair of PEs into a single PE; the success of this step depends on the availability of a PE type in the component library which is powerful enough to run the processes of two existing PEs. After pairwise merging, the load is rebalanced. These three steps are repeatedly applied to the design until no improvement is obtained.

Experiments showed that this heuristic could find optimal results to most of Prakash and Parker's examples and near-optimal results for the remaining examples. It also showed good results on larger examples. The algorithm requires very little run time thanks to the multiple-move strategy used during PE cost minimization.

Wolf [43] extended this algorithm to cover object-oriented specifications. Object-oriented specifications provide useful information for co-synthesis because objects partition the system's functionality: objects themselves provide a coarse granularity of partitioning of data and code; methods which operate on objects provide finer-grained elements of code. Object-oriented heuristics try to keep data together as much as possible as specified in the object-oriented specification, but does allow data to be split across multiple processors if required. Methods may be moved to separate PEs to increase parallelism.

An alternative methodology for system optimization was developed by Kalavade and Lee [21]. Their **global criticality/local phase** algorithm iteratively improves the system performance (global criticality) and cost (local phase). The global criticality measure estimates the criticality of a node in the system schedule; this criterion is used to select which nodes in the behavior should be moved into the ASIC side of the partition. The local phase criterion is used to determine a low-cost implementation for a function. Two major criteria are used to select low-cost implementations for some of the functions. First, some functions are much cheaper to implement in one style than in another; the cheapest style is selected early. These heuristics are used to guide a partitioning algorithm which allocates nodes in the behavior to hardware or software implementations. Second, some nodes consume a large amount of resources (area or time) when implemented in one way.

At the start of an iteration in the GCLP, a global criticality measure is computed based on both mappedd and unmapped nodes. The next node to mapped is selected from the ready nodes—nodes whose predecessors have already been mapped. After calculating local phase change, the mapping for the selected node is found. Experiments show that this algorithm gives near-optimal solutions when compared to ILP solutions, but in a much smaller amount of CPU time.

The TOSCA system [3, 7] provides the designer with a set of design transformations for design space exploration. TOSCA's design representation and transformations target control-dominated systems which can be described as systems of communicating machines. Optimizations include: unfolding entry and exit actions; unfolding FSM enabling conditions; flattening hierarchies; replacing timers with counters; and collapsing a set of communicating processes into a single process.

5.4. SYSTEM PARTITIONING

Some co-synthesis methodologies are adapted from graph partitioning algorithms but do not make strong assumptions about the hardware architecture. An example is the system of Adams and Thomas [1]. They used partitioning algorithms to divide a unified behavioral description, equivalent to a single task, into hardware and software components. Their synthesis process uses operator-level partitioning, but explicitly compares hardware and software implementations for clusters. They implemented their methodology in a tool called Co-SAW (Co-Design System Architect's Workbench).

The first-step in synthesis, preprocessing, includes two parallel phases:

- One phase simulates the complete system model to obtain execution traces for system-level performance analysis.
- Another phase first clusters operators in the behavioral description to form tasks; it then estimates the costs of hardware and software implementations of the clusters.

The data gathered in the preprocessing step is used to drive synthesis. The synthesis algorithm modifies the initial partitioning (*code motion*) to try to expose useful parallelism. They use non-deterministically generated moves to select code motion operations both within and between processes. They showed that they could effectively explore the design space using Co-SAW.

Hou and Wolf [17] developed a partitioning algorithm used to increase the granularity of a task graph before co-synthesis. Partitioning is driven by a cost function which estimates the execution time for subgraphs of a task without explicit scheduling. Their algorithm restricts partitioning moves to create clusters in which data never flows out of the cluster and then back in. This restriction improves the estimation of the total execution time of a cluster, since a new cluster can never be blocked by waiting for data to be returned from an external process. They use a non-greedy clustering algorithm which selects clusters which meet the criteria and determine whether combining the processes into a single process will unnecessarily increase the execution time of the task. They showed that their algorithm could reduce the number of processes in tasks, greatly reducing the time required for sensitivity-driven co-synthesis, without significantly changing the quality of the co-synthesized implementation.

Shin and Choi [37] developed a partitioning and scheduling algorithm for embedded code. Their algorithm partitions a CDFG into a set of threads and then generates a combined static/dynamic scheduler for the system. They partition the CDFG based on accesses to hardware. After generating a large number of small threads, they combine threads in deadlock-free combinations to generate a reasonable number of threads. They then generate a scheduler which statically schedules threads whose initiation times are known and dynamically schedules threads which have data-dependent initiation times.

5.5. REACTIVE SYSTEM CO-SYNTHESIS

Control-dominated systems are often A **reactive system** reacts to external events; such systems usually have control-rich specifications. Several research efforts have targeted reactive real-time systems. both at the process-abstraction and the detailed design levels.

Rowson and Sangiovanni-Vincentelli [36] describe an interface-based design methodology for reactive real-time systems and a simulator to support that methodology.

POLIS [4] is a co-design system including both synthesis and simulation tools for reactive real-time embedded systems. As described in Section 2, POLIS uses the CFSM model to represent behavior. Central to the POLIS synthesis flow is the **zero-delay hypothesis**—events move between communicating CFSMs in zero time. (This hypothesis is also used by Esterel and other reactive system modeling methodologies.) The communication within the system can be analyzed by collapsing the component CFSMs into a single reactive block; the effects of non-zero response time can be taken into account by analyzing and modifying this block. After analysis, a partitioning step allocates functions to hardware or software implementations. Hardware and software components can be synthesized from the state transition graphs of the resulting CFSM models [24, 5]. Elements of the system can be mapped into microcontroller peripherals by modeling them with library CFSMs [18]. Chiodo et al. [8] describe a design case study with the POLIS system.

The Chinook system [9] is another co-synthesis system for reactive real-time systems. The main steps in the Chinook design flow are:

- hardware-software partitioning, which chooses hardware or software implementations and allocates to processors;
- device driver synthesis and low-level scheduling, which synthesizes an appropriate driver for a peripheral based on the results of allocation;
- I/O port allocation and interface synthesis, which assigns I/O operations to physical ports on the available microcontrollers;
- system-level scheduling, which serializes operations based on the more accurate timing estimates available at this point in synthesis; and
- code generation.

Ortega and Borriello [32] developed an algorithm for synthesizing both inter-CPU and intra-CPU communication as part of the Chinook system. Their system can insert queues, synthesize bus protocol implementations, and related operations to implement inter-CPU communication; it uses shared memory to implement intra-CPU communication.

5.6. COMMUNICATION MODELING AND CO-SYNTHESIS

Communication is easy to neglect during design but is often a critical resource in embedded systems. Communication links can be a significant cost of the total system implementation cost. As a result, bandwidth is often at a premium. For example, the automotive industry recently adopted an optical bus standard, but since the optical link is plastic for low cost and ruggedness, its bandwidth is limited. Limited bandwidth makes it important that required communications are properly scheduled and allocated to ensure that the system is both feasible and low-cost.

Daveau *et al.* [11] first formulated communication co-synthesis as an allocation problem in the Cosmos system [20]. Like TOSCA, Cosmos is targeted to communication-rich systems and provides a set of transformations for design exploration; however, Cosmos can map to a wider range of hardware architectures. The givens for the problem are: a set of processes and communication actions between them; and a library of communication modules (busses, etc.). The objective is to allocate the communication ac-

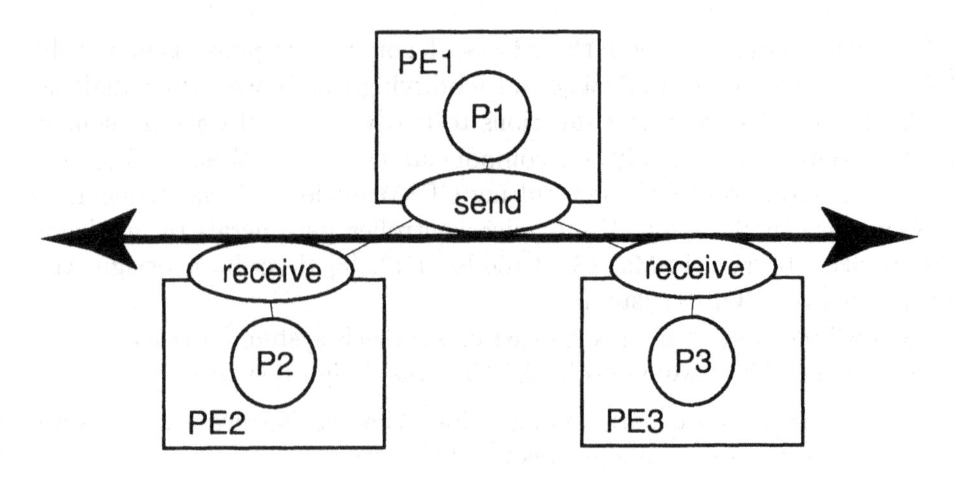

Figure 8. Communication processes.

tions to modules such that all the communications are implemented and feasibly scheduled; some cost objective may also be considered. Cosmos is described in more detail in Chapter [3].

Solar models a communication module by its protocol, its average transfer rate, its peak transfer rate, and its implementation cost. To allocate modules to communication actions, they build a decision tree to map out possible allocations. Nodes in the decision tree (except for the leaf nodes) correspond to the logical channel units which must be implemented. A path through the decision tree therefore enumerates a combination of logical channel units. An edge enumerates an allocation decision for the edge's source node. Each leaf node is labeled with the cost of the path which it terminates. The algorithm performs a depth-first search of the node to evaluate allocation combinations and choose a minimum-cost allocation. Results show that the algorithm can quickly identify a minimum-cost allocation of communication modules for the system.

The CoWare system [41] is a design environment for heterogeneous systems which require significant effort in building interfaces between components. CoWare is based on a process model, with processes communicating via ports using protocols. SHOCK is a synthesis subsystem in CoWare which generates both software (I/O drivers) and hardware (interface logic) interfaces to implement required communication. The interfaces are optimized to meet the requirements of the specification, taking into account the characteristics of the target components.

Yen and Wolf [45] developed a model for communication channels in embedded computing systems and a co-synthesis algorithm which makes use of that model. This model assumes that communication on a bus will be prioritized, with some communications requiring higher priority than others. Communication can be modeled by dummy processes which represent send and receive operations; these communication processes are allocated to the interface between the PE which executes the process doing the sending/receiving and the communication channel on which the communication is performed. In the example of Figure 8, one sending node on PE1 models the send operations from P1; each receiving process has its own receiving node. The total delay incurred by a communication action has two components: the intrinsic delay of the communication and the wait required to obtain the bus from competing processes. These components are similar to the case for the delay of processes on a CPU. However, communication is treated differently because busses cannot usually be preempted in the middle of a transaction. As a result, this work assumes that a communication is **non-preemptable**—a lower-priority communication will finish once it has started, even if another higher-priority communication request arrives after the lower-priority request has started. Non-preemptability complicates the calculation of the worst-case delay for a communication, but that delay can still be solved for by numerical techniques.

5.7. CO-SYNTHESIS FOR LOW POWER

The importance of design for low power is unquestioned today. The recent past has seen the development of models and algorithms for co-synthesis of low-power embedded systems.

Kirovski and Potkonjak [22] developed an synthesis algorithm for low-power real-time systems. Their algorithm uses a template of multiple CPUs communicating via multiple shared bussesbus; the busses also handle input and output data streams. Their algorithm prefers assigning a function to the CPU on which it requires the least power. The algorithm also balances system load across the processors to maximize opportunities for voltage scaling.

Fornaciari et al. [12] developed a methodology for methodology and optimization of control-dominated embedded systems within TOSCA. Their target architecture includes a data path, an embedded core processor, a control unit, a crossbar interconnection network, and memory. They use an analytic model to estimate the power consumption of hardware elements, based on net capacitance and switching activity. They estimate the power consumption of software components by bottom-up analysis from the basic blocks of the specification.

Paleologo et al. [33] developed a stochastic model for power management in electronic systems. The system components are modeled as Markov chains whose states represent various power management states. They show that finding the maximize average performance subject to a bound on maximum power consumption can be solved in polynomial time.

Li and Henkel [27] developed a framework for estimating energy dissipation in embedded systems. Their framework concentrates on the memory components of the system. They formulated models for various sytem components whose parameters can be measured experimentally. They use CPU execution traces to adjust the estimated energy for software components based on the cache behavior.

5.8. TECHNOLOGY MAPPING

Technology mapping—the binding of generic operations to specific types of components chosen from a technology library—has for the most part been given only minimal attention in most co-synthesis algorithms. Since many algorithms concentrate on high-level modeling and design space exploration, they have not considered detailed models for performance, power consumption, and other characteristics of the final components chosen. As co-synthesis matures, however, we see increasing attention paid to technology mapping.

Rhodes and Wolf [35] formulated a technology mapping problem for the arrival of data at an embedded microprocessor. Data may be sensed by a program on the CPU either by polling (reading data within a loop) or interrupts. Reading data by polling generally requires less CPU time than is used by the context switch of an interrupt, but the read may also be delayed by other computations. Microprocessors have a limited number of primitive interrupts available and more can be added only at some cost. Furthermore, some problems have a feasible schedule only when some data arrivals are implemented with polling rather than interrupts. Their algorithm heuristically allocates data arrivals first to polling, then to interrupts if necessary, based on evaluation of the system schedule.

6. Conclusions

Our understanding of co-synthesis has been hugely increased over the past several years. We now have a much better understanding of the types of co-synthesis problems which are of interest, a set of models for those problems, and some strong algorithms for both analysis and co-synthesis.

Of course, there remains a great deal of work to be done. Outstanding challenges range from the development of more accurate models for both hardware and software components to the coupling of co-synthesis algorithms with software and system design methodologies. Beyond this, larger, more realistic, and more examples would be a great benefit in driving co-synthesis research in realistic directions. However, given the enormous amount of progress over the past few years, we can expect the state of the art in co-synthesis to be well advanced over the next five to ten years.

Acknowledgments

Work on this chapter was supported in part by the National Science Foundation under grant CCR-9802661.

References

1. Jay K. Adams and Donald E. Thomas. Multiple-process behavioral synthesis for mixed hardware-software systems. In *Proceedings, 8th International Symposium on System Synthesis*, pages 10–15. IEEE Computer Society Press, 1995.
2. Samir Agrawal and Rajesh K. Gupta. Data-flow assisted behavioral partitioning for embedded systems. In *Proceedings, 1997 Design Automation Conference*, pages 709–712. ACM Press, 1997.
3. S. Antoniazzi, A. Balboni, W. Fornaciari, and D. Sciuto. A methodology for control-dominated systems codesign. In *Third International Workshop on Hardware-Software Codesign*, pages 2–9. IEEE Computer Society Press, 1994.
4. F. Balarin, M. Chiodo, P. Giusto, H. Hsieh, A. Jurecska, L. Lavagno, C. Passerone, A. Sangiovanni-Vincentelli, E. Sentovich, K. Suzuki, and B. Tabarra. *Hardware-Software Co-Design of Embedded Systems: The Polis Approach.* Kluwer Academic Publishers, Norwell, MA, 1997.
5. F. Balarin, M. Chiodo, A. Jurecska, L. Lavagno, B. Tabbara, and A. Sangiovanni-Vincentelli. Automatic generation of a real-time operating system for embedded systems. Presented at the Fifth International Workshop on Hardware/Software Codesign, March 1997.
6. Felice Balarin and Alberto Sangiovanni-Vincentelli. Schedule validation for embedded reactive real-time systems. In *Proceedings, 1997 Design Automation Conference*, pages 52–57. ACM Press, 1997.
7. A. Balboni, W. Fornaciari, and D. Sciuto. Partitioning and exploration strategies in the TOSCA design flow. In *Proceedings, Fourth International Workshop on Hardware/Software Codesign*, pages 62–69. IEEE Computer Society Press, 1996.
8. Massimiliano Chiodo, Daniel Engels, Paolo Giusto, Harry Hsieh, Atilla Jurecska, Luciano Lavagno, Kei Suzuki, and Alberto Sangiovanni-Vincentelli. A case study in computer-aided co-design of embedded controllers. *Design Automation for Embedded Systems*, 1(1-2):51–67, January 1996.
9. Pai Chou, Elizabeth A. Walkup, and Gaetano Borriello. Scheduling for reactive real-time systems. *IEEE Micro*, 14(4):37–47, August 1994.
10. J. G. D'Ambrosio and X. Hu. Configuration-level hardware/software partitioning for real-time embedded systems. In *Third International Workshop on Hardware-Software Codesign*, pages 34–41. IEEE Computer Society Press, 1994.
11. Jean-Marc Daveau, Tarek Ben Ismail, and Ahmed Amine Jerraya. Synthesis of system-level communication by an allocation-based approach. In *Proceedings, 8th International Symposium on System Synthesis*, pages 150–155. IEEE Computer Society Press, 1995.
12. W. Fornaciari, P. Gubian, D. Sciuto, and C. Silvano. Power estimation of embedded systems: a hardware/software codesign approach. *IEEE Transactions on VLSI Systems*, 6(2):266–275, June 1998.
13. Rajesh K. Gupta and Giovanni De Micheli. Specification and analysis of timing constraints for embedded systems. *IEEE Transactions on Computer-Aided Design of Integrated Circuits and Systems*, 16(3):240–256, March 1997.

14. Wolfram Hardt and Wolfgang Rosenstiel. Speed-up estimation for HW/SW-systems. In *Proceedings, Fourth International Workshop on Hardware/Software Codesign*, pages 36–43. IEEE Computer Society Press, 1996.
15. Jorg Henkel and Rolf Ernst. The interplay of run-time estimation and granularity in HW/SW partitioning. In *Proceedings, Fourth International Workshop on Hardware/Software Codesign*, pages 52–58. IEEE Computer Society Press, 1996.
16. Jorg Henkel, Rolf Ernst, Ullrich Holtmann, and Thomas Benner. Adaptation of partitioning and high-level synthesis in hardware/software co-synthesis. In *Proceedings, ICCAD-94*, pages 96–100. IEEE Computer Society Press, 1994.
17. Junwei Hou and Wayne Wolf. Process partitioning for distributed embedded systems. In *Proceedings, Fourth International Workshop on Hardware/Software Codesign*, pages 70–75. IEEE Computer Society Press, 1996.
18. Harry Hsieh, Luciano Lavagno, Claudio Passerone, Claudio Sansoè, and Alberto Sangiovanni-Vincentelli. Modeling micro-controller peripherals for high-level co-simulation and synthesis. In *Fifth International Workshop on Hardware/Software Codesign*, pages 127–130. IEEE Computer Society Press, 1997.
19. Xiaobo (Sharon) Hu and Joseph G. D'Ambrosio. Hardware/software communication and system integration for embedded architectures. *Design Automation for Embedded Systems*, 2(3/4):339–358, May 1997.
20. Tarek Ben Ismail, Mohamed Abid, and Ahmed Jerraya. COSMOS: a codesign approach for communicating systems. In *Third International Workshop on Hardware-Software Codesign*, pages 17–24. IEEE Computer Society Press, 1994.
21. Asawaree Kalavade and Edward A. Lee. The extended partitioning problem: hardware/software mapping, scheduling, and implementatino-bin selection. *Design Automation for Embedded Systems*, 2(2):125–164, March 1997.
22. Darko Kirovski and Miodrag Potkonjak. System-level synthesis of low-power hard real-time systems. In *Proceedings, 1997 Design Automation Conference*, pages 697–702. ACM Press, 1997.
23. David Ku and Giovanni de Micheli. *High-level Synthesis of ASICs under Timing and Synchronization Constraints.* Kluwer Academic Publishers, Boston, 1992.
24. L. Lavagno, J. Cortadella, and A. Sangiovanni-Vincentelli. Embedded code optimization via common control structure detection. Presented at the Fifth International Workshop on Hardware/Software Codesign, March 1997.
25. Edward Ashford Lee and David G. Messerschmitt. Statis scheduling of synchronous data flow programs for digital signal processing. *IEEE Transactions on Computers*, C-36(1):24–35, January 1987.
26. John Lehoczky, Liu Sha, and Ye Ding. The rate monotonic scheduling algorithm: exact characterization and average case behavior. In *Proceedings, Real-Time Systems Symposium*, pages 166–171. IEEE Computer Society Press, 1989.
27. Y. Li and J. Henkel. A framework for estimating and minimizing energy dissipation of embedded HW/SW systems. In *Proceedings, 1998 Design Automation Conference*, pages 188–183. ACM Press, 1998.
28. Yanbing Li and Wayne Wolf. A task-level hierarchical memory model for system synthesis of multiprocessors. In *Proceedings*, 34th *Design Automation Conference*, pages 153–156. ACM Press, 1997.
29. Yanbing Li and Wayne Wolf. Hardware/software co-synthesis with memory hierarchies. In *Proceedings, ICCAD '98*. IEEE Computer Society Press, 1998.
30. Bill Lin, Steven Vercauteren, and Hugo De Man. Embedded architecture co-synthesis and system integration. In *Proceedings, Fourth International Workshop on Hardware/Software Codesign*, pages 2–9. IEEE Computer Society Press, 1996.
31. C. L. Liu and James W. Layland. Scheduling algorithms for multiprogramming in a hard-real-time environment. *Journal of the ACM*, 20(1):46–61, January 1973.
32. Ross B. Ortega and Gaetano Borriello. Communication synthesis for embedded systems with global considerations. In *Fifth International Workshop on Hardware/Software Codesign*, pages 69–73. IEEE Computer Society Press, 1997.

33. G. A. Paleologo, L. Benini, A. Bogiolo, and G De Micheli. Policy optimization for dynamic power management. In *Proceedings, 1998 Design Automation Conference*, pages 182–187. ACM Press, 1998.
34. Shiv Prakash and Alice C. Parker. SOS: Synthesis of application-specific heterogeneous multiprocessor systems. *Journal of Parallel and Distributed Computing*, 16:338–351, 1992.
35. David L. Rhodes and Wayne Wolf. Allocation and data arrival design of hard real-time systems. In *Proceedings, ICCD-97*. IEEE Computer Society Press, 1997.
36. James A. Rowson and Alberto Sangiovanni-Vincentelli. Interface-based design. In *Proceedings, 1997 Design Automation Conference*, pages 178–183. ACM Press, 1997.
37. Youngsoo Shin and Kiyoung Choi. Software synthesis through task decomposition by dependency analysis. In *Digest of Papers, ICCAD '96*, pages 98–102. IEEE Computer Society Press, 1996.
38. Mani B. Srivastava, Trevor I. Blumenau, and Robert W. Brodersen. Design and implementation of a robot control system using a unified hardware-software rapid-prototyping framework. In *Proceedings, ICCD '92*. IEEE Computer Society Press, 1992.
39. John A. Stankovic, Marco Spuri, Marco Di Natale, and Giorgio C. Buttazzo. Implications of classical scheduling results for real-time systems. *IEEE Computer*, 28(6):16–25, June 1995.
40. Steven Vercauteren and Bill Lin. Hardware/software communication and system integration for embedded architectures. *Design Automation for Embedded Systems*, 2(3/4):359–382, May 1997.
41. D. Verkest, K. Van Rompaey, I. Bolshens, and H. De Man. CoWare—a design environment for heterogeneous hardware/software systems. *Design Automation for Embedded Systems*, 1(4):357–386, October 1996.
42. Wayne Wolf. Hardware-software co-design of embedded systems. *Proceedings of the IEEE*, 82(7):967–989, July 1994.
43. Wayne Wolf. Object-oriented co-synthesis of distributed embedded systems. *ACM Transactions on Design Automation of Electronic Systems*, 1(3), July 1996.
44. Wayne Wolf. An architectural co-synthesis algorithm for distributed, embedded computing systems. *IEEE Transactions on VLSI Systems*, 5(2):218–229, June 1997.
45. Ti-Yen Yen and Wayne Wolf. Communication synthesis for distributed embedded systems. In *Proceedings, ICCAD-95*, pages 64–69. IEEE Computer Society Press, 1995.
46. Ti-Yen Yen and Wayne Wolf. Performance estimation for real-time distributed embedded systems. In *Proceedings, ICCD '95*, pages 64–69. IEEE Computer Society Press, 1995.

RAPID PROTOTYPING, EMULATION AND HARDWARE-SOFTWARE CO-DEBUGGING

WOLFGANG ROSENSTIEL
Technische Informatik
Universität Tübingen
Tübingen, Germany

1. Introduction

ASICs will continue to grow increasingly complex. Errors of specification, design and implementation are unavoidable. Consequently, designers need validation methods and tools to ensure a perfect design before the production is started. Errors caught after fabrication incur not only added production costs, but also delay the product, which is an increasingly serious detriment in today's fast-paced international markets. "First-time-right silicon" is, therefore, one of the most important goals of chip-design projects.

Figure 1 shows loss due to relatively late marketing. In order to get "first-time-right silicon," a variety of approaches is needed. Figure 2 describes the necessary steps of synthesis and validation for the design of integrated circuits. There are four methods to reach the goal of "first-time-right silicon":

- specification on high levels of abstraction followed by automatic synthesis;
- simulation on various levels;
- formal verification;
- prototyping and emulation.

In the last few years the growing significance of synthesis has become apparent, that is synthesis in a general sense and—more specifically—synthesis on higher levels of abstraction, such as RT level synthesis and the so-called high-level synthesis, i.e. synthesis from behavioral descriptions. The increasing number of available commercial tools for RT and high-level synthesis indicates that the abstraction level of the design entry will increase to higher levels in the near future. Especially with respect to

A. A. Jerraya and J. Mermet (eds.), System-Level Synthesis, 219–262.

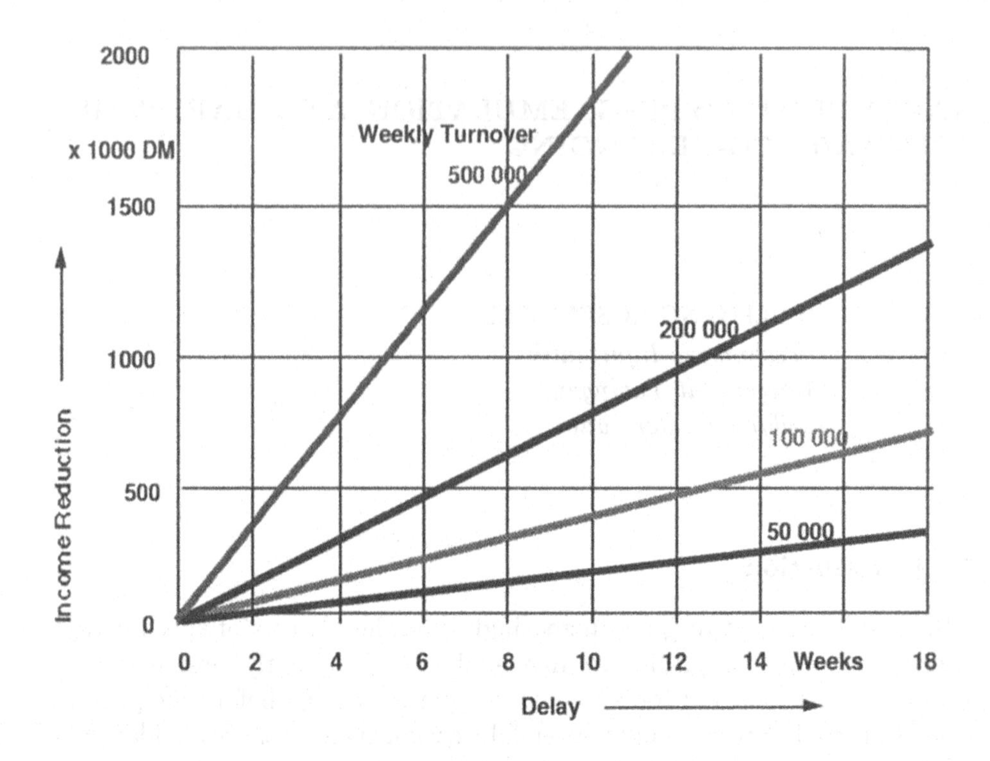

Figure 1. Cost of product delays.

hardware/software co-design high level synthesis gathers momentum. Only by integrating high-level synthesis in the hardware software co-design cycle real hardware/software co-design will be possible. The investigation of various possibilities with different hardware/software trade-offs is only sensible if the different hardware partitions can be implemented as fast as software can be compiled into machine code. High-level synthesis is, therefore, the enabling technology for hardware software co-design.

Another advantage of high-level synthesis is its high level of abstraction and as a consequence of it a shorter formal specification time of the circuit to be implemented. This more compact description also results in a much faster simulation. High-level synthesis combined with hardware/software co-design also supports the investigation of many different design alternatives resulting in higher quality designs. Design exploration compensates for the reduced efficiency and low optimality of synthesized designs when compared to manually optimized solutions, which may only be done for very few design alternatives.

The CAD vendor Synopsys reports from user experience that, through the application of behavioral compilation, design effort may be reduced by a factor of 10 and that, despite this shorter design time, area and delay could

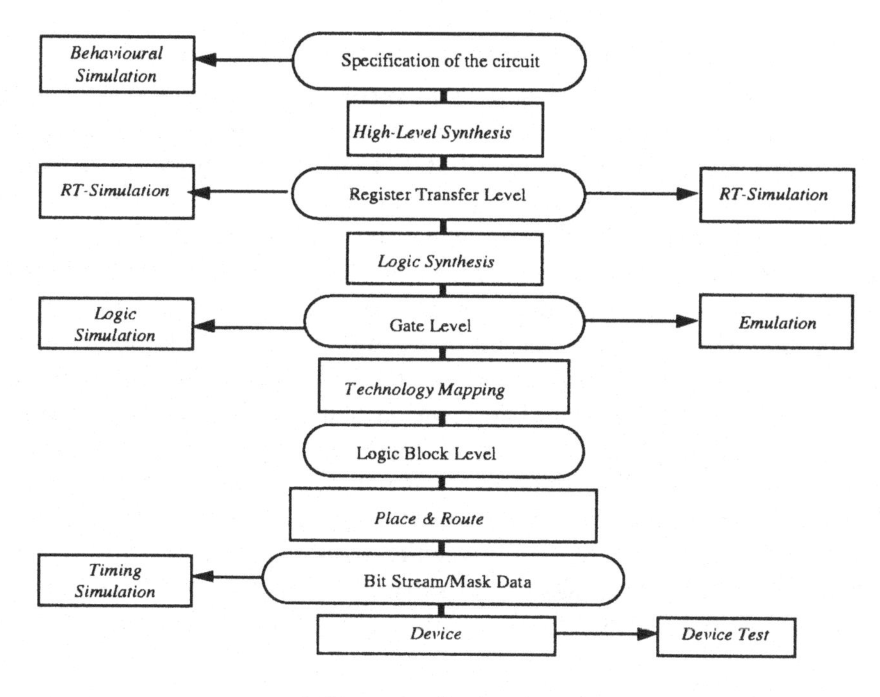

Figure 2. Design flow.

be reduced by 10-20% compared to manual designs. The higher quality of the implementation is mainly due to the investigation of further design alternatives and more sophisticated pipelining schemas, which automatically optimize clock frequency and latency.

Simulation of different design levels constitutes the standard technique for the validation of designs on various design levels. Nobody denies that simulation has many advantages and it is very likely to remain the standard technique for design validation. On the other hand, it is also indisputable that simulation is extremely time-consuming—a simulation of a large digital system may run a million times slower than the system itself. Moreover, simulation requires an explicit model of its test bench. The expenditure for the construction of a test bench often can be found to equal the expenditure of the actual circuit design.

Much effort goes into the development of new systems, which allow the integrated simulation of both large hardware/software systems and of peripheral devices (mechanical, mechatronic, etc.). One problem of simulation is the very difficult consideration existing interface components and custom components. They may best be treated with "hardware in the loop" concepts, even though simulation can be speeded up considerably by means of newly introduced cycle based simulators. But even cycle based simulation is much slower than its real-time execution so that the problem of an integration of real existing hardware cannot be solved this way.

As a result of these problems, formal verification becomes increasingly important. There are two types of formal verification to be distinguished: **spec debugging** ("Did I specify what I want?") and **implementation verification** ("Did I implement what I specified?"). Despite the growing importance of formal verification and despite the introduction of some early verification tools, it is limited in many ways. One of its major restrictions is the high complexity of verification procedures, which limit an application to comparably simple design modules. It is difficult to include time-related constraints into the verification process. The difficulty of including an environment also occurs for formal verification. Another important technique for the "first -time-right silicon" goal is therefore an increasing use of emulation and prototyping. This contribution is devoted to this issue. Emulation and prototyping support hardware/software co-design by means of simple coupling with real hardware— in particular with standard microprocessors which run parts of a total system in software—and are closely related to other contributions in this course. This paper is a shorter and updated version of [1].

2. Prototyping and Emulation Techniques

Validation in hardware/software co-design has to solve two major problems. First, appropriate means for software and hardware validation are needed; second, these means must be combined for integrated system validation. Since validation methods for software are well known, the main effort is spent on hardware validation methods and the integration of both techniques. Today, the main approaches for hardware validation are simulation, emulation [6], [19], [8] and formal verification Simulation provides all abstraction levels from the layout level up to the behavioral level. Formal verification of behavioral circuit specifications has up to now only limited relevance, since the specifications which can be processed are only small and can only use a very simple subset of VHDL. All these validation and verification are harder to implement during hardware/software co-design: low-level validation methods are not really useful, since hardware/software co-design aims at automatic hardware and software design starting from behavioral specifications. This would be like writing programs in C++ code and debugging the generated assembly code. This even holds for debugging at the RT level , since HW/SW co-design usually implies high-level synthesis for the hardware parts. The RT description of the hardware is produced automatically and may therefore be very difficult to interpret for a designer. Though simulation can operate at the algorithmic level, there are some problems with it, too. First, simulation is very time-consuming and thus it may be impossible to simulate bigger systems in the same way as systems consisting of hardware and software parts. Second, simulation is too slow to bring simulated components to work with other existing components. Thus, it is necessary to model the outside world within the simulator, which requires at least an enormous effort.

One great advantage of emulation and prototyping in contrast to simulation is higher speed. Emulation, for instance, is only 100 times slower than real time. Furthermore, the integration of the environment is much easier. Likewise high expenditure for the generation of test benches can be avoided.

Today, emulation is the standard technique CPU design and is increasingly used for ASIC design. Emulation can be expected to become an essential part of co-design methodologies since—compared with simulation—the speed advantage of emulation with respect to co-emulation is even more apparent. One of the main disadvantages of emulation in contrast to simulation is that timing errors are hard or impossible to detect. Therefore, emulation mostly serves the testing of functional correctness. There are further disadvantages: slow compilation once the circuit changes, different design flows for implementation and emulation, and the high expenditure, which may be up to one Dollar per emulated gate. However, it is possible to overcome many of these disadvantages by not restricting emulation to the gate level as is normally done, but by aiming at emulation on higher abstraction levels, which then may be more viable in real time.

First steps in this direction have been taken by Quickturn [9] with the HDL-ICE system, which allows to relate the probed signals to a register-transfer specification. Synopsys has introduced the Arkos system which also performs a cycle based emulation of an RT-level design. A similar approach is taken by Quickturn in the CoBalt system. Potkonjak et. al. [17] have described a method to do design for debugging of an RT level circuit based on emulation which multiplexes output pins of a design and register contents to the same pins, whenever they are not used by the design itself. This approach concentrates rather on a minimization of output pins needed for debugging then on source code based debugging. Although solutions for the RT level start to evolve on the market there is a big gap between the algorithmic level and the RT level where emulation is available today. In this contribution we will therefore first present classical gate level emulation techniques based on FPGA structures. Later we will explain how these classical gate level emulation systems are currently evolving to support the RT level. Still in the research state is emulation support at higher levels together with high-level synthesis and hardware/software co-design. This outlook on current state of the art in emulation research will then be described by explaining our own approach which we call **source-level**

emulation (SLE).

As mentioned before, emulation nowadays forms part of usual validation techniques in microprocessor design, but also is applied to ASICs. SUN Microsystems, for instance, reports at length on the use of emulation in connection with the design of the UltraSparc-architecture. With a speed of 500,000 cycles per second it was possible to test the boot sequence of Solaris or to emulate with the same speed the start and use of window systems, such as Openwin, etc. Problems mentioned in this context are related to the integration of memory but also to the simulation of pre-charging and other dynamic effects of circuit design. An interesting side effect which was reported, is that the return on investment costs for emulation sytems, which are still very expensive, are distributed by 50% each on return on investment before first silicon and 50% after first silicon. Thus, the availability of a complete emulation system allows also the improvement of fault diagnosis in the case that there are still errors in the implemented chip.

In addition, it is possible to test further developments of this chip with respect to what-if design alternatives. Bug fixes are conceivable as well. Tandem has also given a detailed account of the results derived from the use of emulation. They pointed out that without emulation it would have been impossible to make complete operating system tests, to simulate input/output behavior and to integrate finished ASICs step by step into the total system. The emulated components are gradually substituted by real chips. This allows a complete validation at any point of the development time. Yet, various users of emulation systems record unanimously that emulation is generally only used after thorough simulations so that there is only a low expectation of design errors anyway. Simulation mainly aims at the reduction of the number of functional and timing errors, whereas emulation in general serves realistic loads in demonstrating functional correctness.

Some of the problems which have been mentioned in connection with emulation systems are caused by the different design flows required for implementation and emulation. Figure 3 demonstrates these varying design flows. Alternative 1 is today still very common. The cycle between simulation and emulation is passed for so long until no errors occur anymore. This is followed by a translation into the final ASIC implementation. The disadvantage of this alternative is that the migration which is required by emulation results in an implementation which is different from the subsequent translation into an ASIC. The ensuing validation problem is to be solved by an increased application of alternative 2. Here, a synthesis for emulation is implemented, for example on the basis of a gate level or a future RT level VHDL description. Accordingly, errors bring on changes in the VHDL source. This identical VHDL source file is then also converted by synthesis into a corresponding ASIC. Postulating an identical VHDL

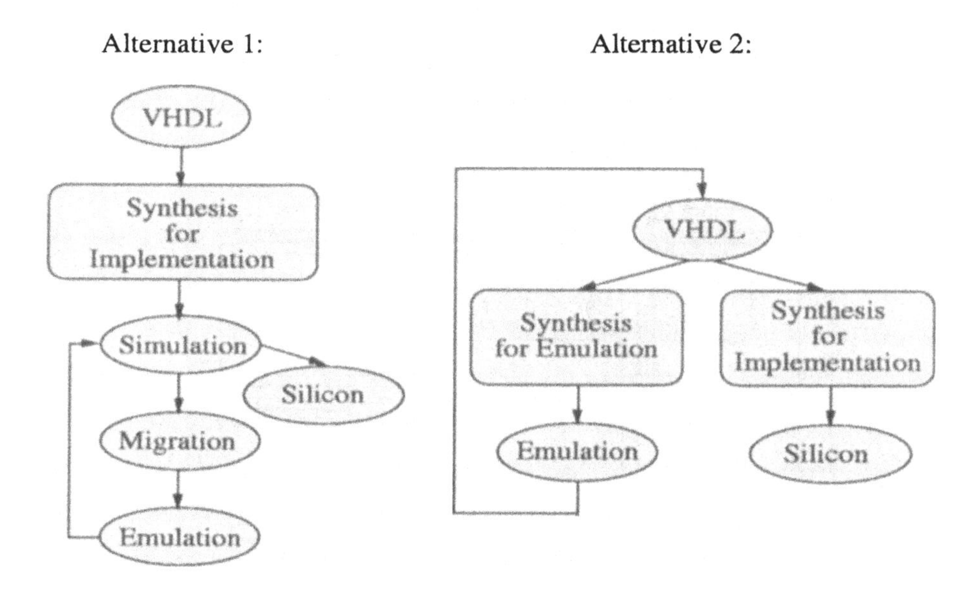

Figure 3. Different design flow integrations for emulation.

subset with identical semantics enables a sign-off on the basis of the VHDL source code. In the future, alternative 2 will support a RT level sign-off by means of a more abstract description on RT level.

Today's existing emulation systems are distinguished into FPGA based and custom based solutions. Additionally, the FPGA based approaches may be categorized as off-the-shelf FPGAs and FPGAs which have been developed especially for emulation. Essentially, off-the- shelf FPGAs go back to the FPGAs of Xilinx. Currently, XC4000 families are the predominant FPGAs for this application. Their superiority comes from their static RAMs-based implementation, which allows unlimited reprogrammability. Figures 4, 5 and 6 show respectively the basic set-up of the Xilinx device architecture, the general function of the look-up tables as well as the internal architecture of a CLB of the Xilinx XC4000 series.

3. Prototyping and Emulation Environments

This section describes the functionality and characteristics of today's existing prototyping environments. There are many different approaches for prototyping architectures. Most of them address the prototyping of hardware but not of embedded systems. Even if their gate complexity is sufficient for total systems, they display various disadvantages with respect to hardware/software co-design and prototyping of whole systems. First, until now there is no integration of microprocessors for the execution of the software parts supported. Integration facilities are not flexible enough

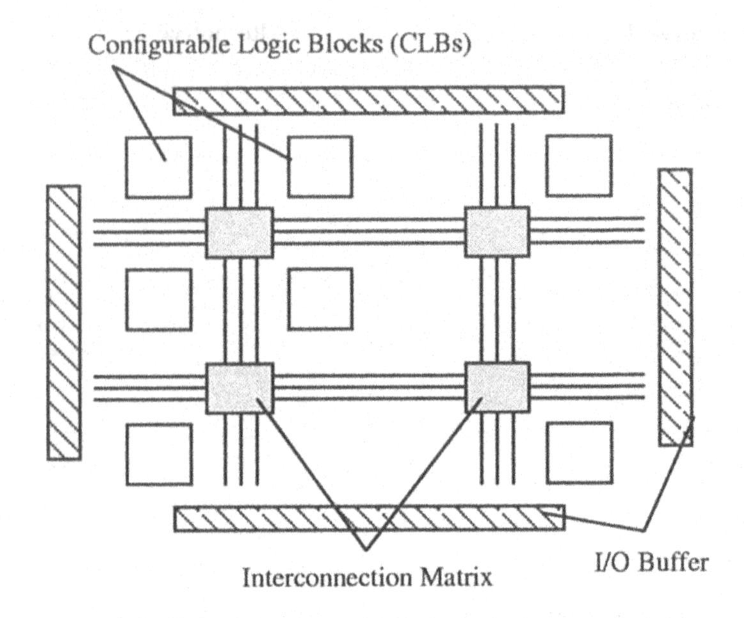

Figure 4. **Xilinx interconnection structure.**

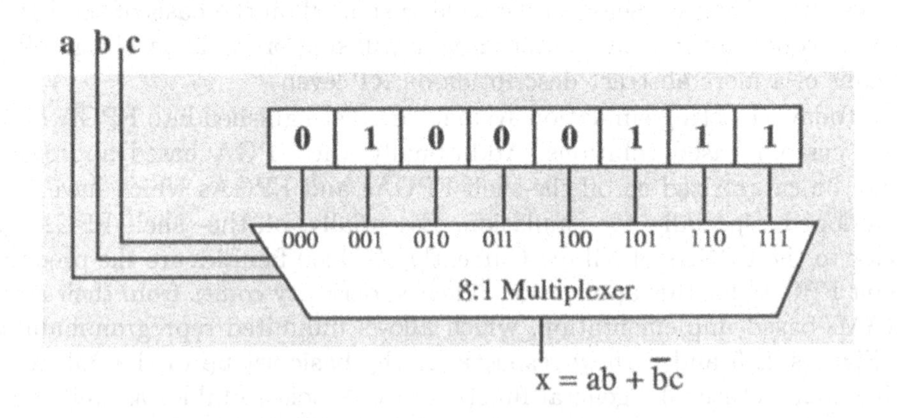

Figure 5. **CLB principle.**

to adequately support prototyping in hardware/software co-design; in particular, it is very difficult for the emulator's synthesis software to route buses which are usually introduced by the integration of microprocessors. Second, hardware partitioning is too inefficient, leading to inadequate utilization of the emulator's hardware resources. Third, debugging facilities are not sufficient.

Improvements in programmable logic devices especially by means of field programmable gate arrays support their use in emulation. FPGAs are

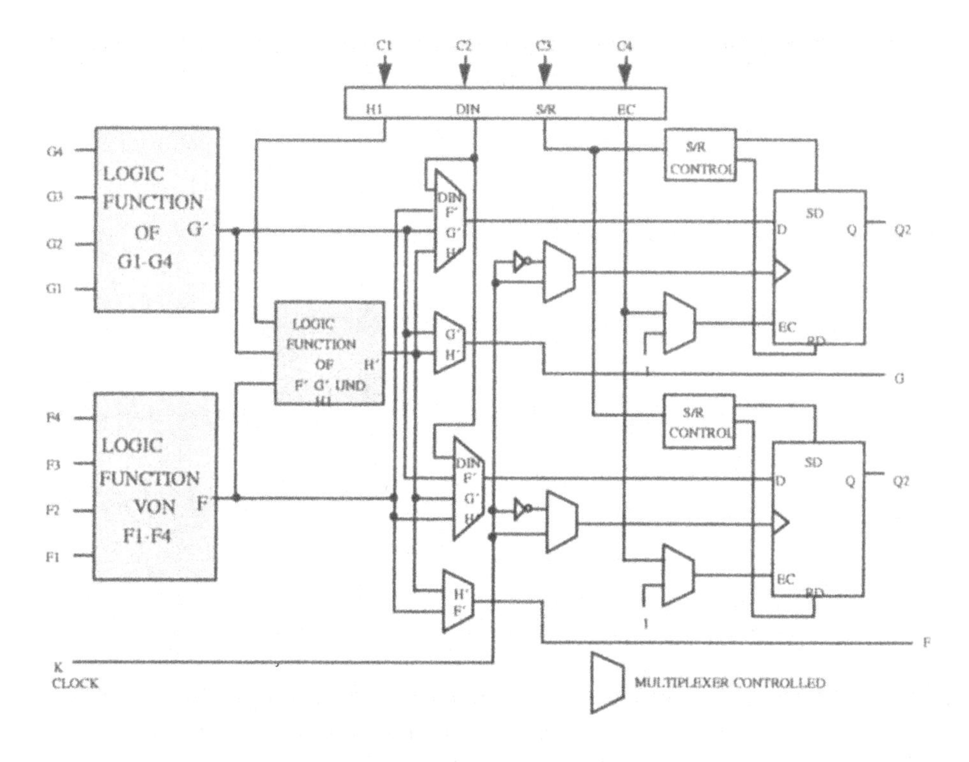

Figure 6. CLB structure.

growing faster and they are capable of integrating more and more gate functions and of supporting flexible emulation systems.

Figure 7 explains the principle of emulation. A logic design is mapped on an emulation board with multiple FPGAs. Different interconnection schemes support the interconnection of these multiple FPGAs.

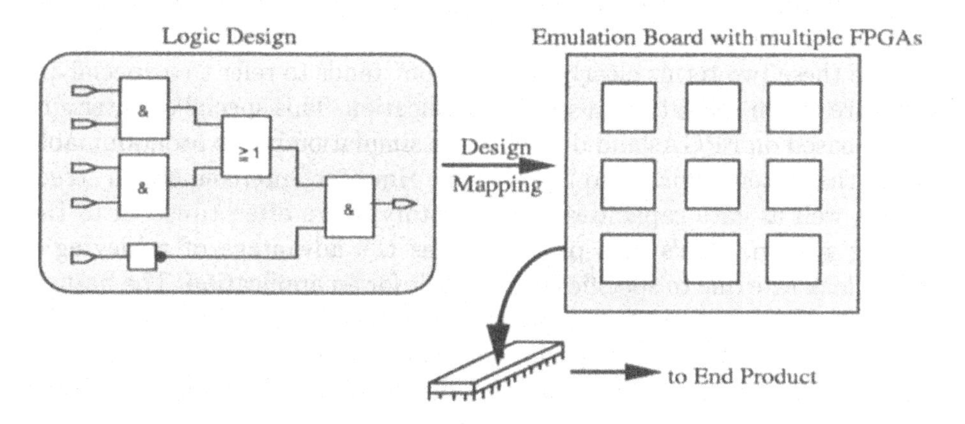

Figure 7. FPGA mapping.

Figure 8 explains the general structure of such an emulation system. Just as individual emulation systems differ with respect to their use of

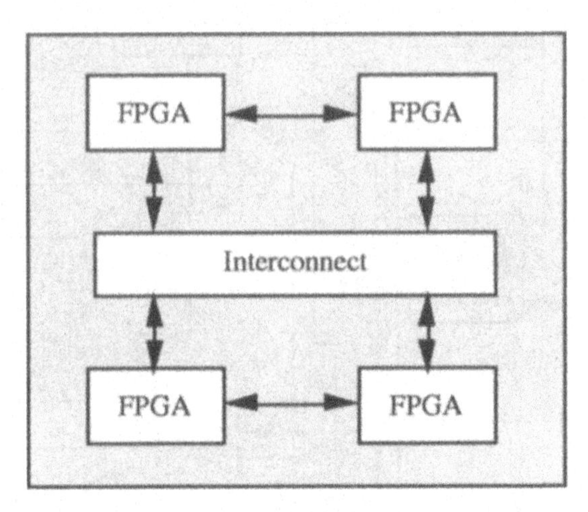

Figure 8. Principal structure of an FPGA-based emulation.

FPGAs, there are differences between emulation systems on the market concerning their interconnection structure. The systems which are available on FPGAs may be arranged in three classes. Below concrete examples for each class are provided:

Some systems also use programmable gate arrays for their interconnection structure, others take recourse to special interconnect switches (ex. Aptix), and there are some which apply especially developed custom interconnect chips for the connection of individual FPGAs.

Until now, we have not suggested any distinctions between prototyping and emulation. In fact, the dividing-line between the two terms is blurred. Many manufacturers of emulation systems claim that these may be used in particular for prototyping. In my opinion, it is actually impossible to separate these two terms clearly. "Prototype" tends to refer to a special architecture which is cut to fit a specific application. This special architecture is often based on FPGAs and describes the simulation in the programmable logic of the system which is to be designed. However, interconnection structures as well as gate capacities of the prototype are often tuned in to the evolving system. One's own prototype has the advantage of achieving a higher clock rate due to specific cutting to fit for an application. The best elements for the corresponding application can be selected, and thus, RAMs, microcontroller, and special peripheral components etc. can be used. Its weak point is, however, that new expenditure for the development of such a special prototype occurs for every new prototype. Therefore, it is appropriate to ask to what extent this disadvantage overthrows the above mentioned advantages. Some kind of compromise between general purpose emulation systems, which will be elaborated later on, and special prototypes may be seen in the Weaver system. It was developed by the University of Tübingen and Siemens Munich under the auspices of the Basic Research ESPRIT project COBRA (Co-Design Basic Research Action).

3.1. THE WEAVER PROTOTYPING ENVIRONMENT

The Weaver prototyping environment is designed especially for prototyping in the domain of hardware/software co-design. It is a modular and extensible system with high gate complexity. Since it fulfills the requirements on a conventional prototyping system it is also useful for hardware prototyping. This section describes the hardware and supporting software for this environment.

Weaver uses a hardwired regular interconnection scheme. This way fewer signals have to be routed through programmable devices, which results in a better performance. Nevertheless, it will not always be possible to avoid routing of signal through the FPGAs. This must be minimized by the supporting software. For the interconnection of modules bus modules are provided. These offer 90-bit-wide datapaths. This is enough even if 32 bit processors are integrated. Depending on the type of Xilinx FPGAs used, it provides 20K to 100K gates per board. Application specific components and standard processors can be integrated via their own boards which must have the same interface as the other modules. For debugging it is possible to trace all signals at runtime, to read the shadow registers of the FPGAs and to read back the configuration data of any FPGA in the system. A very interesting feature is the ability to reconfigure particular FPGAs at the runtime of the system without affecting the other parts of the architecture. The base module of the Weaver board carries four Xilinx FPGAs for the configurable logic. Each side of the quadratic base module has a connector with 90 pins. Each FPGA is connected with one of these connectors. Every also FPGA has a 75 bit link to two of its neighbours. A control unit is located on the base module. It does the programming and readback of the particular FPGAs. A separate bus comes to the control unit of each base module in the system. The programming data comes via this bus serially. This data is associated with address information for the base module and the FPGA on the base module. This way the control unit can find out if the programming data on the bus is relevant for its own base module and if so if it can forward the programming data to the FPGA for which it is intended. The readback of configuration data and shadow registers is done in the same way. The ROM is used to store the configuration data for each FPGA of a base module. While the startup of the control unit reads out the ROM and programs each FPGA with its configuration data. Figure 9 shows the base module equipped with Xilinx 4025 FPGAs.

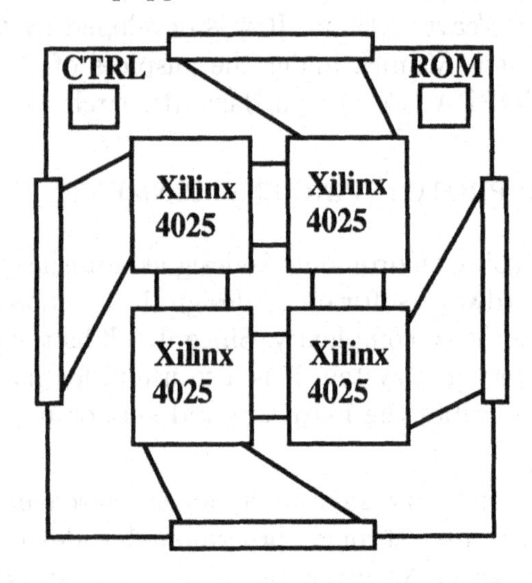

Figure 9. The base module.

An I/O module provides a connection to a host. Several interfaces are available. Examples are serial and parallel standard interfaces as well as processor interfaces to PowerPC and Hyperstone. With such a module the host works as I/O preprocessor which offers an interface for interaction to the user. A RAM module with 4 MB static RAM can be plugged in for the storage of global or local data. It can be plugged in a bus module so that several modules have access to it in the same way, or it may be connected directly to a base module. Then this module has exclusive access to the memory. Requests of other modules must be routed through the FPGAs of the directly connected base module. This is possible but time-consuming. The bus module allows the plugging together of modules in a bus-oriented way. Given the option of having a bus on each side of the base module, this architecture can be used to build multiprocessing systems with an arbitrary structure. For the integration of standard processors they must be located on their own modules which have to meet the connection conventions of the other modules. A standard processor module for the Hyperstone 32 bit processor is already included in the set of module types. Even though this list of module types is sufficient for many applications, there may be even more applications which require other types of modules like networking modules or DSP modules. One may think of any other type of module, and certainly the list of existing module types will grow in the future. With these modules arbitrary structures can be built. The structure depends on the application which is prototyped. Thus a running system contains all modules needed but not one more. It is important to reduce overhead and to keep the price low. An example of a more complex structure is depicted in figure 10.

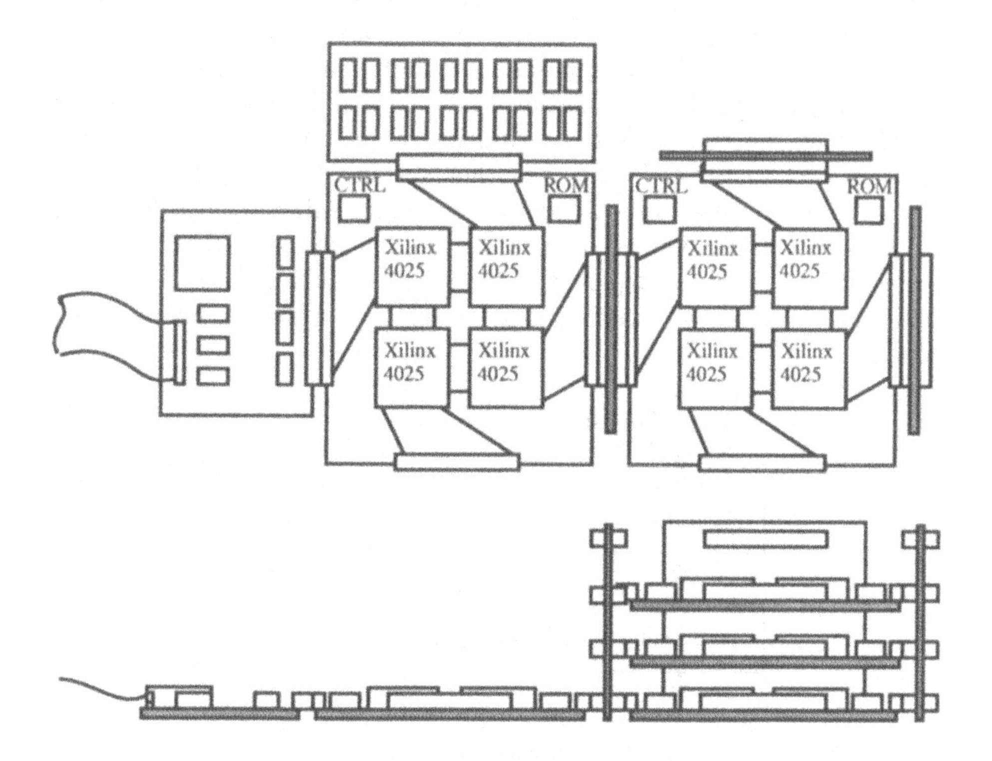

Figure 10. A more complex structure.

This picture shows an architecture in top view and in side view. As can be seen, the architecture is built in three dimensions and consists of four basic modules, a RAM module and an I/O module. On the right side is a tower of three basic modules which are connected via three bus modules. On the left side is another basic module with a local RAM module and an I/O module. The gross amount of gates in this example is about 400K. Thus it is an example which would be sufficient for many applications. One may assume that the left basis carries an application specific software executing processor, which works with the RAM on the RAM module and which communicates with software on a host for providing a user interface via the I/O module. The tower on the right side may carry some ASICs.

After the hardware/software partitioning the resulting behavioral hardware description has to be synthesized via existing high-level synthesis tools. The hardware debugging software provides a user interface for the readback of configuration data, for signal tracing and for shadow register reading. A special partitioning tool was developed to partition design on several FPGAs [24].

This example of a prototyping system is used as experimental environment for the source-level emulation which will be elaborated lateron. In the following some examples for emulation systems are given.

These example emulation systems are manufactured by Quickturn, Mentor and Aptix. The new simulation system from Quickturn (CoBalt) is elaborated at the end of this contribution since it utilizes very different emulation architectures. While the other systems from Quickturn, Mentor and Aptix are based on off-the-shelf or in-house-developed FPGAs and since they essentially depend on a rather complicated and lengthly configuration process, such as depicted in figure 11, CoBalt is based on special processors, on which an existing logic or RT level design is mapped by a compiler.

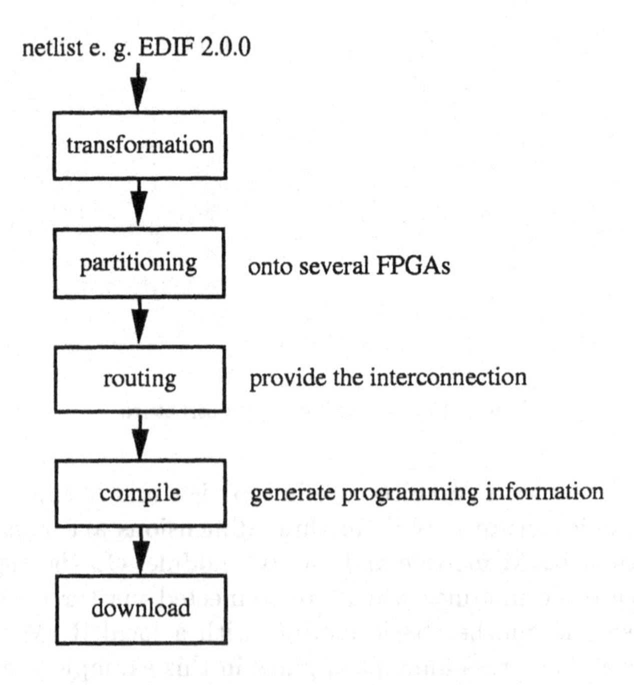

Figure 11. The FPGA configuration process.

All these emulation systems can operate in two operation modes, shown in figure 12. By means of a host system individual stimuli are applied to an emulator in a test vector mode (a). The resulting test answers are evaluated this way as well.

The emulated system in the target environment is operated in a dynamic mode (b). A logic analyzer is controlled by the emulator and connected to the communication signals between the emulator and the target environ-

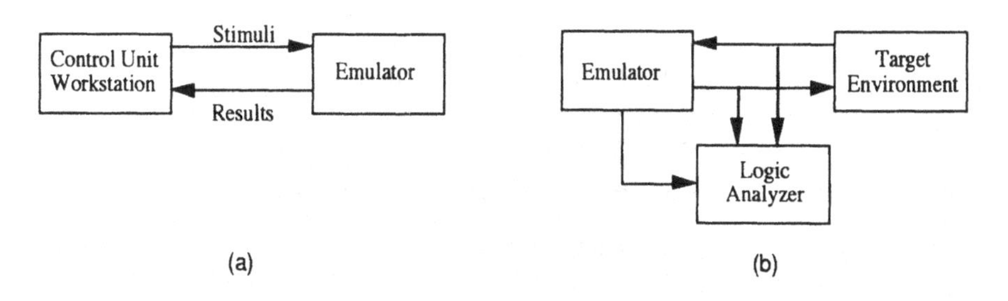

Figure 12. Operation modes of emulation systems.

ment. The logic emulator traces the signals and allows the later analysis of this trace. The size of the trace buffer is very critical and distinguishes the different emulation systems. Important requirements for all these prototyping or emulation environments are their flexibility in terms of a flexible integration of microcontrollers, microprocessors or DSP cores.

Flexibility is also required in several other areas. First, it must be possible to implement large memory blocks, which are required for both program and data elements of the software; this is particularly important in co-design applications. The clock must run as fast as possible to allow for dynamic testing and for validation in real time within the system-under-design. High gate capacity, high observability of internal nodes, the ability to read back information out of the FPGAs, and the ability to create long signal traces are also important.

Most of the information on the systems described in the following sections is taken from product information of the mentioned companies. It is collected from various sources, including advertising, technical reports, customer experience reports and publically available information on the world wide web.

3.2. SYSTEM REALIZER (QUICKTURN)

Quickturn offers different kinds of emulation system. A smaller version for less complex designs, the so-called logic animator, an emulation system for designs up to 3 Million gates, namely the System Realizer with the HDL-ICE software support, an accelerator architecture CoBalt which is similar to the Synopsys Arkos system, and the 1998 introduced new Mercury system. Table 1 gives an overview of the Logic Animator, which is not described here in much detail. Tab. 2 describes two versions of the System Realizer: the basic M250 and the more sophisticated M3000. Figure 13 gives an overview of the hardware architecture in the M3000.

The System Realizer allows a maximum capacity of 3 million gates. The architecture allows a modular system of up to twelve 250,000 emulation

TABLE 1. Logic Animator overview.

Model	Animator Model 550
Architecture	Single board, reprogrammable system
Capacity	50K gates typical
Memory support	Memory compiler for synchronous RAM (1 port and 2 port) and ROM
Prototyping speed	8 - 16 MHz typical
I/O connections	448 bi-directional pins (for cores, logic analysis and target interface)
Software	Single-pass automatic compilation, interactive speed optimization, incremental probe changes
Design entry	EDIF, TDL, NDL, Verilog netlist formats, over 40 ASIC libraries available
EDA compatibility	Verilog, ViewSim, or VSS simulation environments
Debug	Connects to commercial logic analyzers (HP 1650 and 16500 families, Tektronix)

(Source: Quickturn Product Description)

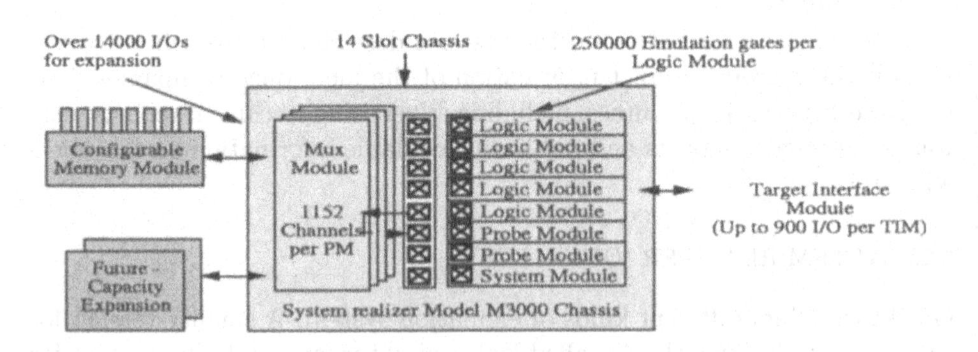

Figure 13. Realizer hardware architecture.

gate logic modules with probe modules through a programmable backplane. Stand alone 250,000 emulation gates integrated system for mid-range designs (basic model of the System Realizer: M250) is also possible. The software support includes partitioning and mapping on the different FPGAs. There is also an integrated instrumentation available to capture and process real-time data for design debugging. The Quickturn System Realizer model

TABLE 2. System Realizer overview.

Features and Specifications	**Model M3000**	**Model M250**
Architecture	Hierarchical multiplexing architecture Custom interconnect chip Xilinx 4013 FPGA	Hierarchical multiplexing architecture Custom interconnect chip Xilinx 4013 FPGA
Design entry	Netlist formats: Verilog structural EDIF, TDL, NDL Over 55 ASIC libraries available	Netlist formats: Verilog structural EDIF TDL NDL Over 55 ASIC libraries available
Emulation	Single-pass automatic compilation	Single-pass automatic compilation
Debug Software	Functional test Vector debug Interactive readback	Functional test Vector debug Interactive readback
Capacity	250,000 - 3,000,000 emulation gates	250,000 emulation gates
Memory support	Memory compiler for RAM (single or multi-port) and ROM Configurable Memory Module up to 14MB each single port up to 875k, 4 write ports, 16 read port	Memory compiler for RAM (single or multi-port) and ROM
Emulation speed	1-4MHz typical	4-8MHz typical
I/O connections	Up to 7200 signal pins 4,000 expansion pins (for CMM or capacity expansion)	600 signal pins
Debug	Integrated logic analyzer 1152 probes (thousands possible through multiple probe modules) 128k memory depth 8-event trigger and acquire 16MHz operation	Integrated logic analyzer 1152 probes 128k memory depth 8 events, complex trigger and acquire 16MHz operation

(Source: Quickturn Product Description)

especially supports a very flexible memory emulation. Small scattered memories can be compiled into the FPGAs internal RAMs to emulate single, dual, and triple-port memories. For larger memories configurable memory modules (CMMs) are offered. They are specificly designed for memory applications, such as large multiport cache arrays, multi-port register files, asynchronous FIFOs, microcode ROMs and large on-chip RAM arrays. In addition, special target interface modules (TIM) interface the emulator with the target system. Standard adapters route system I/O signals to signals in the emulated design. Furthermore, complex ASIC functions including microcontrollers, CPUs, peripheral controllers etc. are integrated into the emulation environment by plugging real chips into this TIM. The System Realizer currently supports 3 and 5 Volt designs. Like all other emulation systems it allows debug in the test vector mode and in the dynamic mode. In the test vector mode test vectors are applied and results are stored. This vector debug mode supports breakpoints, query signals, read-back of internal nodes, single step and continue. A special functional test mode supports regression testing where 128 K of vectors can be applied at up to 4 MHz in an IC tester-like functional validation environment. An automatic vector compare logic for quick go/no go testing is built-in. In the dynamic mode a logic analyzer can be connected to the emulation environment. This logic analyzer includes a state machine based trigger and acquire capability with up to 8 states, 8 event trigger support, 128K channel depth and 1152 channels. Moreover, further 1152 channel capacity modules may be added in place of emulation modules. The trigger-and-acquire conditions are very similar to advanced commercial logic analyzers.

3.3. MERCURY SYSTEM (QUICKTURN)

Quickturn introduced its latest prototyping and emulation system at DAC 1998. This system not only provides the successor version for the FPGA based System Realizer (see above), but also includes many new features, especially towards an integration of simulation and emulation. Like the System Realizer, also the Mercury emulation system is based on Xilinx FPGA's. These FPGA's are of course based on more recent Xilinx chips supporting about three times as many gates per chip as the 4013 type which have been used in the System Realizer. By these bigger chips and the special use of Xilinx programming software, which has been especially adapted to Quickturn's emulation requirements, more complex designs can be translated in shorter compilation time, and also the complexity of emulated systems can be increased. The second change with respect to the system realizer is the development of a new special purpose interconnection ASIC, developed by Quickturn to connect the different FPGA's. This advanced hardware architecture is called IMPX (Inherent Multiplexed Partial Crossbar) and contains a special custom MUX chip. The third big change with respect to the system realizer is that the Mercury emulation cards do not only contain FPGA's and interconnect chips, but also RISC processors to support a combined simulation and emulation of behavioural and RTL descriptions. In the following, some more details will be given in order to describe this new Mercury emulation system. Most of this information is extracted from product information received from Quickturn.

Mercury is available in two different sizes. The smaller so-called SI series can handle up to 2.000.000 gates and 12 MB of memory in increments of 500.000 gates and 3 MB of memory. The larger so-called E series can handle up to 10.000.000 gates plus up to 60 MB of memory in increments of 1.000.000 gates and 6 MB of memory.

Mercury also supports three different levels of memory. On the FPGA chips themselves, 40 kbits of memory can be implemented. Per FPGA board another 3 MB of memory can be implemented. This on-board memory is SRAM-based and allows the implementation of complex multi-port memories. In addition, special add-on memory boards for very large memory of up to one GB will be offered.

IP-integration is supported by integrating emulatable IP models as well as simulatable IP models. Mercury also offers a kind of trade-off between visibility and depth of traces. In principle, 100% visibility at full emulation speed is offered, but then only shorter traces are supported. With less than 100% visibility traces can be much longer. With respect to the generation of traces, a special language has been developed to describe breakpoints and trigger conditions.

Finally, as already mentioned, the emulation cards contain not only FPGA's, but also RISC processors. This combination supports mixed level simulation and emulation. Behavioural models can be simulated on the RISC processors, RTL models can be emulated on the FPGA's. By a close integration, a combined simulation and emulation is possible. In the first version of Mercury, which is already available on the market, Verilog behavioural and RTL descriptions are supported. VHDL will be available in the future.

3.4. MENTOR SIMEXPRESS EMULATION SYSTEM

Like the Quickturn system, the Mentor emulation system also consists of FPGAs. The difference is that the Mentor system is based on a special purpose full custom FPGA architecture developed by the French company Meta Systems, which has been taken over by Mentor. Based on Mentor product information the compilation for this special purpose FPGA emulation architecture is 100 times faster than for off-the-shelf FPGA based emulators. The special purpose FPGAs especially support on-chip logic analyzers with access to every net in the design. It also in particular supports high speed fault simulation. The special purpose FPGA also allows higher emulation clock speed up to a maximum of 20 MHz. Typical values are 500 KHz to 4 MHz. In a full version the capacity allows up to 1.5 million user gates plus memory mapping. The same as in the Quickturn system the system configuration consists of a basic cabinet with 23 universal slots. One, two, four and six cabinet configurations are available. For these universal slots different card types are available. All these card types can be plugged into these universal slots. There are actually four different card types available. Namely logic card (10K-12K user gates), I/O card (336 pins), memory card (64 Mbytes) and stimulus card (12 Mbytes).

The custom FPGA architecture which is used in the Mentor emulation system also uses reprogrammable logic blocks (comparable to Xilinx). These programmable logic blocks contain flexible four-input/one-output modules together with a configurable latch or flip-flop. The logic cards themselves allow 24 programmable 32 Kbyte RAM chips. Furthermore, as mentioned before, special dedicated 64 Mbyte memory cards are available. The custom FPGA chips moreover support an automatic mapping of clock distribution trees to avoid timing races and hazards. A special purpose custom interconnection chip supports a scalable network architecture and improves the efficiency of the compilation process and the design access.

As all other systems a test vector and a dynamic mode are supported. The test vector mode supports an interactive debugging, taking advantage of the logic analyzer build into the custom FPGA chip. This custom FPGA chip also supports accesses to all design modes without recompiling the design. Additionally, a signal name cross-reference supports debugging in the context of the signal names and design hierarchy of the original netlist. In the dynamic mode traces and triggers are used. There are four system wide trigger conditions and a 62-state sequencer to generate breakpoints and to perform complex event analysis.

3.5. APTIX PROTOTYPING SYSTEM

Aptix prototyping environment is based on the Aptix switches, which are called **field programmable interconnect components** (FPICs). They are special switching elements which can be programmed and which support the routing of signals between the chips of boards. By means of these field programmable switches a fast breadboarding of system designs is possible, allowing fast prototyping of PCBs and ASICs. The field programmable switches have 1024 I/0 pins arranged in a 32x32 matrix. Figure

14 shows how the different board components are interconnected to these programmable switches. Figure 15 shows such a field programmable circuit board.

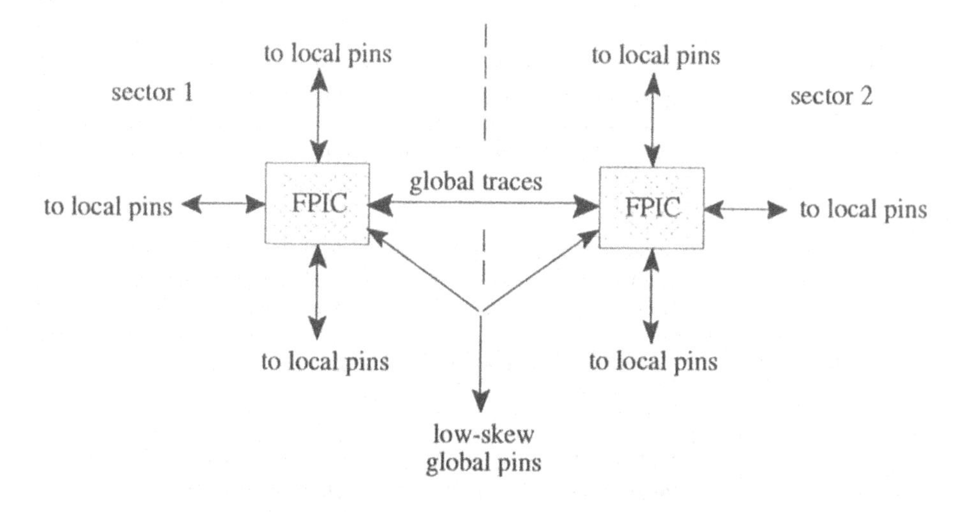

Figure 14. Programmable interconnection.

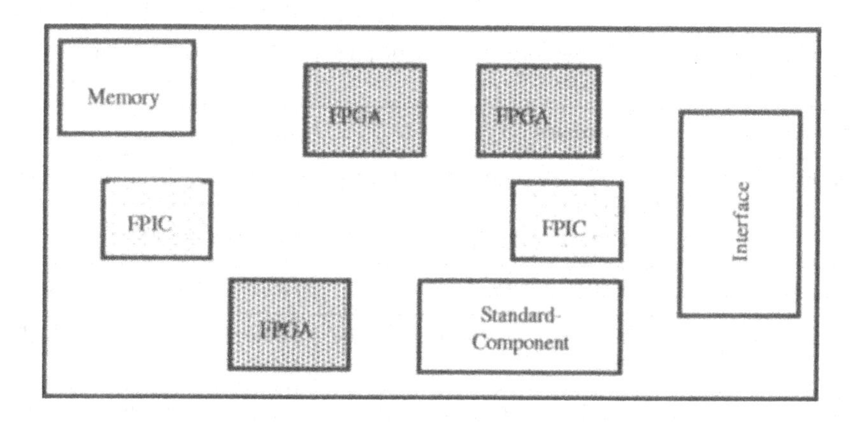

Figure 15. Programmable circuit board.

New versions of the Aptix prototyping system (MP4 system explorer, MP4Pro Explorer) support the integration of the shelf components and FPGAs of any type. This environment also supports test vector and dynamic mode. In the test vector mode up to 544 probe points for debugging are available. An additional stimulus response interface module is available with up to 288 pattern generation channels. The host system compares the uploaded results with original simulation results. Tab. 3 gives an overview on the MP4 and the MP4Pro architectures.

TABLE 3. MP4 architecture overview.

Feature	**MP4**	**MP4Pro**
Architecture	Open architecture Accepts all components	Open architecture Accept all components
FPGAs	Support for all popular families	Support for all popular families
Interconnection	Fully re-programmable 4 FPIC with 1024 pins each 4ns propagation delay through FPICs	Fully re-programmable 4 FPIC with 1024 pins each 4ns propagation delay through FPICs
Emulation Speed	20 - 35 MHz typical 50MHz high speed electronic bus	10 - 20 MHz typical
Debug	256 probes interface to HP logic analyzers and pattern generators (HP 1660 and 16500 families supported)	256 probes interface to HP logic analyzers and pattern generators (HP 1660 and 16500 families supported), Stimulus Response Interface with 544 probes and 288 pattern generation channels
Pattern generation	All popular simulators plus ASCII	All popular simulators plus ASCII
vector formats	interface	interface
Capacity	4 FPICs 2,880 free holes Up to 20 FPGAs (208 QFP)	4 FPICs 2,880 free holes Up to 20 FPGAs (208 QFP)
Design input	EDIF, Xilinx XNF	EDIF, Xilinx XNF
System programming	Ethernet access through the Host	Ethernet access through the Host
Interface Module	On-board flash memory for standalone operation Controlled by on-board microcontroller	Interface Module On-board flash memory for standalone operation Controlled by on-board microcontroller
Partitioning	Manual partitioning	Automatic partitioning software
I/O connections	448 I/O plus 160 bus pins	448 I/O plus 160 bus pins
Software platforms	Sun SPARC, HP 700, PC	Sun SPARC, HP 700, PC

3.6. PROCESSOR BASED EMULATION

These systems are completely different from all other FPGA based emulation and prototyping environments. They are not FPGA- but processor-based. They contain processors specialized for logic computation and use a special interconnection architecture. In contrast to the FPGA based emulation and prototyping environments for processor based systems the design is compiled into a program that is loaded and executed on these logic processors. Compiling the hardware description into code for these specialized logic processors especially supports debugging. A compilation is of course also much faster than mapping on FPGAs. In addition, special custom logic processors are fast enough for the emulation speed to be compared to FPGA based emulators. Arkos and CoBalt, recent examples for processor based emulation systems, support emulation speed in the range of 1 MHz.

Quickturn and Synopsys each announced their new generation of custom processor-based emulation systems in late 1996. These new generations of emulation systems offer certain unique capabilities in design capacity, compile time and debug capability. Although the Synopsys Arkos emulator and the Quickturn CoBalt (Concurrent Broadcast Array Logic Technology) System have different architectures, they provide similar benefits to the user including fast compile times, enhanced runtime and debug performance, and improved design visibility.

Just recently Quickturn and IBM entered into a multi-year agreement to co-develop and market the CoBalt compiled-code emulation system based on a 0.25-micron custom processor and an innovative broadcast interconnect architecture developed at IBM, which is coupled with Quickturn's Quest II emulation software. CoBalt provides verification capabilities for designs between 500,000 and 8,000,000 logic gates. The Synopsys Arkos emulation system is also based on a custom processor and was originally introduced by Arkos Design Systems in 1994 and was re-engineered after Synopsys acquired that company in 1995. Recently Quickturn acquired Arkos, i.e. only CoBalt will be supported in the future.

4. Future Developments in Emulation and Prototyping

Advances in design automation enable synthesis of designs from higher levels of abstraction. Advances in synthesis have natural consequences for emulation, which may lead to problems, in particular in connection with debugging. Interactive debugging by means of a designer requires that in mapping the description on the emulation environment the designer is enabled to reconstruct the transformations which have been implemented by the synthesis system. In the case of implementation of emulation starting off from a description on RT level, the resulting necessity of keeping up the pace may mostly be reconstructed by cross reference lists. However, it is considerably harder to keep up the interrelation between the description done by the designer and the final emulated circuit in emulation from a design created by high-level synthesis. A particular complication ensues from the fact that the time response of the circuit is determined or optimized only through high-level synthesis.

As a solution to this problem we have suggested the so-called source-level emulation (SLE), which aims at closing the gap between description, or simulation on behavioral level and hardware emulation. The basic idea of SLE is the retention of a relation between hardware elements and the source program (which may be given as behavioral VHDL) during the execution of the application on an emulation hardware. In analogy to a software debugger it aims at a potential debugging on the basis of the source code. Consequently, it should be possible to set up breakpoints based on the VHDL source text, to select the active variable values and to continue the program. This kind of source emulation is of particular relevance in connection with hardware/software co-design and co-emulation so that in the end debugging implementation will be possible independently of the implementation in hardware or software—debugging on the the same level and with the same user interface.

Our approach supports symbolic debugging of running hardware analogous to software debugging, including examining variables, setting breakpoints, and single step execution. All this is possible with the application running as real hardware implementation on a hardware emulator. Back annotating the values read from the circuit debugging can be done at the source code level. For certain applications it is not even necessary to capture the environment of the application in test vectors. It is possible to connect the emulator to the environment of the application. There is no need to develop simulation environments in VHDL. Avoidance of the test bench development is another advanced side effect of this approach, since these simulation environments are at least as error-prone as the application itself. Figure 16 shows the steps of high level synthesis (HLS) together with the data that these produce. Both actions, the sharing of components and the sharing of registers add multiplexers to the circuit.

4.1. TARGET ARCHITECTURE

Figure 17 shows the target architecture for HLS. To get a better view on what SLE is, it is necessary to take a closer look on the relation between the behavioral VHDL source code and the generated RT level circuit. Figure 18 shows different possible implementations of variables by nets at the RT level. In case 1) the requested variable (bold line) is an input net of the

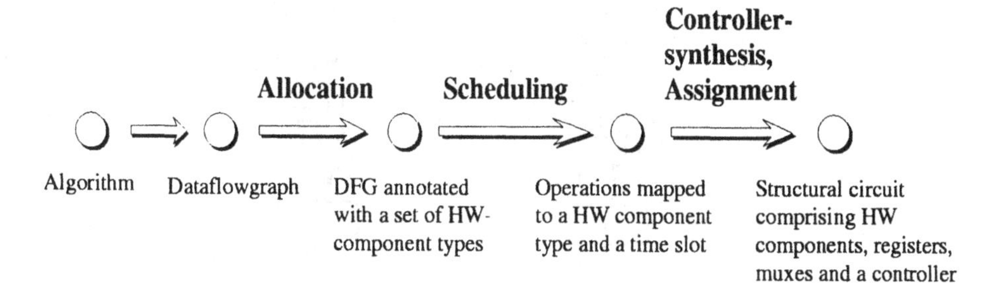

Figure 16. Steps in high-level synthesis.

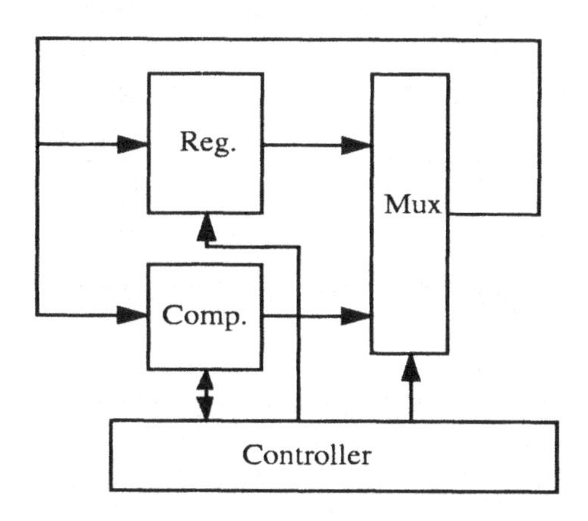

Figure 17. Target architecture.

component and connected to register via several other components.

Only the register has to be read for retrieving the value of this net. In case 2) it is backtracked from the component which is connected to a register via some copying components. The corresponding value may be retrievable by reading the register in the next clock cycle. In case 3) the input net is connected to another computing component. It is impossible to retrieve the value of the net directly. In this case the input nets of all other components have to be recursively backtracked. Until registers are reached at all inputs of the resulting combinational circuit, the value of the requested net must be computed from the values of these registers by the debugging software. Therefore, the debugging software maintains a model of the circuit and the components. In case 4), where the requested output net is connected to another computing component, the input nets of the component of which the output is requested must be backtracked as in case 3).

The time overhead for the software computation does not affect the running hardware since such computations occur only at breakpoints when

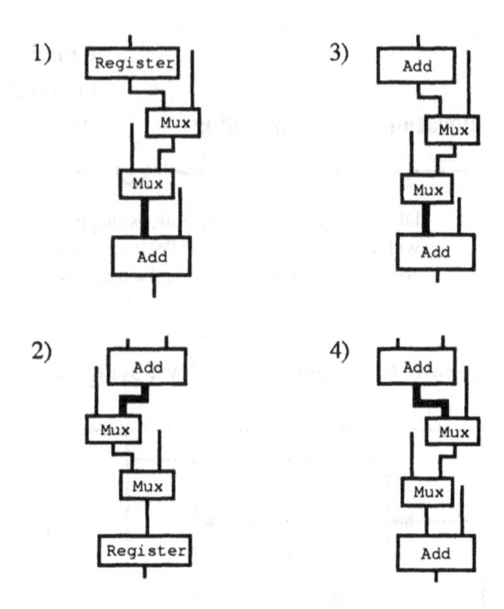

Figure 18. Possible net connections.

the hardware stops. It only requires a library where a behavioral model of each component is stored.

High-level synthesis maps operations to components in the circuit. Thereby it can happen that an operation corresponds to several components as well as several operations corresponding to the same component. For SLE it is necessary to keep track of the correspondence between operations and components after high-level synthesis, and to resolve such dependencies. This correlation is used to set breakpoints at particular steps in the algorithm. Therefore the corresponding components are identified together with the state of the controller at which this components execute the requested operation. The state of the controller is known from scheduling and the component is known from the assignment.

When the correspondence between the behavioral specification and the synthesized RT level is known, a special synthesis step is needed, which does not destroy the structure for debugging through retiming optimization etc. Also some additional hardware is required for the retrieval of register values from outside and for setting breakpoints and which is to interrupt the hardware when a breakpoint is reached. A special debugging controller has been developed to provide these additional debugging support operations. A special target hardware system has been developed as a back-end environment for this source level emulation system. This hardware is based on the Weaver environment presented in section 3.1 and allows the access of the requested debugging structure inside the circuit. As mentioned before, the programming and the controlling of breakpoints is done by the debugging controller. This component allows the host to control the circuit operation. Figure 19 shows the scheme of this debugging controller.

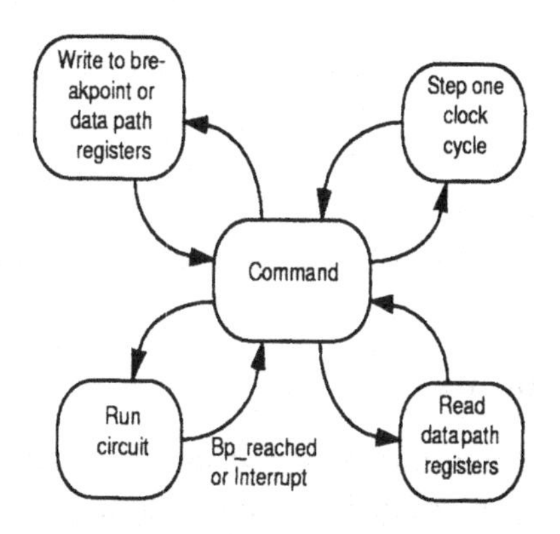

Figure 19. Simplified state diagram of the debugging controller.

The command state is the initial state of the controller. In this state the controller waits for a command and the circuit stops. The following commands are implemented:

- Write a breakpoint.
- Write values to the data path registers.
- Read the contents of the data path registers.
- Step one clock cycle.
- Run the circuit.

Each command is decoded and the controller switches to the corresponding macro state, which controls the required action. In each macro state, the controller returns to the command state after the action is finished. Only the *run circuit* state is kept until a breakpoint is reached or a interrupt command is issued by the host. The circuit is interrupted by setting a clock control signal to 0, which then prevents the data path registers and the circuit controller from operation.

Each register in the data path is exchanged according to figure 20 by a register which supports the required operation control. The inputs added to the original data path register are used for programming or to read back of the registers via a scan path and for enabling normal data path operation. These additional inputs are controlled by the debugging controller. The debugging controller itself is not application dependent. It is implemented together with the application on the Weaver system for reasons of simplicity, but it can also be implemented outside the programmable logic.

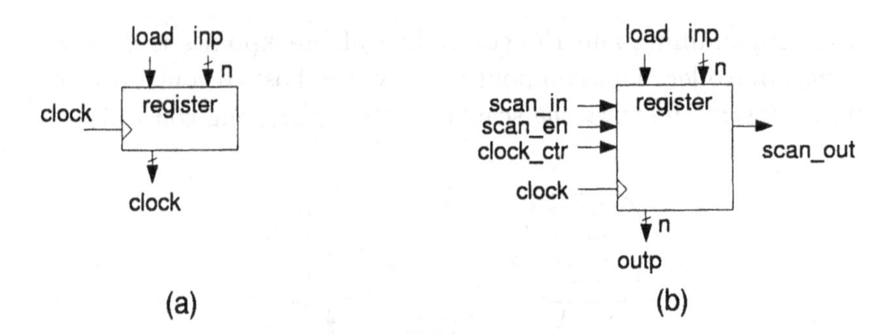

Figure 20. **(a) Original data path register and (b) inserted register for debug operations.**

5. Breakpoint Encoding

Since we want to debug a running circuit at the source code level, a breakpoint must be defined as some language construct that is visible in the source code. On the other hand, a breakpoint must also be visible in the circuit to enable detection. Thus, a breakpoint is defined as an operation like +, *, etc. If such an operation is executed by a component, it can be detected in the running hardware.

Identifying an operation as a breakpoint is not really sufficient if the operation is implemented by a synchronous component which reads different inputs at different times (i.e, controller states) and produces the outputs a few cycles later. Therefore, the point of interest is actually the state at which an input or output port of a component is active. This is determined from the component library, where all components are listed with their behavior and their timing specification. This library provides for each port the clock cycle, at which the port transports data, relative to the starting time of the component.

Figure 21 illustrates the definitions. Part a) shows all inputs and outputs of a '+' operation as visible in the data path. Part b) shows an example, where the input/output X is not visible due to an optimization during synthesis.

In the following, we will use the term "breakpoint" to refer to detectable breakpoints. The use of nondetectable breakpoints is not allowed since these cannot be detected by hardware means.

One possibility to detect the control step T of a breakpoint in the hardware is by looking at the controller states, since the controller manages the sequential behavior of the data path. The relation between control steps, and a controller state is static and known from the synthesis process. In the data path, there are operations which are always executed in a certain controller state. We call breakpoints corresponding to this type of operations Moore breakpoints. Such a breakpoint is reached, if the controller

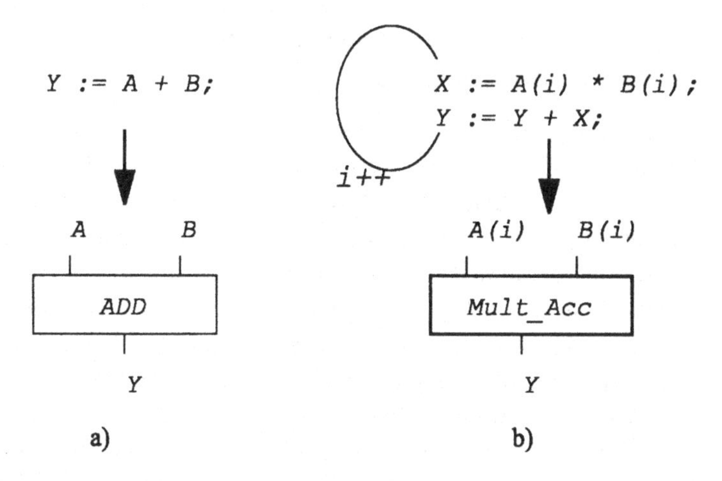

Figure 21. a) A, B and Y as detectable breakpoints and b) X as non detectable breakpoint

state matches the control step T.

There are also operations which are executed at transitions of controller states. To detect breakpoints at those operations, we need to know about the current state and the next state of the controller. These breakpoints are called Mealy breakpoints. A Mealy breakpoint is reached if the controller state matches the control step T and if the next state is the right one for the breakpoint.

Both, Mealy and Moore breakpoints, can be extended to allow for the setting of conditional breakpoints. For instance, a request could be "stop at B in the expression $Y := A * B$; if $B = 5$." More sophisticated conditions are possible, but require more hardware overhead. Naturally, it is possible to stop at a breakpoint whenever it is reached and perform the evaluation of the conditional expression in software. Since this would slow down the emulation speed significantly, it is not acceptable. Thus, the conditions have to be very simple and they have to be hardwired to allow for breakpoint detection at full emulation speed.

The remainder of this paper will concentrate on Moore breakpoints only, since the methods used for Mealy breakpoints are very similar.

Since the goal is to stop the application clock immediately, when a breakpoint is reached, the breakpoints have to be detected purelly by hardware means. To do that, the breakpoints need to be encoded in a way that allows for very simple testing as to whether a controller state corresponds to a breakpoint or not. This code of the breakpoint (breakpoint ID) has to satisfy the following constraints:

- Different breakpoints must have different IDs.

– The state code must allow for breakpoint detection at full emulation speed by simple hardware means.

Above, we have implicitly mentioned that a breakpoint corresponds to one controller state. In this case, simple state code would also serve as a breakpoint code for hardware breakpoint detection. In reality, it is not this simple. A breakpoint can be represented by a whole set of controller states. This can be formulated as the problem of finding a binary code for the elements of an arbitrary set structure in such a way, that, given a particular subset code, one can see whether or not an element is part of that subset just by ignoring some bit positions of the element code (which are fixed for a subset). As many other problems in synthesis, this is an NP-hard problem. Thus, for the general case, a heuristic has to be used, which is explained by the following example. The example in figure 22 shows a number of intersecting breakpoints to which our algorithm was applied to clarify the operation.

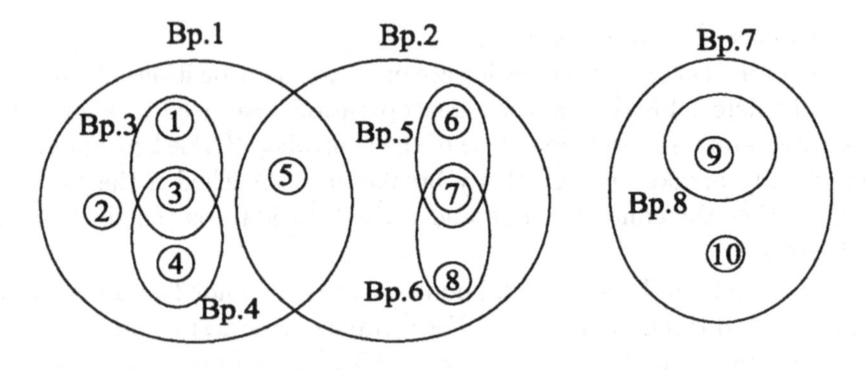

Figure 22. Example for the encoding algorithm

First, we identify two clusters at the top level of hierarchy:
C1: Bp. 1, 2
C2: Bp. 7
The first bit of the code is then '0' for C1 and '1' for C2. The next bits depend on the clusters and may be shared for different clusters. For C2 we get one bit for Bp. 7 on the highest level and after going down one level, we get an additional bit for Bp. 8.

For cluster C1, we get 2 bits, one for Bp.1 and one for Bp. 2 at the highest level. Going down one level, we get another 2 bits for Bp. 3 and Bp. 4, which share the positions with the 2 bits for Bp. 5 and Bp. 6. The breakpoint code is given in table 4 and the resulting code of the states is shown in table 5.

TABLE 4. Breakpoint code

Breakpoint	Code
1	01----
2	0-1---
3	01001-
4	0100-1
5	0011--
6	001-1-
7	11----
8	1101--

TABLE 5. State code

State	Code
1	010010
2	0101--
3	010011
4	010001
5	011---
6	00110-
7	00111-
8	00101-
9	1101--
10	111---

Bit 1 identifies the cluster, bit 2 = '1' indicates that a state belongs to Bp 1, if we are in cluster C1. The meaning of the other bits is similar. The code uses 6 bits for each state, which is not optimal. In fact, the implemented algorithm results in a 4 bit code by using local optimizations. For example,

```
fsmlogic: process(current_state, input_list)
begin
  default_assignments;
  case current_state is
    WHEN "01010" =>
      output_assignments;
      next_state_assignment;
    WHEN "01000" =>
      output_assignments;
      next_state_assignment;
    ......
  end case;
end process fsmlogic;
```

Figure 23. VHDL style of breakpoint IDs in the controller

since Bp. 7 is a cluster on its own, it is not necessary to assign a 1-hot code to it as done in loop 6. The breakpoints Bp.1 and Bp. 2 are 1-hot encoded, thus the value "00" is still free and can be assigned to the cluster with Bp. 7 as cluster identity, making it unnecessary to binary encode the clusters {Bp. 1, Bp. 2} and {Bp. 7}. For the same reason, the cluster {Bp.3, Bp. 4} and the free state 2 do not need to be enumerated.

As mentioned above, breakpoints can be detected by comparing the encoded controller state identifier with a given breakpoint ID. Depending on the breakpoint of interest, certain bits of the state code need to be ignored. An example of the VHDL style of the generated controller together with the state code is shown in figure 23. It is a process with a case statement, in which the outputs and the next state are computed based on the current state.

The breakpoint detection logic that is added to the generated circuit is shown in figure 24. A register to write in the desired breakpoint and a mask register are necessary. The mask register tells the comparator which bits of the state code and the breakpoint register it should ignore. The first bit of the breakpoint register tells the comparator to constantly output '0' if it is set, i.e. it indicates continuous operation with no breakpoint set. The comparator tests the nonmasked bits of the breakpoint identifier and the breakpoint register for equality. If equal, the BP-reached signal is raised to '1'. The signal "BP-scanin" represents a scan path for setting and changing breakpoints in the circuit. By doing this, setting breakpoints or changing breakpoint settings can be done online without doing a re-synthesis of the circuit every time. All this is done dynamically in the implemented circuit.

The detection of Mealy breakpoints is done exactly in the same way as the detection of Moore breakpoints. As opposed to Moore breakpoints,

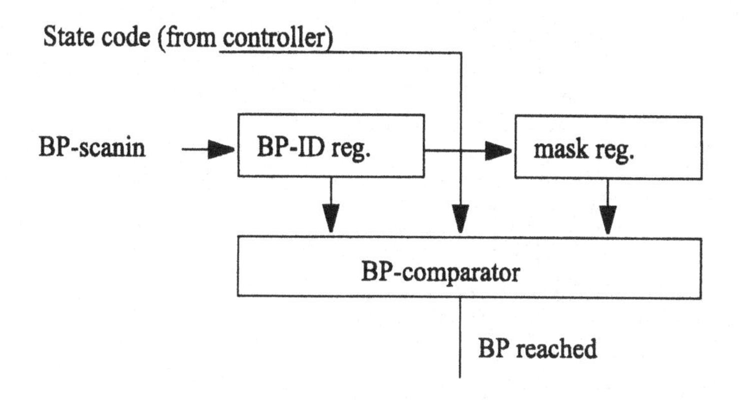

Figure 24. Breakpoint detection logic for Moore breakpoints

the logic for Mealy detection produces a constant '1' if Mealy detection is disabled. There must also be an additional identifier as controller output, which is set in each path of the IF-clause within the WHEN-clause in the VHDL code of the controller. The encoding of the Mealy identifier is much less complex than the encoding of the Moore identifiers and it is done using the same algorithm. It requires less bits and has fewer intersections. Thus, it is not described more detailed here.

The detection of data dependent breakpoints is also very similar to Moore breakpoints. All that has to be done is to replace each register in the data path by a component which contains a data register, a programmable register for breakpoint detection and a comparator.

More details can be found in [3].

6. Example

As example we have chosen Euclid's gcd algorithm. The VHDL code is shown in figure 25.

It is a very simple algorithm but it shows the principles of the approach. Figure 26 shows the intermediate flowgraph, the generated circuit and the controller graph of the gcd algorithm.

The lines between the flow graph and circuit show the relationships between operations and components. Both subtractions are executed by the same component, and both comparisons are performed by the same comparator. Thus the while-condition and the if-condition of the VHDL source are evaluated in parallel at the same time. The conditional transitions in the controller graph are marked with their conditions. Table 6 shows the full relationship between operations, components and controller states. The operations and controller states are numbered according to their numbers in figure 26, and the components are also named according to figure 26.

```
entity gcd is
  port (I1, I2: in Bit_Vector (15 downto 0);
        Ou: out Bit_Vector (15 downto 0));
end gcd;

architecture beh of gcd is
begin
process
  variable X1, X2: Bit_Vector(15 downto 0);
begin
  while (true) loop
    wait for t1;
    read(X1,I1);
    read(X2,I2);
    while X1 /= X2 loop
      if X1 < X2 then
        X2 := X2 - X1;
      else
        X1 := X1 - X2;
      end if;
      wait for t2;
    end loop;
    wait for t3;
    write(OU,X1);
  end loop;
end process;
end beh;
```

Figure 25. The example VHDL code.

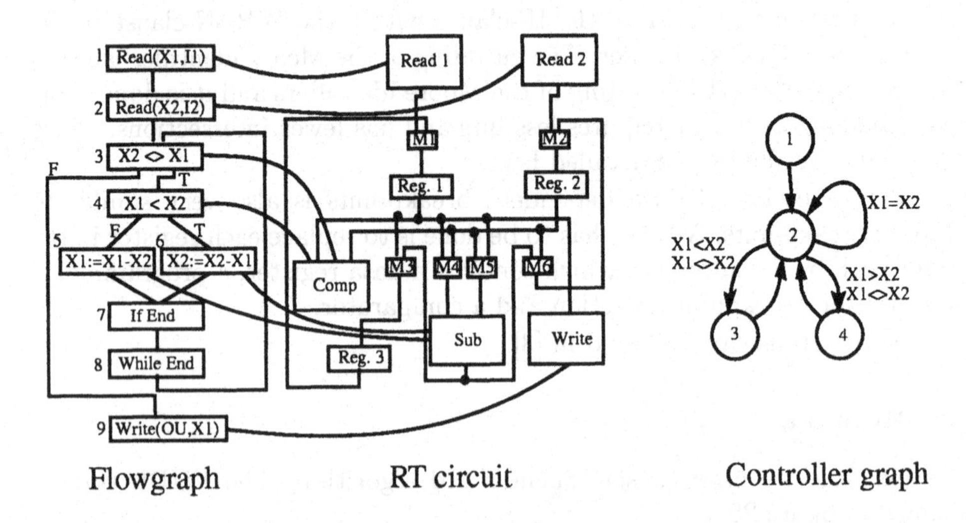

Figure 26. FG, circuit, and controller of the gcd algorithm.

The column Mealy/Moore is used to describe the behavior of the controller output which controls the component. For synchronous components this is of type Mealy, and for others it is of type Moore. If it is Mealy, then the component belongs to a state transition rather than to a state. In this case the outputs of the comparator determine together with the controller state whether the component is executed or not.

For instance, operation No. 1 is executed by the component Read 1 in the controller states 1 and 2. It is controlled by a Mealy output of the FSM and therefore also the comparator outputs are relevant to decide whether

TABLE 6. Relation operation-component-FSM state.

Operation	Component	FSM state	Mealy/Moore
1	Read 1	1→2, 2→2	Mealy
2	Read 1	1→2, 2→2	Mealy
3	Comp	2	Moore
4	Comp	2	Moore
5	Sub	2	Moore
6	Sub	3	Moore
9	Write	2→2	Mealy
- -	Reg 1	1→2, 2→2, 3→2, 4→2	Mealy
- -	Reg 2	1→2, 2→2, 3→2, 4→2	Mealy
- -	Reg 3	2→3, 2→4	Mealy

TABLE 7. Multiplexer controls.

Mux	State 1	State 2	State 3	State 4	Type
M1	0, 1→2	0, 2→2	1, 3→2	1, 4→2	Mealy
M2	0, 1→2	0, 2→2	1, 3→2	1, 4→2	Mealy
M3	-	0, 2→3 1, 2→4	0, 3→2	1, 4→2	Mealy
M4	-	0	1	-	Moore
M5	-	0	1	-	Moore
M6	-	0, 2→3 1, 2→4	0, 3→2	1, 4→2	Mealy

the component is actually executed or not. In state 1 it is executed anyway, since there is no relevant condition. But in state 2 it is only executed if the following state is also 2, that is if the condition X1=X2 is true.

Table 7 shows how the multiplexers are controlled in each controller state. Some of the multiplexers are controlled by controller outputs with Mealy behavior, others with Moore behavior. Those with Mealy behavior have a state transition after the controlling value in the table. Also these may have different values for different transitions in the same state. For instance, in state 2 the multiplexer M3 puts its first input through (control = 0) if the following state is state 3, and the second input is put through (control = 1) if the following state is state 4. In our example we request the input variable X1 of the operation No. 5. Table 6 shows that this operation is executed by the component sub and that the clock must be interrupted in controller state No. 2. An evaluation of the multiplexer controls given in table 7 shows that the requested input of this component is connected directly via a put-through component to register Reg 1. Thus, this register holds the requested value. Doing the same for Reg 2 that at the same time Reg 2 holds the value of X2 as the other input variable of operation No. 5. The values correspond also to the inputs of the operations 3 and 4. The value of Reg 3 cannot be assigned to a variable since the history of the circuit is not known.

7. Hardware-Software Co-Debugging

Systems may contain several software and hardware processes. Each process runs in its own debugger while the presented approach provides the necessary synchronization between the processes that precludes loss of data at the communication interfaces due to breakpoints. Therefore it allows to continue the operation after a breakpoint which is essential for interactive debugging.

Hardware-Software Co-Debugging concentrates on systems containing multiple processes. These processes may be software processes, running on a microprocessor, or hardware processes, designed as ASICs. All processes are synchronous processes and each of them is running on its own clock. Thus, all the processes run asynchronously. This is a typical constellation for embedded systems. Hardware-software communication will be treated as horizontal communication between a coprocessor and the software. We do not consider vertical communication, which denotes the software being executed by a processor. The idea is that for each sequential component in the component library the designer also provides a debug model of the component in the debug library. In most cases doing this is trivial. However, it is not trivial for interface components

Unfortunately we cannot provide a general rule how to derive a debug model from an arbitrary interface component. But we can provide a rule for a certain type of interface component and we can show that all common communication schemes can be implemented with components of that type. A typical protocol for reading from a standard parallel port as shown in figure 27 may serve as an illustration. It uses an out of band data synchro-

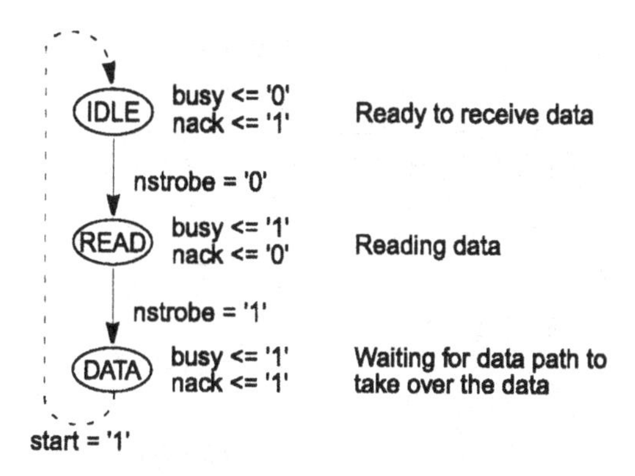

Figure 27. Simple protocol

nisation via the busy signal. If busy is '1', the partner cannot send data because the component is not ready to receive it. The rest of the protocol is a 4-way handshake involving the nstrobe and nack signals. The behavior in a breakpoint is that the states IDLE and READ are processed as in normal operation, allowing the component to receive data even in a breakpoint, if it is ready to receive. In the state DATA, the transition triggered by "start = '1'" is delayed until the next active clock edge after the breakpoint. Since the value "busy"-signal is constant '1' in that state, the sender cannot send any data and has to wait until the end of the breakpoint. Thus no data is overridden or lost.

We can identify three important elements in that example. First, there is a data synchronisation through the "busy"-signal. This allows the component to prevent the sender from sending data that cannot be received during a breakpoint. Second, there is a state, where the component is waiting for the data path to act via the "start"-signal. In that state, the communication partner is held back by the data synchronisation. Third, there is a set of states, in which the component runs through the protocol. In these states, requests from the data path ("start = '1'") are answered with "done = '0'", meaning that there is no data waiting. The general cases for sending and receiving data are depicted in figure 28. In both directions there is a state, where the component is waiting for the data path to give the start trigger. In this state, data synchronisation prevents the partner process from becoming active at the interface.

A component of this type can easily be transformed into a debug model by delaying the state transition from the state, where the component is waiting for the data path, until the next active clock edge after a breakpoint. Additionally, all outputs to the data path have to remain constant during a breakpoint. Since this cannot be guaranteed due to the possible operation that is going on in a breakpoint, some additional hardware

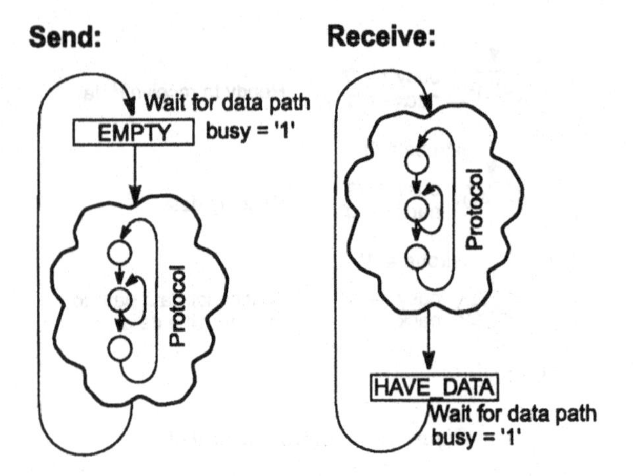

Figure 28. Behavior of interface components

is needed to hold these outputs at a constant value during a breakpoint. Bidirectional interface components can easily be implemented by combining the state machines for sending and receiving as two independent state machines in one component. These components still follow the same rules and are transformed in the same way.

The protocol part of the state machines can implement any type of protocol. It is easily possible to implement interrupts or DMA protocols. In case of an interrupt protocol the state machine would first set an interrupt signal and then wait for the interrupt procedure to handle the data transfer.

If a DMA is implemented, the component would only handle the protocol with the DMA controller of the processor, not the data communication itself. This would be implemented via a multiport RAM or an addressable register file.

It is obvious that there are limitations to the presented approach. In general, it cannot be applied to two types of systems. First, if the environment of the design imposes hard realtime constraints on the design, which cannot be softened for the prototype. E.g. an ethernet controller which has to listen at the bus with a certain speed cannot be interrupted at a breakpoint and still continue to work after a while. Second, if the design communicates with external devices without any data synchronisation and without a handshake protocol. In these cases, the external device cannot be manipulated by a special synthesis for debugging and thus, the presented method is not applicable. However, if two processes communicate with a protocol that uses neither data synchronisation nor handshake, but both processes are part of the synthesized design, the protocol can be transparently exchanged by a suitable protocol during synthesis as to make the prototype debuggable.

If the approach is not applicable, then there are two possible debugging strategies. One can operate with a logic analyser as it is done today, and one can run until a breakpoint occurred, evaluate the state of the system and restart from the beginning with the same parameters, but stop at another breakpoint.

More details can be found in [28].

8. Results

We have applied the approach to some example systems. For hardware synthesis, we have used the CADDY High Level Synthesis tool [10][11][12], which was developed at FZI. The mechanism for transparent component exchange for debugging purposes has been implemented as a "-g" switch in this tool.

The first example is a circuit that computes a 2-dimensional DCT. Data is transferred from and to a software process via DMA accesses. Figure 29 shows the structure of the example. It uses two DMA channels of the PPC403GA [29] embedded processor, one for receiving the source data and the other for sending the transformed data. A DMA component which negotiates with the DMA controller of the processor, and an internal RAM which holds the data are associated with each channel.

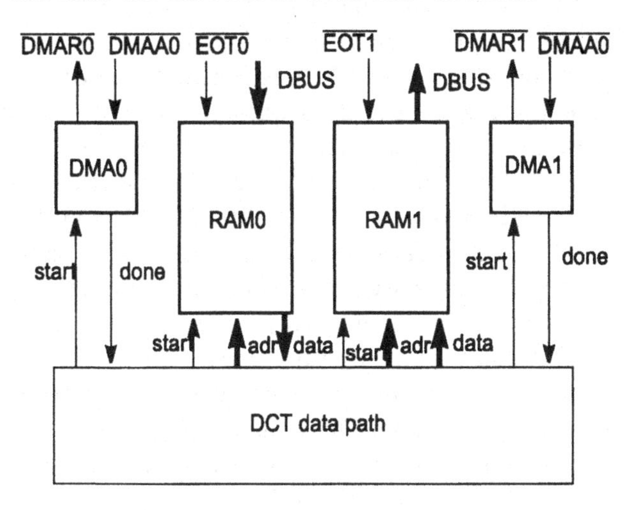

Figure 29. Structure of DMA example

Figure 30 shows an excerpt from the behavioral VHDL code of the design. The first inner loop initiates the DMA request and waits until this has been acknowledged by the component. The second inner loop waits until the DMA has been finished and the data is available. Then the DCT Algorithm can start to work on the data. Writing the result back to the processor is done analogously. Both, the DMAx()-procedure and the syncx()-procedure are implemented by the DMA-component.

```
loop
   while DMA0() = '0' loop
      wait until clk'event and clk = '1';
   end loop;
   while sync0() = '0' loop
      wait until clk'event and clk = '1';
   end loop;

   -- compute the DCT

   while DMA1() = '0' loop
      wait until clk'event and clk = '1';
   end loop;
   while sync1() = '0' loop
      wait until clk'event and clk = '1';
   end loop;
end loop;
```

Figure 30. Behavioral code

The behavior of the debug model of the DMA component is shown in figure 31. In the IDLE state, the component waits for the data path to initiate the DMA request via a call of the procedure DMAx(). It then works through the handshake protocol with the DMA controller and ends up in the SYNC state. Now the whole DMA is finished, the data is transferred and the component waits for the data path to call the syncx()-procedure and to begin working on the data. The two transitions from the IDLE state and from the SYNC state, which are triggered by the start-signal from the data path, are delayed until after the breakpoint, if the circuit is stopped.

The version of the design that uses an interrupt to trigger the data transfer has exactly the same structure. The only difference is in the state machine of the DMAx() component. So, from a hardware point of view, an interrupt protocol doesn't infer any problems.

Two other protocols have been implemented and tested. One is a bidirectional RegisterIO communication with the Hyperstone [14] processor, the other is a communication via a standard parallel interface. The RegisterIO protocol uses a status register for data synchronisation and two 32bit registers for data transfer, one for each direction. Each direction of the communication is guarded by an own state machine, which handles the data synchronisation.

The parallel interface protocol was implemented for both directions, each direction in an own component. On the software side, the communication is associated with a timeout. This timeout is implemented in the operating system. When the hardware is stopped due to a breakpoint and

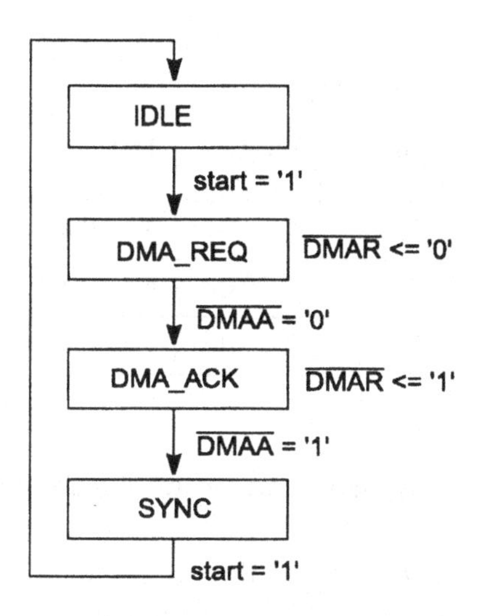

Figure 31. DMA component

TABLE 8. Area values

Circuit	**Orig.**	**Debug**	**Protocol**
GCD	60	84	Hyperstone RegIO
SIRGD	99	146	Sparc10 parallel Port
DCT	387	430	PPC403GA DMA

the software wants to transfer data, then the timeout will occur. To allow debugging, the software must catch the timeout and retry to initiate the transfer in a loop. A similar protocol encapsulation in software as it is proposed for hardware in this paper allows to do that in a simple way.

We can run these designs in our debugger and do single stepping, set breakpoints and evaluate the contents of the data path registers without loosing any data at the communication channel or causing a failure of the protocol.

In table 8, the area values of the different designs, which use the different protocols, are listed. All values are given in CLBs for the Xilinx XC4000 series.

- **Orig.** refers to the original circuit without debug overhead.
- **Debug** includes the overhead for debugging without the possibility of data dependent breakpoints.
- **Protocol** denotes the communication protocol that was used to communicate with a software process.

TABLE 9. Clock frequencies

Circuit	Orig.	Debug
GCD	5.7 MHz	4.7 Mhz
SIRGD	7.7 Mhz	4.8 Mhz
DCT	3.5 Mhz	3.5 Mhz

Table 9 shows the delay that is added by the additional logic for debugging.

The values are obtained by the Xilinx tool 'xdelay'. This tool provides a quite pessimistic estimation. All designs run at a significantly higher clock speed on our Weaver [13] board. Nevertheless, it shows the relation between the different implementations. Our approach mainly adds controller delay. Therefore we do not add any delay to the DCT, which is mainly determined by the combinational multiplier. To this path, we do no add any delay. Therefore, the debug versions can be clocked with equal frequency.

9. Conclusions

This contribution has given a survey on the state of technique in the fields of prototyping and emulation. Both existing commercial prototyping and emulation systems have been presented as well as actual research activities. The existing commercial systems are featured by different starting points. Most of the systems take user programmable gate arrays like off-the-shelf FPGAs, especially those from Xilinx. Mentor uses FPGA structures which are especially optimized for emulation. These systems differ regarding the interconnection structure among these numerous FPGAs where a logic design is mapped onto. Some systems use again programmable gate arrays as interconnection structure. Others take special user programmable switches for interconnect. Another alternative would be to use especially developed VLSI custom switch circuits. A new starting-point has been developed by Synopsys and Quickturn. The Arkos and CoBalt emulation systems of Synopsys and Quickturn do not use FPGAs but closely coupled special processors with an instruction set whose instructions are optimized for execution of logic operators. A circuit design will then be translated into Boolean equations which on their part are compiled into machine code for these special logic processors. Recently Arkos was bought by Quickturn and therefore CoBalt is currently the only commercial processor based emulation system.

Future work aims especially to support emulation at higher levels of abstraction. These approaches will take into account that automatic synthesis systems will be used more intensively, as these are necessary for the increase of design productivity. Increased design productivity and short design times are necessary for the "first-time-right silicon" concept. Emulation and prototyping are an important contribution to reach this aim. Source-level emulation, which is presented in the last chapter, does moreover help to support hardware/software co-design by an improved hardware/software co-simulation, respectively hardware/software co-emulation.

Acknowledgments

This work has been partially supported by grants from the European Commission (Project EC-US-045) and Deutsche Forschungsgemeinschaft (DFG Research Programme: Rapid Prototyping of Embedded Systems with Hard Time Constraints). I would also like to thank my research assistants Julia Wunner and Gernot Koch for helping me to put this chapter together.

References

1. W. Rosenstiel, "Prototyping and Emulation" in: J. Staunstrup, W. Wolf (Eds.), "Hardware/Software Co-Design" Kluwer Academic Publishers, 1997, pp. 75 -112
2. U. Weinmann, O. Bringmann, W. Rosenstiel, "Device Selection for System Partitioning" EURODAC '95, GB- Brighton, September 18-22, 1995
3. G. Koch, U. Kebschull, W. Rosenstiel, "Breakpoints and Breakpoint Detection in Source Level Emulation" 9th Intern. Symposium on System Synthesis (ISSS 96), La Jolla, USA, Nov. 1996
4. Y. Tanurhan, S. Schmerler, K. Mueller-Glaser, "A Backplane Approach for Cosimulation in High-Level System Specification Environments" European Design Automation Conference, EURODAC'95, GB-Brighton, September 18-22, 1995
5. P. Eles, Z. Peng, A. Doboli, "VHDL System-Level Specification and Partitioning in a Hardware/Software Co-Synthesis Environment" Proceedings of 3rd International Workshop on Hardware/Software Codesign Codes/CASHE'94, F-Grenoble, 1994
6. S. Note, J. Van Ginderdeuren, P. Van Lierop, R. Lauwereins, M. Engels, B. Almond, B. Kiani, "Paradigm RP: A System for the Rapid Prototyping of Real-Time DSP Applications" DSP Applications, Vol. 3, No. 1, 1. 1994
7. J. Soininen, T. Huttunen, K. Tiensyrjae, H. Heusala, "Cosimulation of Real-Time Control Systems" European Design Automation Conference EURODAC'95, GB-Brighton, September 18-22, 1995
8. H. Owen, U. Kahn, J. Hughes, "FPGA based ASIC Hardware Emulator Architectures" School of Electrical and Computer Engineering, Georgia Institute of Technology, 1993
9. S. Sawant, "RTL Emulation: The Next Leap in System Verification" DAC'96, USA-Las Vegas (CA), 1996
10. R. Camposano, W. Rosenstiel, "Synthesizing Circuits from Behavioral Descriptions" IEEE Transactions on CAD, Vol. 8, 2-1989
11. P. Gutberlet, J. Müller, H. Krämer, W. Rosenstiel, "Automatic Module Allocation in High Level Synthesis" European Design Automation Conference EURODAC'92, Hamburg, 1992
12. P. Gutberlet, W. Rosenstiel, "Scheduling Between Basic Blocks in the CADDY Synthesis System" European Conference on Design Automation EDAC'92, B-Brussels,

1992
13. G. Koch, U. Kebschull, W. Rosenstiel, "A Prototyping Architecture for Hardware/Software Codesign in the COBRA Project" Proceedings of 3rd International Workshop on Hardware/Software Codesign Codes/CASHE'94, F-Grenoble, 1994
14. Hyperstone electronics, Hyperstone E1 32-Bit-Microprocessor User's Manual, 1990
15. G. Koch, U. Kebschull, W. Rosenstiel, "Debugging of Behavioral VHDL Specifications by Source Level Emulation" European Design Automation Conference EURO-DAC'95, GB-Brighton, 1995
16. H. Krämer, "Automatische Synthese von parallelen Prozessor-Strukturen für VLSI-Schaltungen" PhD thesis, University of Karlsruhe, 1992
17. M. Potkonjak, S. Dey, K. Wakabayashi, "Design-For-Debugging of Application Specific Designs" ICCAD'95, USA-San Jose (CA), 1995
18. S. Note, J. Van Ginderdeuren, P. Van Lierop, R. Lauwereins, M. Engels, B. Almond, B. Kiani, "Paradigm RP: A System for the Rapid Prototyping of Real-Time DSP Applications" DSP Applications, Vol. 3, No. 1, 1. 1994
19. D. Bittruf, Y. Tanurhan "A Survey of Hardware Emulators" Technical Note, ESPRIT Basic Research Project No. 8135, 1994
20. J. Staunstrup "A Formal Approach to Hardware Design" Kluwer Academic Publishers, 1994
21. X. Ling, H. Amano, WASMII "A Data Driven Computer on a Virtual Hardware" IEEE Workshop on FPGAs for custom Computing Machines, 1993
22. P. Lysaght, J. Dunlop, "Dynamic Reconfiguration of Field Programmable Gate Arrays" Field Programmable Logic Workshop, GB-Oxford, 1993
23. P. French, R. Taylor, "A self-reconfiguring processor" IEEE Workshop on FPGAs for Custom Computing Machines, 1993
24. U. Weinmann, "FPGA Partitioning under Timing Constraints" Field Programmable Logic Workshop, GB-Oxford, 1993
25. C.A. Valderrama, F. Naçabal, P. Paulin, A. A. Jerraya, "Automatic Generation of Interfaces for Distributed C-VHDL Cosimulation of Embedded Systems: an Industrial Experience" Proceedings of 7th International Workshop on Rapid System Prototyping, Thessaloniki, 1996
26. "ASIC-Emulation auf RTL-Level" Markt&Technik - Wochenzeitung für Elektronik Nr. 42, 1994
27. T. Benner, R. Ernst, I. Könenkamp, P. Schüler, H. Schaub, "A Prototyping System for Verification and Evaluation in Hardware-Software Cosynthesis" Proceedings of 6th International Workshop on Rapid System Prototyping, Chapel Hill, 1995
28. G. Koch, U. Kebschull, W. Rosenstiel, "Co-Emulation and Debugging of HW/SW-Systems" 10th Intern. Symposium on System Synthesis (ISSS 97), Antwerpen, Belgium, Sept. 1997
29. IBM, PPC403GA Embedded Controller User's Manual, 1995

DYNAMIC POWER MANAGEMENT OF ELECTRONIC SYSTEMS

GIOVANNI DE MICHELI
Stanford University
Stanford, CA 94305

LUCA BENINI
DEIS - Università di Bologna
Bologna, Italy

AND

ALESSANDRO BOGLIOLO
DEIS - Università di Bologna
Bologna, Italy

Abstract— Dynamic power management is a design methodology aiming at controlling performance and power levels of digital circuits and systems, with the goal of extending the autonomous operation time of battery-powered systems, providing graceful performance degradation when supply energy is limited, and adapting power dissipation to satisfy environmental constraints.
We survey dynamic power management applied at the system level. We analyze first idleness detection and shutdown mechanisms for idle hardware resources. We review industrial standards for operating system-based power management, such as the Advanced Configuration and Power Interface (ACPI) standard proposed by Intel, Microsoft and Toshiba. Next, we review system-level modeling techniques, and describe stochastic models for the power/performance behavior of systems. We analyze different modeling assumptions and we discuss their validity. Last, we describe a method for determining optimum policies and validation methods, via simulation at different abstraction levels, for power managed systems.

A. A. Jerraya and J. Mermet (eds.), System-Level Synthesis, 263–292.

1. Introduction

Design methodologies for energy-efficient system-level design are receiving an increasingly larger attention. The motivations for such interest are rooted in the widespread use of portable electronic appliances (e.g., cellular phones, laptop computers, etc.) and in the concerns about the environmental impact of electronic systems (whether mobile or not). In the former case, low-power circuit and system design are required to provide a reasonable operation time to battery-operated devices. In the latter case, heat dissipation may pose a practical limitation to the design and use of high-performance processors. Moreover, studies [19] have shown that present computers consume a significant amount of electric energy, and a corresponding amount of non-renewable energy resources. Thus system-level design must strike the balance between providing high service levels to the users while curtailing power dissipation. In other words, we need to increase the energetic efficiency of electronic systems, as it has been done, by other means, with other types of engines.

Electronic systems are *heterogeneous* in nature, by combining digital with analog circuitry, using semiconductor (e.g., RAM, FLASH memories) and electro-mechanical (e.g., disks) storage resources, as well as electro-optical (e.g., displays) human interfaces. Power management must address all types of resources in a system. The power breakdown for a well-known laptop computer [32] shows that, on average, 36% of the total power is consumed by the display, 18% by the *hard-disk drive* (HDD), 18% by the wireless *local area network* (LAN) interface, 7% by non-critical components (keyboard, mouse etc.), and only 21% by digital VLSI circuitry, mainly memory and *central processing unit* (CPU). Reducing the power in the digital components of this laptop by *10X* would reduce the overall power consumption by less than 19%.

Lowering system-level power consumption, while preserving adequate service and performance levels, is a difficult task. Indeed, reducing system performance (e.g., by using lower clock rates) is not a desirable option when considering the increasingly more elaborate software application programs for computers and features of portable electronic devices. On the other hand, present systems have several components which are not utilized at all times. When such components are *idle*, they can be put in sleep states with reduced (or null) power consumption, with a limited (or null) impact on performance.

Dynamic power management is a design methodology aiming at controlling performance and power levels of digital circuits and systems, by exploiting the idleness of their components. A system is provided with a *power manager* that monitors the overall system and component states and controls the state transitions. The control procedure is called power management *policy*. We believe that dynamic power management is the most appropriate approach to reduce power consumption under performance constraints, because significant power waste is associated with idle resources and because of its general applicability. Note that support for dynamic power management must be provided by the overall system organization, and system architects often envision system partitions that enable power management. Despite the fact that some systems are designed with power management schemes, *computer-aided design* (CAD) support for this task is limited, if any at all exists. Thus, designing and implementing dynamic power management schemes are usually manual tasks.

The power management policy plays a key role in determining the efficacy of a power managed system. This chapter describes some computational methods for the determination of policies that are optimum under some system modeling assumptions. Management policies can be implemented in hardware or in software. In the case of simple systems, it may be best to implement policies as specialized hardware control units that can be modeled as *finite-state machines* (FSMs). Such units monitor the system components' states to determine their evolution in time.

When considering programmable digital systems, or generic systems with a programmable digital component, it is possible to migrate to software the task of controlling the power states. In particular, the *operating system* (OS) is the software layer where the policy can be implemented best, for those systems, like computers, that have an OS. *OS-based power management* (OSPM) has the advantage that the power/performance dynamic control is performed by the software layer (the OS) that manages the computational, storage and I/O tasks of the system. Implementing OSPM is a *hardware/software co-design* problem, because the hardware resources need to be interfaced with the OS-based software power manager, and because both the hardware resources and the software application programs need to be designed so that they cooperate with OSPM.

Recent initiatives to handle system-level power management include Microsoft's *OnNow* initiative [36] and the *Advanced Configuration and Power Interface* (ACPI) standard proposed by Intel, Microsoft and Toshiba [34]. The former supports the implementation of OSPM and targets the design of personal computers with improved usability through innovative OS design. The latter simplifies the co-design of OSPM by providing an interface standard to control system resources. On the other hand, the aforementioned standards do not provide procedures for optimal control of power-managed system.

System-level dynamic power management can be complemented by specific chip-level design techniques for power reduction. Low-power consumption in integrated circuits can be achieved through the combination of different techniques, including architectural design choices [7], logic and physical design [18, 23], choice of circuit families and implementation technology [26]. In particular, dynamic power management can be applied to digital circuits by specific techniques, such as supply voltage, frequency and activity control. Examples of activity control are clock gating of control units [2] and signal gating of data-path [16] units. Such approaches are based on the principle of exploiting idleness of circuits or portions thereof. They involve both detection of idle conditions and the freezing of power-consuming activities in the idle components. We refer the interested reader to [1] for a comprehensive and comparative description of dynamic power management methods at the chip and system level.

In this chapter, we survey system-level dynamic power management. We consider first system-level design issues, such as idleness detection and shutdown mechanisms for idle resources. We review the OnNow and ACPI standards, as well as previous work in the area of power management.

Next, we review system-level modeling techniques, and introduce stochastic models for the power/performance behavior of systems. We analyze different modeling assumptions and we discuss their validity. We consider then a working model, for which optimal policies can be computed. We present next how policies can be implemented in electronic systems.

Last but not least, we describe several methods for validating the policies, based on simulation at different abstraction levels. We conclude by stressing the need for CAD tools to support model identification, policy optimization and validation for dynamically power-managed systems.

2. System design

In this section we consider issues related to system-level design. We view the system hardware as a collection of resources, we characterize their idleness and present methods for their shutdown. We consider then the interface standards that support resource monitoring and control from the operating system, and we review current related work on dynamic power management.

2.1. IDLENESS AND SHUTDOWN MECHANISMS

The basic principle of a dynamic power manager is to detect inactivity of a resource and shut it down. A fundamental premise is that the idleness detection and power management circuit consumes a negligible fraction of the total power.

We classify idleness as *external* or *internal.* The former is strongly tight to the concept of observability of a resource's outputs, while the latter can be related to the notion of internal state, when the resource has one. A circuit is externally idle if its outputs are not required during a period of time. During such period, the resource is functionally redundant and can be shut down, thus reducing power consumption. A resource is internally idle when it produces the same output over a period of time. Thus, the outputs can be stored and the resource shut down.

While external idleness is a general concept applicable to all types of resources (e.g., digital, analog, memories, hard-disks, displays), internal idleness is typical of digital circuits. Thus, we will be concerned with external idleness detection and exploitation, since we address here system-level design.

There are several mechanisms for shutting down a resource. Digital circuits can be "frozen" by disabling registers (by lowering the enable input) or by gating the clock. By freezing the information on registers, data propagation through combinational logic is halted, with a corresponding power saving. (This saving may be significant in CMOS static technologies, where power is consumed mainly during transitions).

A radical approach to shutdown is to turn off power to a resource. While this mechanism is conceptually simple and applicable in general, it usually involves a non-negligible time to restore operation. Note that in some cases the context must be saved before shutdown (e.g., in non-volatile memory) and restored at restart.

Some components can be shut down at different levels, each one corresponding to a power consumption level and to a delay to restore operation. As a first example, consider a backlit display. When the display is used, both the LCD array and the backlighting are on. When the user is idle, the backlighting and/or the LCD array can be turned off with different power savings.

As a second example, a hard-disk drive [37] may have an operational state, in addition to an idle, a low-power idle, a standby, and a sleep state. In the idle states the disk is spinning, but some of the electronic components of the drive are turned off. The transition from idle to active is extremely fast, but only 50-70% of the power is saved in these states. In the standby and sleep states, the disk is spun down, thus reducing power consumption by 90-95%. On the other hand, the transition to the active state is not only slow, but it causes additional power consumption, because of the acceleration of the disk motor.

This example shows the trade-off of power versus performance in dynamic power management. The lower the power associated with a system state, the longer the delay in restoring an operational state. Dynamic power management strategies need to take advantage of the low-power states while minimizing the impact on performance.

2.2. INDUSTRIAL DESIGN STANDARDS

Industrial standards have been proposed to facilitate the development of operating system-based power management. A precursor standard is *Advanced Power Management* (APM) [35], which provides several layers of software to support power management on computers with compliant resources. APM defines the software interface between an OS power management policy driver and software for hardware-specific power management. APM partitions the power management functionality into a hierarchy of cooperating layers and standardizes the flow of information and control among them. With this scheme, APM-compliant application software issues commands to an APM driver which control the resources through the BIOS layer [1].

More recently, Intel, Microsoft and Toshiba proposed the *Advanced Configuration and Power Interface* (ACPI) [34], described in detail in the next section, which superseeds APM. In contrast to APM, which hinges upon the BIOS code layer, ACPI provides an OS-independent power management and configuration standard. It provides for an orderly transition from *legacy* hardware to ACPI-compliant hardware. Although this initiative targets *personal computers* (PCs), it contains useful guidelines for a more general class of systems. The characterizing feature of ACPI is that it recognizes dynamic power management as the key to reducing overall system power consumption, and it focuses on making the implementation of dynamic power management schemes in personal computers as straightforward as possible.

The ACPI specification forms the foundation of the *OnNow initiative* [36] launched by the Microsoft Corporation. The OnNow initiative is specific to the design of personal computers (PCs) and proposes the migration of power management algorithms and policies into the computer's operating system (OS). The scope of OnNow goes beyond dynamic power management.

An OnNow-compliant PC platform must conform to a set of requirements [36], including:

- The PC is ready for use as soon as the user turns it on.
- The PC appears as off when not in use, but it must be capable of responding to wake-up events (originated by the user or by a resource, such as a modem sensing an incoming call).
- Software tracks hardware status changes and adjusts accordingly. OS and software applications cooperate in the frame of dynamic power

[1] The *basic input output system* (BIOS) is the lowest layer of the OS that is often customized to the hardware.

management: applications are aware that resources may be in different service states and release resources when they are unneeded.

- All hardware devices participate in the power management scheme, by responding to the OS commands.

Personal computers are just beginning to meet the requirements of OnNow in 1998. The migration of power management to the operating system level will yield a profound improvement of the performance, power consumption and quality of service of personal computers, because it will give the control of the system to the component (i.e., the OS) that can make the most informed decisions. OnNow relies on the ACPI infrastructure to interface the software to the hardware components to be managed.

2.2.1. *ACPI*

ACPI [34] is an OS-independent, general specification that applies to desktop, mobile and home computers as well as to high-performance servers. The specification has emerged as an evolution of previous initiatives that attempted to integrate power management features in the low-level routines that directly interact with hardware devices (firmware and BIOS). It also provides some form of *backward compatibility* since it allows ACPI-compliant hardware resources to co-exist with legacy non-ACPI-compliant hardware.

ACPI is the key element for implementing operating system power management strategies, such as *OnNow.* It is an open standard that is made available for adoption by hardware vendors and operating system developers. The main goals of ACPI are to:

- Enable all PCs to implement motherboard dynamic configuration and power management.
- Enhance power management features and the robustness of power managed systems.
- Accelerate implementation of power-managed computers, reduce costs and time to market.

The ACPI specification defines most interfaces between OS software and hardware. The software and hardware components relevant to ACPI are shown in Figure 1. Applications interact with the OS kernel through *application programming interfaces* (APIs). A module of the OS implements the power management policies. The power management module interacts with the hardware through kernel services (system calls). The kernel interacts with the hardware using device drivers. The front-end of the ACPI interface is the *ACPI driver.* The driver is OS-specific, it maps kernel requests to ACPI commands, and ACPI responses/messages to kernel signals/interrupts. Notice that the kernel may also interact with non-ACPI-compliant hardware through other device drivers.

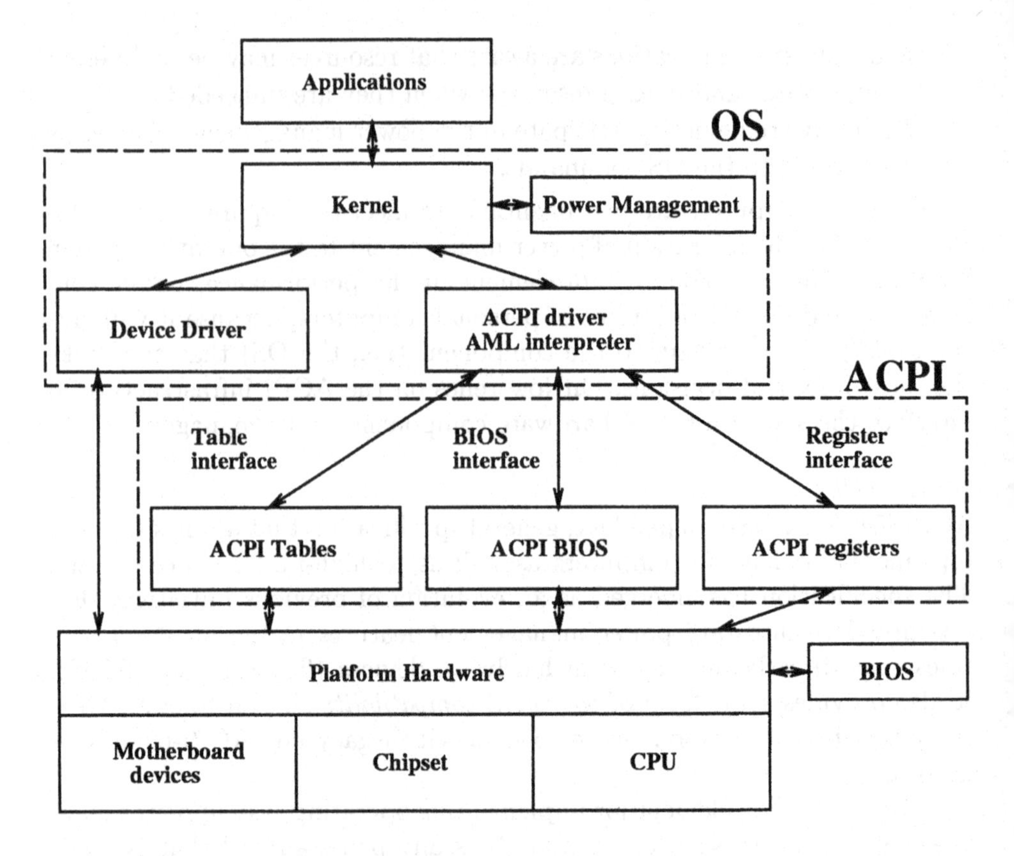

Figure 1. ACPI interface and PC platform

At the bottom of Figure 1 the hardware platform is shown. Although it is represented as a monolithic block, it is useful to distinguish three types of hardware components. First, hardware resources (or *devices*) are the system components that provide some kind of specialized functionality (e.g., video controllers, modems, bus controllers). Second, the *CPU* can be seen as a specialized resource that need to be active for the OS (and the ACPI interface layer) to run. Finally, the *chipset* (also called core logic) is the motherboard logic that controls the most basic hardware functionalities (such as real-time clocks, interrupt signals, processor busses) and interfaces the CPU with all other devices. Although the CPU runs the OS, no system activity could be performed without the chipset. From the power management standpoint, the chipset, or a critical part of it, should always be active, because the system relies on it to exit from sleep states.

It is important to notice that ACPI specifies neither how to implement hardware devices nor how to realize power management in the operating system. No constraints are imposed on implementation styles for hardware and on power management policies. Implementation of ACPI-compliant

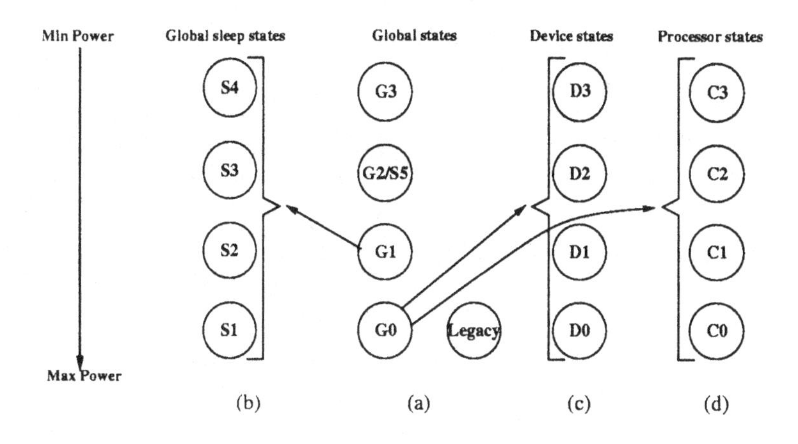

Figure 2. State definitions for ACPI

hardware can leverage any technology or architectural optimization as long as the power-managed device is controllable by the standard interface specified by ACPI.

In ACPI, the system has five *global power states*. Namely:

- *Mechanical off* state $G3$, with no power consumption.
- *Soft off* state $G2$ (also called $S5$). A full OS reboot is needed to restore the working state.
- *Sleeping* state $G1$. The system appears to be off and power consumption is reduced. The system returns to the working state in an amount of time which grows with the inverse of the power consumption.
- *Working* state $G0$, where the system is ON and fully usable.
- *Legacy* state, which is entered when the system does not comply with ACPI.

The global states are shown in Figure 2 (a). They are ordered from top to bottom by increasing power dissipation.

The ACPI specification refines the classification of global system states by defining four sleeping states within state $G1$, as shown in Figure 2 (b):

- $S1$ is a sleeping state with low wake-up latency. No system context is lost in the CPU or the chipset.
- $S2$ is a low wake-up latency sleeping state. This state is similar to the $S1$ sleeping state with the exception that the CPU and system cache context is lost.
- $S3$ is another low wake-up latency sleeping state where all system context is lost except system memory.
- $S4$ is the sleeping state with the lowest power and longest wake-up latency. To reduce power to a minimum, all devices are powered off.

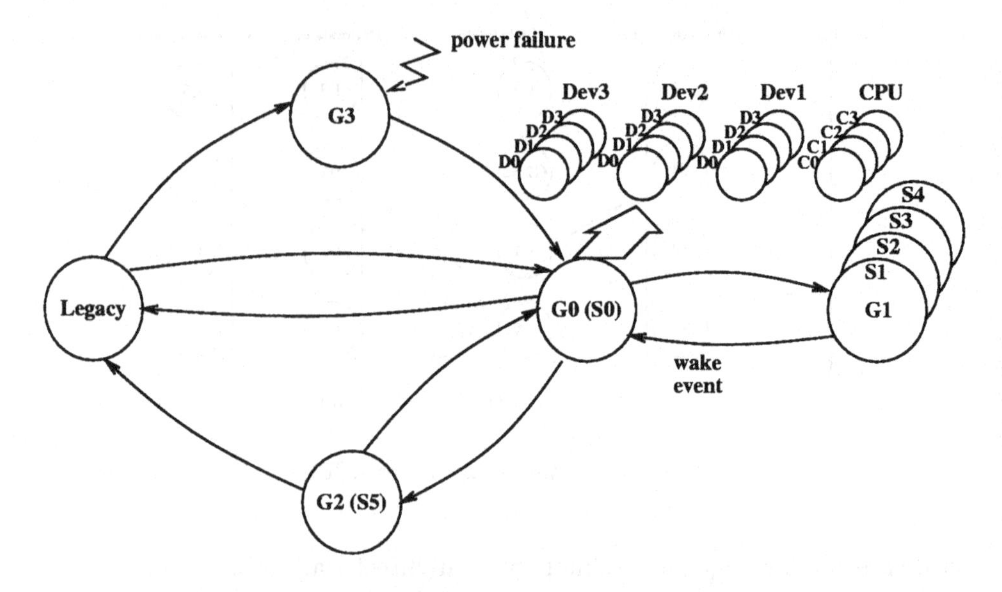

Figure 3. Global and power states and substates

Additionally, the ACPI specification defines states for system components. There are two types of system components, *devices* and *processor*, for which power states are specified. Devices are abstract representations of the hardware resources in the system. Four states are defined for devices, as shown in Figure 2 (c). In contrast with global power states, device power states are not visible to the user. For instance, some devices can be in an inactive state, but the system appears to be in a working state. Furthermore, state transitions for different devices can be controlled by different power management schemes.

The processor is the central processing unit that controls the entire PC platform. The processor has its own power states, as shown in Figure 2 (d). Notice the intrinsic asymmetry of the ACPI model. The central role of the CPU is recognized, and the processor is not treated as a simple resource.

Special devices are *embedded controllers*, that function as resources for the main CPU. ACPI defines a specialized interface for embedded controllers. Although from a power management point of view embedded controllers are treated as normal resources, they have specialized drivers because they may be used to monitor power-related system characteristics, perform low-level complex calculations, and they may provide data that is required to implement power management policies. For example, an embedded controller can be used to control board temperature sensors and provide valuable data for thermal management.

States and transitions for an ACPI-compliant system are shown in Figure 3. Usually the system alternates between the working ($G0$) and the sleeping ($G1$) states. When the entire system is idle or the user has pressed the power-off button, the OS will drive the computer into one of the states on the left side of Figure 3. From the user's viewpoint, no computation occurs. The sleeping sub-states differ in which *wake* events can force a transition into a working state, and how long the transition should take. If the only wake-up event of interest is the activation of the user turn-on button and a latency of a few minutes can be tolerated, the OS could save the entire system context into non-volatile storage and transition the hardware into a soft-off state ($G2$). In this state, power dissipation is almost null and context is retained (in non-volatile memory) for an arbitrary period of time. The mechanical off state ($G3$) is entered in the case of power failure or mechanical disconnection of power supply. Complete OS boot is required to exit the mechanical off state. Finally, the *legacy* state is entered in case the hardware does not support OSPM.

It is important to note that ACPI provides only a framework for designers to implement power management strategies, while the the choice of power management *policy* is left to the engineer.

2.3. RELATED WORK

We consider now work in different areas related to dynamic power management. The common theme is the search of methods for power/performance management. Techniques and application domains vary widely.

Chip-level power management features have been implemented in mainstream commercial microprocessors [9, 10, 11, 14, 29]. Microprocessor power management has two main flavors. First, the entire chip can be put in one of several sleep states through external signals or software control. Second, chip units can be shut down by stopping their local clock distribution. This is done automatically by dedicated on-chip control logic, without user control. Techniques for the automatic synthesis of chip-level power management logic are thoroughly surveyed in [1].

At a much higher level of abstraction, energy-conscious communication protocols based on power management have been extensively studied [17, 25, 27, 33]. The main purpose of these protocols is to regulate the access of several communication devices to a shared medium trying to obtain maximum power efficiency for a given throughput requirement.

Power efficiency is a stringent constraint for mobile communication devices. Pagers are probably the first example of mobile device for personal communication. In [17], communication protocols for pagers are surveyed. These protocols have been designed for maximum power efficiency. Protocol power efficiency is achieved essentially by increasing the fraction of time in which a single pager is idle and can operate in a low-power sleep state without the risk of loosing messages.

Golding et al. considered HDD sub-systems [12, 13], and presented an extensive study of the performance of various disk spin-down policies. The problem of deciding when to spin down a hard disk to reduce its power dissipation is presented as a variation of the general problem of predicting idleness for a system or a system component. This problem has been extensively studied in the past by computer architects and operating system designers (reference [13] contains numerous pointers to work in this field), because idleness prediction can be exploited to optimize performance (for instance by exploiting long idle period to perform work that will probably be useful in the future). When low power dissipation is the target, idleness prediction is employed to decide when it is convenient to spin down a disk to save power (if a long idle period is predicted), and to decide when to turn it on (if the predictor estimates that the end of the idle period is approaching).

The studies presented in [28, 15] target hypothetical "interactive terminals". A common conclusions in these works is that future workloads can be predicted by examining the past history. The prediction results can then be used to decide when and how transitioning the system to a sleep state. In [28], the distribution of idle and busy periods for the interactive terminal is represented as a time series, and approximated with a least-squares regression model. The regression model is used for predicting the duration of future idle periods. A simplified power management policy is also introduced, that predicts the duration of an idle period based on the duration of the last activity period. The authors of [28] claim that the simple policy performs almost as well as the complex regression model, and it is much easier to implement. In [15], an improvement over the simple prediction algorithm of [28] is presented, where idleness prediction is based on a weighted sum of the duration of past idle periods, with geometrically decaying weights. The weighted sum policy is augmented by a technique that reduces the likelihood of multiple mispredictions.

A common feature of these power management approaches is that policies are formulated heuristically, then tested with simulations or measurements to assess their effectiveness.

3. Modeling

In the sequel we consider the hardware part of the system as a set of resources. We model the resources at a very-high level of abstraction, i.e., we view them as units that perform or request specific services and that communicate by requesting and acknowledging such services. Resources of interest are, of course, those that can be power managed, i.e., those that can be set in different states, as in the ACPI scheme.

From a power management standpoint, we model the hardware behavior as a *finite-state system*, where each resource is associated with a set of states and can be in one of the corresponding states. Power and service levels are associated with the different states and transitions among states. In this modeling style, we abstract away the functionality of the resource, and we are concerned only with the ability of the resource to provide and/or request a service.

Because of the high-level of abstraction in resource modeling, it is difficult, if not impossible, to have precise information about power and performance levels of each resource. This uncertainty can be modeled by using random variables for the observable quantities of interest (eg., power, performance), and by considering average values as well as their statistical distributions [1]. This stochastic approach is required to capture both the non-determinism due to lack of detailed information in the abstract resource models as well as the fluctuations of the observed variables due to environmental factors.

With this modeling style, computing optimum dynamic power management policies becomes a *stochastic optimum control* problem [6]. The problem solution, and its accuracy in modeling reality, depend highly on the assumptions we use in modeling. We will discuss next the impact of some modeling assumptions, and then consider is detail a system model under some specific assumption that enables us to compute optimum policies, as shown in Section 4.

3.1. MODELING ASSUMPTIONS

A system model can be characterized by the ensemble of its components, their mode of interaction and their statistical properties.

In general, we can view resources both as providers and requesters of services to other resources. In practice, some resources will be limited to providing or requesting services. We call *system structure* the system abstraction where resources are vertices of a directed graph and where resource interaction is shown by edges. The interaction is the request of a service and/or its delivery. Special resources, such as *queues*, can be used to model the accumulation or requests waiting for services [31].

A simple example of a CPU requesting data to a hard-disk drive is shown in Figure 4 (a). A more complex example is reported in Figure 4 (b): it shows a CPU interacting with a LAN interface, a HDD, a display, a keyboard and a mouse. Requests to the CPU can be originated from the keyboard, mouse and LAN interface. Requests to the display come from the CPU (which also forwards requests from the keyboard and mouse). The

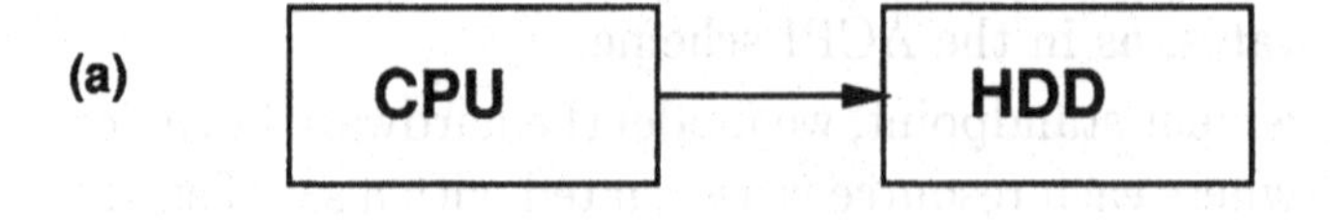

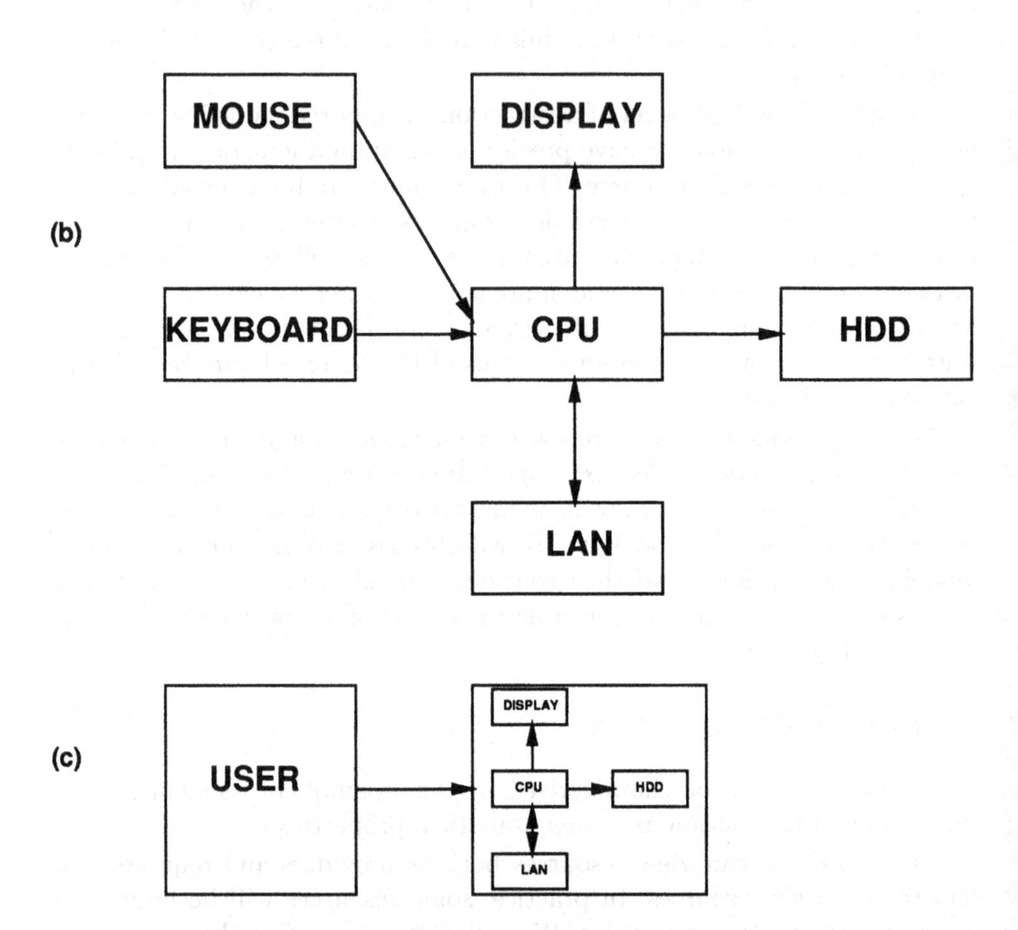

Figure 4. (a) CPU requesting data to a HDD. (b) Simple model of some resources of a personal computer and their interaction. (c) User-PC model where the requests sent by the keboard and by the mouse are lumped as a single requester and the CPU, HDD, LAN and display are lumped as a single provider.

CPU can request services to the HDD and LAN. Note that the keyboard and mouse models can express also the behavior of the human user who hits their keys and buttons.

A modeling trade-off exists in the *granularity* of the resources, i.e., between the number and average complexity of the resources. Whereas a system model with several resources and an associated structure can capture the interaction of the system components in a detailed way, most researchers view systems with a very coarse granularity. Namely, systems are identified by one resource providing a service, called *service provider*, and one unit requesting a service, called *service requester*. The requester models the *workload source*. This granularity can be used to model systems like user-PC, as shown in Figure 4 (c), where the keyboard and mouse are lumped as a single requester and the CPU, HDD and LAN are seen as a single provider.

Let us consider now the statistical properties of the components of a system. *Stationarity* of a stochastic process means that its statistical properties are invariant to a shift of the time origin [22]. When resources are viewed as providers of services in response to input stimuli, it is conceivable to model their behavior as stationary. Conversely, when resources act as workload sources, and when we model users' requests as such, the stationarity assumption may not hold in general. For example, patterns of human behavior may change with time, especially when considering the fact that an electronic system may have different users. On the other hand, observations of workload sources over a wide time interval may lead to stationary models that are adequately accurate. An advantage of using stationary models is the relative ease of solving the corresponding stochastic optimization problems.

The statistical properties of each component are captured by their distributions. An important aspect is the statistical independence (or dependence) of the resources' statistical models from each others. When a system structure can be captured by disjoint graphs corresponding to statistically-independent resources, the system decomposition allows us to consider and solve independent subproblems. In practice, weak dependencies can sometimes be neglected. Conversely, system structures with many dependencies correspond to complex models requiring a large computational effort to solve the related optimization problems. As a result, the identification of the system resources, interactions and statistics is a crucial step in modeling real systems.

3.2. A WORKING MODEL

We consider here a working model, with one provider that receives requests through a queue, and that is controlled by a power manager (PM), as shown in Figure 5. This model is described in more detailed in [21]. We summarize here the salient features of the model.

We assume stationary stochastic models for a service provider (SP), a service requester (SR), and a queue (Q). We assume also that the service

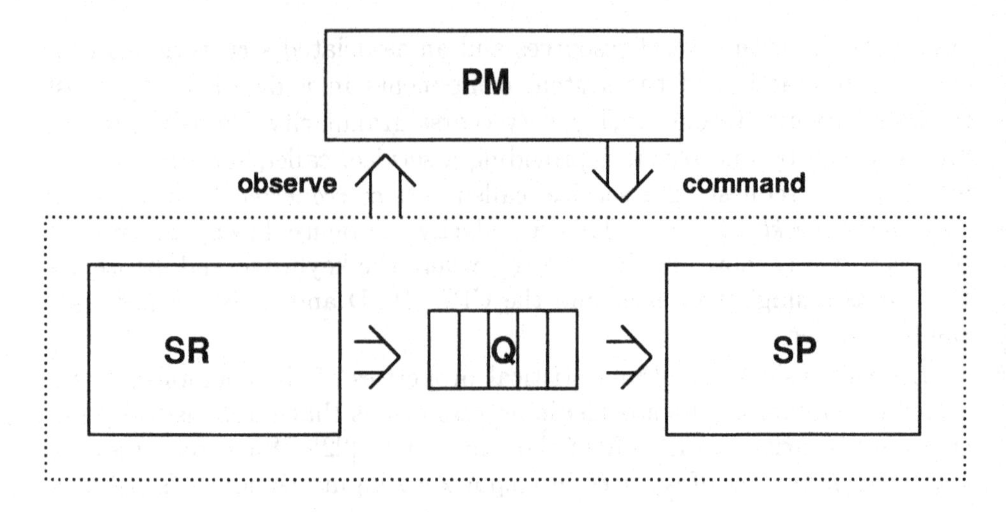

Figure 5. Coarse-grained system model

requester is statistically independent from the other components. We consider a discrete-time setting, i.e., we divide time into equally-spaced time slices. We use a parametrized Markov chain model to represent the statistical properties of the system resources. By using the Markov assumption, transition probabilities depend only on the current state and not on the previous history. Moreover, we assume that transition probabilities depend on a parameter, that models the command issued by the power manager. We consider next the system components in detail.

Service provider. It is a device (e.g., HDD) which services incoming requests from a workload source. In each time interval, it can be in only one *state*. Each state $s_p \in \{1, 2, \ldots, S_p\}$ is characterized by a performance level and by a power-consumption level. In the simplest case, we could have two states ($S_p = 2$): *on* and *off*. Otherwise, the states may be more, and in particular match states (and substates) as defined by the ACPI standard. At each timepoint, transitions between states are controlled by a power manager through *commands* $a \in A = \{1, 2, \ldots, N_a\}$. For example, we can define two simple commands: switch on (s_on) and switch off (s_off). When a specific command is issued, the SP will move to a new state at the next timepoint with a fixed probability dependent only on the command a itself, and on the departure and arrival states. In other terms, after being given a transition command by the power manager, the SP can remain in its current state during the next time slice with a non-zero probability. This aspect of the model takes into account the uncertainty in the transition time between states caused by the abstraction of functional information. Our probabilistic model is equivalent to the assumption that the evolution in time of states is modeled by a Markov process that depends on the commands issued by the power manager. Each state has a specific *power consumption rate*, which is function both of the state and the command issued. The SP provides service in one state only, that we call active state.

Service requester. It sends requests to the SP. The SR is modeled as a Markov chain, whose state corresponds to the number of requests s_r (with $s_r \in \{0, 1, \ldots, S_r - 1\}$) sent to the SR during time slice of interest.

Queue. It buffers incoming service requests. We define its length to be $(S_q - 1)$. The queue length is also Markov process with state $s_q \in \{0, 1, \ldots, S_q\}$. The state of the queue depends on the state of the provider and requester, as well as on the command issued by the power manager in the time slice of interest.

Power manager. It communicates with the service provider and attempts to set its state at each timepoint, by issuing commands chosen among a finite set A. For example, the commands can be *s_on*, and *s_off*. The power manager contains all proper specifications and collects all relevant information (by observing SP and SR) needed for implementing a power management policy. The consumption of the power manager is assumed to be much smaller than the consumption of the subsystems it controls and it is not a concern here.

The state of the system consisting of {SP,SR,Q} and managed by PM is a triple $s = (s_r, s_p, s_q)$. Being the composition of three Markov chains, s is a Markov chain (with $S = S_r \times S_p \times S_q$ states), whose transition matrix depends on the command a issued by the PM.

Let us consider a simple example, as shown in Figure 6, representing a power-managed HDD. The service requester has only two states, 0 and 1, representing the number of requests per time slice sent to the provider. The queue of the service provider has two states, 0 and 1, representing the number of requests to be serviced. The service provider has two states, *on* and *off*, representing its functional state. When *on*, it services up to one request per time slice taken from the queue. The corresponding power consumption is of 3W. When *off* it does not service any request and it consumes no power. However, a power consumption of 4W is associated with any transition between the two states. SR evolves independently, while the transition probabilities of SP depend on the command issued by the power manager (*s_on*, *s_off*) and those of the queue depend on the states of both SP and SR, as well as on the command. For example, consider the SP in state *on*. (Center-left of Figure 6.) When command *s_on* is issued,

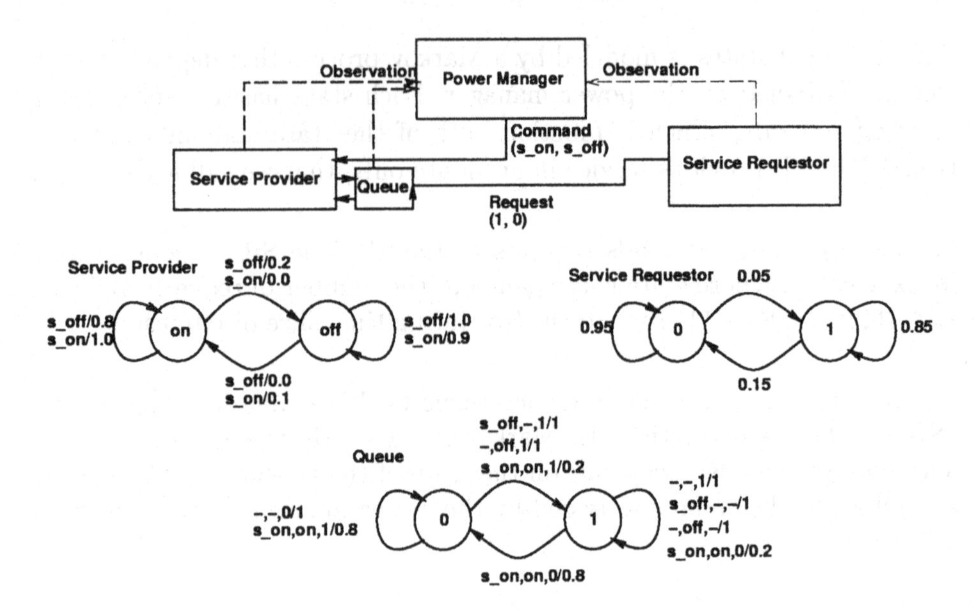

Figure 6. An example of a system model with one service provider, one service requester and one queue,with corresponding Markov chains.

the SP will stay in state *on* with probability 1, and transit to state *off* with probability 0. Conversely, when command *s_off* is issued, it will stay in state *on* with probability 0.8, and transit to state *off* with probability 0.2.

3.3. EXTENSIONS AND LIMITATIONS

System providers, requesters and queues with several internal states can be modeled in a straightforward way. Power costs and performance penalties can be associated with states and transitions of the Markov models. Thus, the simple model exemplified by Figure 6 can be made more detailed, to capture subtle differences among resource states (e.g., discriminating *soft off* states from *sleeping* states).

Similarly, more complex system structures (with multiple providers, requesters and queues) can be modeled by considering the combined effect of the resources' models. This can be easily done under the hypothesis of statistical independence of the resources' behavior, as in the case of several independent providers responding to a single workload source. In this particular case the overall system model can be derived by composing the Markov chains associated with each resource.

Unfortunately, in the general case, the system model is not amenable to a simple decomposition. Consider for example systems such as the one depicted in Figure 4 (b). The interaction among components causes statistical dependence. Most requests to the display from the CPU are triggered

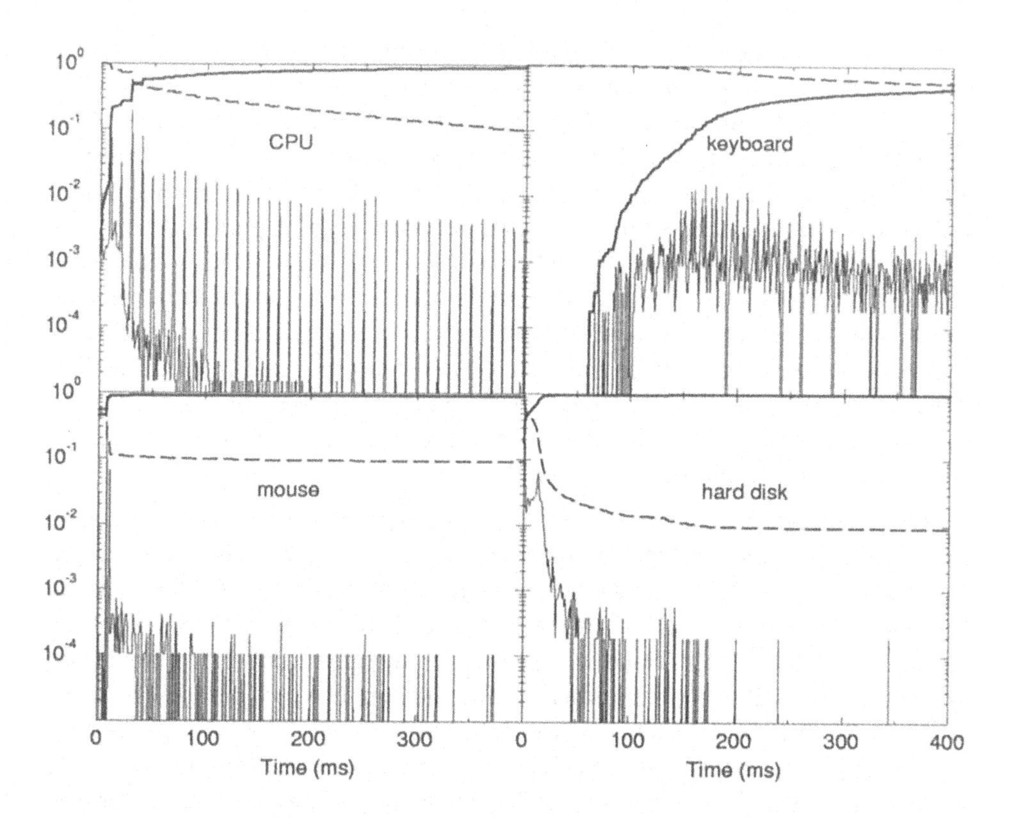

Figure 7. Statistical analysis of the inter-arrival times between service requests for CPU, keyboard, mouse and HDD of a personal computer during software development. For each device, three curves are plotted in lin-log scale: the probability density (solid line), the probability distribution (bold line) and its complement to 1 (dashed line).

by the mouse and keyboard. Thus it is not possible to view the resources as having an independent behavior.

Even when considering systems with simple structures, the identification of the statistical distributions is not a simple matter. The use of stationary Markov models corresponds to use geometric distributions for requests and service times. Such a model may deviate from reality. For example, resources may have known, deterministic service delays compounded with non-deterministic delays depending on the environment.

3.4. EXTRACTING MODELS FOR THE USER

System users can be viewed as workload sources and modeled as service requesters. An approach to model the user behavior consists of *monitoring* the system during a user session and then *extracting* a statistical model of his/her behavior.

System monitoring has to be sufficiently accurate to provide time-stamped traces of service requests. The cumulative counts provided by the system utilities of many computer systems are not sufficient to steer power management. In addition, monitoring has to be non-perturbative in order to affect usage patterns as little as possible. A monitoring system specifically designed for supporting dynamic power management in personal computers is described in reference [4]: the prototype implementation is conceived as an extension of the Linux operating system [30]. The monitoring tool can be configured to collect information about many resources at the same time. Measured overhead for data collection is quite small (around 0.4%). Figure 7 shows usage statistics simultaneously extracted for the CPU, the keyboard, the mouse and the HDD of a personal computer during one-hour of software development.

Once time-stamped request traces have been collected, they are used to characterize the abstract model for the SR. If a discrete-time setting is assumed for modeling, the trace need to be discretized first. For a given time step T, that is usually of the same order of the minimum time constant of the SP, a discretized trace is a stream of integer numbers representing request counts. The k-th number in the stream (i.e., n_k) is the number of requests with time stamps in the interval $[(k-1) \cdot T, k \cdot T]$. According to the definition of SR proposed in Section 3.2, n_k represents the state of the SR at the k-th time step. Characterizing a Markov model for the user consists of tuning the state transition probabilities in order to make the statistical properties of the model as similar as possible to those of the stream. To this purpose, state transition probabilities are directly computed from the discretized trace. For instance, the probability associated with the transition from state $s_r = 0$ to state $s_r = 1$ is obtained as the ratio between the number of $0, 1$ sequences in the stream and the total number of 0's. This procedure extracts the most accurate Markov model of any trace. If the trace cannot be seen as the output of a Markov process, then the SR Markov model needs to be validated by simulation, as described in Section 5.

4. Policy optimization

We consider now the policy optimization problem, for the working model described in Section 3.2. Policy optimization strives at minimizing the average power consumption under performance constraints. Similarly, we can define the complementary optimization of maximizing system performance under a bound on the average power consumption. With the working model of Section 3.2, performance relates to the average delay in servicing a request (i.e., wait time on a hard-disk access). Due to space limitation, we describe only the major steps toward solving the problem. The interested reader is referred to [21] for details.

We need to analyze first how the PM controls the system, to define formally the notion of policy, which is the unknown of the problem to optimize.

At each time point, the power manager observes the history of the system and controls the SP by taking a *decision*. A *deterministic decision* consists in issuing a single command. A *randomized decision* consists of specifying the probability of issuing a command. Randomized decisions include deterministic decisions as special cases (i.e., the probability of a command is 1).

A *policy* is a finite sequence of decisions. A *stationary policy* is one where the same decision (as a function of the system state) is taken at each time point. Note that stationarity means that the functional dependency of the decision on the state does not change over time. Obviously, as the state evolves, the decisions change. *Markov stationary policies* are policies where decisions depend only on the present system state.

The importance of Markov stationary policies stems from two facts: they are easy to implement and it is possible to show that optimum policies belong to this class. Namely, it is possible to prove formally that the aforementioned policy optimization problems have an optimum solution that is a unique randomized Markov stationary policy. In the particular case that either the problem is unconstrained or the constraints are inactive, then the solution is also deterministic [6, 21]. It is possible to show that the policy optimization problem can be cast as a linear program. An intuitive formulation is described here in an informal way. Consider the PM, that observes the system state and issues commands. For each possible pair (state,command), we can compute its *frequency*, i.e., the expected number of times that a system is in that state and issues that command. The frequency is a non-negative number subject to the following *conservation law.* The expected number of times state x is the current state is equal to the expected initial population of x plus the expected number of times x is reached from any other state. Moreover, average power and performance loss can be expressed as linear functions of the $(state, command)$ frequencies. Thus, minimizing power consumption can be expressed as minimizing a linear function of the $(state, command)$ frequencies, under linear constraints.

Overall, linear programs modeling policy optimization can be efficiently solved by standard software packages, for simple topologies and a reasonable number of commands. The policy optimization tool described in [21] is built around PCx, an advanced LP solver based on an interior point algorithm [8].

Figure 8 shows the power-performance trade-off curve obtained for the

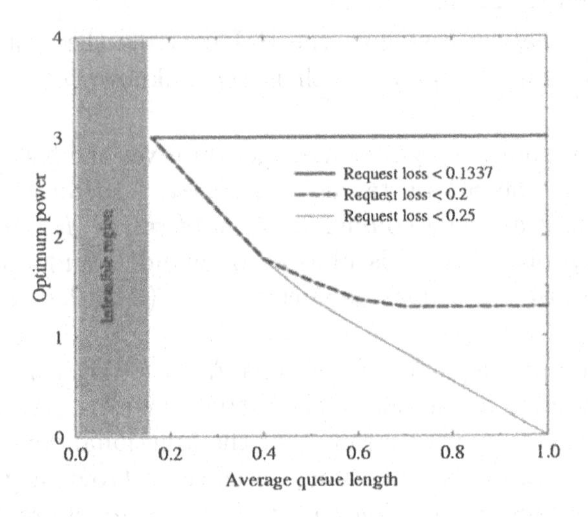

Figure 8. **Power-performance trade-off curves for the example system of Figure 6.**

example system of Figure 6 by iteratively solving the policy optimization problem for different performance constraints. Performance is expressed in term of average queue length, that is the average waiting time for a request. An additional constraint is used, called *request loss*, to represent the maximum probability of loosing a request because of a queue-full condition. It is worth noting how the power-performance trade-off is affected by the additional constraint. In particular, if a request-loss lower than 0.1337 has to be guaranteed, the SP can never be shut down. In this case, no power savings can be achieved regardless of the performance constraint.

The trade-off curve for a more complex system is reported in Figure 9. The SP is a commercially-available power-manageable HDD with one active state and four inactive states, spanning the trade off between power consumption and shut-down/wake-up times [39]. The average power consumption of the disk when in the active state is of 2.5W. The SR model was extracted as described in Section 3.4 from the time-stamped traces of disk accesses provided in [38]. A queue of length 2 was used.

Points associated with several heuristic policies are also plotted in the power-performance plane for comparison. Although we cannot claim that our heuristic policies are the best that any experienced designer can formulate, some of them provide power-performance points not far from the trade-off curve. Note that heuristic solutions do not allow the designer to automatically take constraints into account. On the other hand, trial and error approaches may be highly expensive due to the large number of pa-

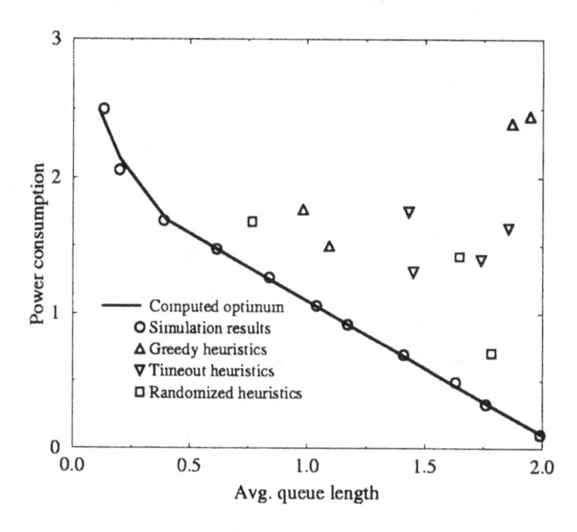

Figure 9. Power-performance feasible trade-off's for a commercially-available power-manageable hard disk.

rameters (in our case study the policy is represented by a 66x5 matrix with 330 entries). Moreover, even if it is possible to produce heuristic policies that produce "reasonable" results, there is no way for the designer to estimate if the results can be improved. For these reasons, computer-aided design tools for policy optimization can be of great help to system designers.

4.1. POWER MANAGER IMPLEMENTATION

Power management policies can be computed *off-line* or *on-line*. In the former case, a policy is computed once for all for the system being designed, and implemented in hardware or software as described in this section. Alternatively, several policies can be computed off-line and stored, each corresponding to a different environmental factor, such as a workload source. The power manager can switch among the policies at run time. On-line policy computation is also possible. Once the power manager has identified a change of the environmental conditions that make the current policy no longer effective, a new policy can be computed which takes into account the new environmental parameters (e.g., request arrival rate). Once the policy is computed, it can be executed until the power manager deems it appropriate.

In the case of simple systems, it may be practical to implement the dynamic power management policy as a hardware control circuit. Since circuit synthesis methods are currently used for hardware design, policy implementation consists of representing the policy in a synthesizable *hardware description language* (HDL) model for the power manager. In general, the circuit input is the system state and the output are the commands.

Deterministic policies can be implemented by table look-up schemes. Randomized policies require storing the conditional probabilities of issuing a command in any given state and comparing them with a pseudo-random number, which can be generated by using a linear feedback shift register (LFSR). The command probabilities should be normalized to the length of the LFSR. In particular, when only two commands are possible (e.g., *s_on* and *s_off*), their conditional probabilities sum up to 1 and thus only one probability needs to be stored. The binary outcome of the comparison with a pseudo-random number corresponds to the chosen command. This scheme can be easily extended to handle N_a commands by means of a table with $N_a - 1$ entries per state and $N_a - 1$ comparisons with the pseudo-random number, which can be executed in parallel.

The implementation of policies in software requires the software synthesis of the power manager (e.g., the generation of a C program that issues the commands as a function of the system state) as well as its embedding in the operating system. In the case of randomized policies, the program should make use of a pseudo-random number generator for deciding which command should be issued. The power manager may be executed in kernel mode and be synchronized and/or merged with the OS task scheduler to reduce the performance penalty due to context switch.

5. Validation

In this section we address the problem of bridging the gap between the high level of abstraction at which policy optimization is performed and the real-world systems, where optimal policies have to be applied. In Section 3 we have described a general approach for modeling power-manageable systems as interacting Markov processes. In Section 4 we have shown that such an abstract model allows us to cast the policy optimization problem as a linear program that can be solved in polynomial time. All modeling assumptions made to formulate and solve the policy optimization task need to be tested in order to validate its results. We briefly describe validation techniques based on simulation and emulation at different abstraction levels, ranging from the direct simulation of the Markov models used for optimization to the actual implementation of the optimal policies in the target systems. We discuss the main strengths and the inherent limitations of each approach.

Discrete-time simulation of Markov processes. Discrete-time simulation is performed at the same abstraction level used for optimization. The simulator takes the policy and the Markov models of the components and iteratively performs the following steps: *i*) take a decision (based on the current state), *ii*) evaluate cost/performance metrics, *iii*) evaluate the next state of all components, *iv*) increment time, update the state and iterate. Notice that both the policy and the next-state functions of the Markov chains are *non-deterministic discrete functions* (NDFs): inputs are present-state variables and commands, whereas outputs are the outcomes of random processes. NDFs can be represented as matrices having as many rows as input configurations and as many columns as output values. Entries represent the conditional probabilities of all possible outcomes for all given input configurations. To evaluate a function, the row associated with the current input configuration is selected and a pseudo-random number (uniformly distributed between 0 and 1) is generated and compared to the entries in the row to select the actual command.

Needless to say, this simulation paradigm cannot be used to validate the policy against the modeling assumptions of Section 3, since it relies on them as well. However, it provides valuable information about the time-domain system behavior. Constraints and objective functions used for optimization are average expected values of the performance/cost metrics of interest. Simulation allows us to monitor the instantaneous values of such parameters (to detect, for instance, the temporary violations of performance constraints) and to measure their variance.

Discrete-time simulation with actual user traces. The simulation paradigm is the same described in the previous paragraph. The only difference is that the model of the service requester is now replaced by a trace taken from a real-world application. At each time step, the present state of the SR is read from the trace, instead of being non-deterministically computed from the previous one.

Though the abstraction level is still very high, trace simulation allows us to remove all assumptions on the time distribution of service requests. As a result, it can be used to check the validity of the Markov model used for the SR during optimization.

Discrete-time simulation with real request traces was performed to validate the trade-off curve of Figure 9. Simulation results are denoted by circles in figure. The small distance of the circles from the solid-line curve is a measure of the quality of the SR Markov model extracted from the user traces and used for optimization.

Event-driven stochastic simulation. In event-driven simulation, model evaluation is no longer periodic. The model of each component is re-evaluated only when an event (i.e., a change) occurs on some of the state/command variables it depends on. The evaluation of a component may produce new events to be scheduled at a future time. Both the output events and their scheduling times may be non-deterministic. For instance, the command issued by a randomized policy can be modeled as an instantaneous non-deterministic event, while the transition between two states of the SP can be viewed as a deterministic event (if the next state is uniquely determined once a command has been issued) to be scheduled at a non-deterministic time (if the transition time is a random variable). The scheduling time is pseudo-randomly chosen according to a given probability distribution. An event-driven stochastic simulator is described in [5].

The main advantage of the event-driven paradigm is that it can easily handle stochastic processes with arbitrary distributions. Conversely, discrete-time simulation is implicitly based on the memory-less assumption that is behind Markov models, that allows us to represent and simulate only geometrically-distributed random variables. Adding memory information to a Markov model in order to represent different distributions is not a practical solution since it causes the exponential increase of the number of states. Event-driven simulation provides a more practical way of applying optimal policies to arbitrary SP models in order to check the validity of the Markov model used for optimization.

Fully-functional simulation. The functionality of a system can be described at many levels of abstraction. Functional simulation can be performed at any level. Here we focus on *cycle-accurate* simulation, that is the most accurate simulation paradigm that can be used to handle systems as complex as a personal computer. Cycle-accurate simulation matches the behavior of the real system at clock boundaries. When the system is a computer, cycle-accurate simulation provides enough detail to boot an operating system and run an actual workload on top of it. A fully-functional simulator specifically designed to study computer systems is *SimOS* [24], that can handle multi-processor architectures and provides models for simulating commercial microprocessors, peripherals and operating systems.

When system functionality comes into the picture, most of the simplifying modeling assumptions can be eliminated. In particular, stochastic models for SP and SR are no longer required, since even their functionality can be exactly simulated. Performance penalties can be realistically estimated and accurate cost metrics (i.e., power consumptions) can be associated with the operating states of the resources. In addition, functional simulation realizes a unique trade-off between realism and flexibility. On one hand, it provides a means of validating the policies against the real world and gives the designer a direct hands-on experience of most of the implementation issues involved in OS-directed power management. On the other hand, it allows the designer to explore the entire design space, balancing hardware and software solutions.

The main drawback of functional simulation is performance: simulation times may be more than three orders of magnitude slower than the run times on the corresponding real system, making the approach impractical to study complex workloads.

Emulation. We use the term *emulation* to denote a validation approach that uses functionally-equivalent real hardware components to exercise the behavior of part of the system. In particular, we are interested in using a computer without power-management features as the hardware platform to emulate a power-managed functionally-equivalent one. As an example, suppose that we are designing a power-management policy for the HDD of a laptop computer, having one active state and several inactive states (with different power consumptions and wake-up times). If such a HDD is not available for validation, the power-managed system can be emulated on an equivalent computer (with the same workload of the target one) with a non-power-manageable HDD. As long as the device used for emulation has the same performance of the target one, it can be employed to emulate the active-state functionality, while inactive states (and transitions between them) can be simulated by the software device driver. The code of the original device driver needs a few changes: *i*) an additional state variable representing the power state, *ii*) a routine for updating the power state according to power-management commands, *iii*) a timer to simulate state transition times, *iv*) a routine to provide power consumption estimates and *v*) a request-blocking mechanism that enables actual accesses to the disk only when in the active state. In general, emulation of power-managed systems is based on the observation that dynamic power management can only reduce system performances. Hence, if a functionally-equivalent real system is available for exercising the active-state performance, lower-performance states can be emulated as well.

Emulation has two desirable features. First, it runs at the same speed of the actual system, thus enabling policy validation against realistic workloads of any complexity and real-time interactive user sessions. This gives the user a direct experience of performance degradations possibly induced by power management. Second, it enables the software specification of the low-power states of the SP. The possibility of easily changing the SP model can be exploited both during the design of a power-manageable resource, to verify the effectiveness of a given low-power state, and during system-level design, to select among equivalent power-manageable components. The main drawback with respect to simulative approaches is that the system architecture is assigned once for all: no architectural choices can be explored.

Implementation. Policies can be validated by testing their implementations. Since the policy is directly applied to the target system, its actual impact on the cost metrics of interest can be measured accurately. Thus experimentation at this level is useful as a final step in validating a given policy.

6. Conclusions

Dynamic power management is an effective means for system-level design of low-power electronic systems. Dynamic power management is already widely applied to system design, but today most electronic products rely on ad-hoc implementation frameworks (e.g., firmware code) and on heuristic management policies (e.g., timeout policies). We expect that the use of industrial standards, such as OnNow and ACPI, will soon facilitate the clean implementation of operating system based power management.

This survey has shown how systems can be modeled so that optimal management policies can be computed, validated and implemented. The computation of optimal policies is a new problem for system-level design. In particular, we have shown a working model for which the optimal stochastic power-management control problem can be efficiently and exactly solved. The solution method we have analyzed relies on a modeling abstraction of system resources in terms of Markov processes. Several extension can be made to the model, at the price of complicating the solution procedure, by considering more detailed system models. As in many design problems, good engineering judgment is key in determining the right balance among model accuracy, exactness of the solution (for the given model), and computational effort.

Due to the proliferation of handheld electronic systems, and due to increasingly stringent environmental constraints on non-mobile systems, we believe that designers will be very often confronted with the challenge of deriving optimal, or near optimal, dynamic power management solutions. As a result, computer-aided design tools for power management will be extremely useful is system-level design for model identification, policy optimization and validation.

7. Acknowledgments

We acknowledge support from NSF under contract MIP-942119. We would like to thank Eui-Young Chung, Yung Hsiang Lu, Giuseppe Paleologo and Tajana Simunic for several discussions on this topic.

References

1. L. Benini and G. De Micheli, *Dynamic Power Management of Circuits and Systems: Design Techniques and CAD Tools*, Kluwer, 1997.
2. L. Benini and G. De Micheli, "Automatic Synthesis of Low-Power Gated-Clock Finite-State Machines," *IEEE Transactions on Computer-Aided Design of Integrated Circuits and Systems*, vol. 15, no. 6, pp. 630–643, June. 1996.
3. L. Benini, P. Siegel and G. De Micheli, "Automatic Synthesis of Gated Clocks for Power Reduction in Sequential Circuits," *IEEE Design and Test of Computers*, pp. 32–40, Dec. 1994.
4. L. Benini, A.Bogliolo, S.Cavallucci and B. Riccò, "Monitoring System Activity for OS-directed Dynamic Power Management," *Proceedings International Symposium on Low-Power Design*, 1998.
5. L. Benini, R. Hodgson and P. Siegel, "System-Level Power Estimation and Optimization," to appear in *Proceedings of Int.l Symposium of Low-Power Electronics and Design*, 1998.
6. D. Bertsekas, "Stochastic Optimal Control," Academic Press, 1978.
7. A. Chandrakasan and R. Brodersen, *Low-Power Digital CMOS Design*. Kluwer, 1995.
8. J. Czyzyk, S. Mehrotra, and S. Wright, "PCx User Guide", *Technical Report OTC 96/01*, Optimization Technology Center, May, 1996.
9. G. Debnath, K. Debnath and R. Fernando, "The Pentium Processor-90/100, Microarchitecture and Low-Power Circuit Design," in *International conference on VLSI design*, pp. 185–190, Jan. 1995.
10. S. Furber, *ARM System Architecture* Addison-Wesley, 1997.
11. S. Gary, P. Ippolito et al., "PowerPC 603, a Microprocessor for Portable Computers," *IEEE Design & Test of Computers* vol. 11, no. 4, pp. 14-23, Win. 1994.
12. R. Golding, P. Bosch and J. Wilkes "Idleness is not Sloth", in *Proceedings of Winter USENIX Technical Conference*, pp. 201–212, 1995.
13. R Golding, P. Bosh and J. Wilkes, "Idleness is not Sloth" *HP Laboratories Technical Report* HPL-96-140, 1996.
14. M. Gowan, L. Biro, D. Jackson, "Power Considerations in the Design of the Alpha 21264 Microprocessor," *DAC - Proceedings of the Design Automation Conference*, 1998, pp. 726-731.
15. C.-H. Hwang and A. Wu, "A Predictive System Shutdown Method for Energy Saving of Event-Driven Computation", in *Proceedings of the Int.l Conference on Computer Aided Design*, pp. 28-32, 1997.
16. H. Kapadia, G. De Micheli and L. Benini, "Reducing Switching Activity on Datapath Buses with Control-Signal Gating," *CICC - Proceedings of the Custom Integrated Circuit Conference*, pp. 589-592, 1998.
17. B. Mangione-Smith, "Low-Power Communication Protocols: Paging and Beyond," *IEEE Symposium on Low-Power Electronics*, pp. 8–11, 1995.
18. J. Monteiro and S. Devadas, *Computer-Aided Techniques for Low-Power Sequential Logic Circuits*. Kluwer 1997.
19. B.Nadel, "The Green Machine," *PC Magazine*, Vol 12, No. 10, p.110, May 25, 1993.
20. W. Nebel and J. Mermet (Eds.), *Low-Power Design in Deep Submicron Electronics*. Kluwer, 1997.

21. G. Paleologo, L. Benini, A. Bogliolo and G. De Micheli, "Policy Optimization for Dynamic Power Management," *DAC - Proceedings of the Design Automation Conference*, 1998, pp. 182-187.
22. A. Papoulis, *Probability, Random Variables, and Stochastic Processes*, McGraw-Hill, 1984.
23. J. M. Rabaey and M. Pedram (editors), *Low-Power Design Methodologies*. Kluwer, 1996.
24. M. Rosenblum, E. Bugnion, S. Devine and S. A. Herrod, "Using the SimOS Machine Simulator to Study Complex Computer Systems," *ACM Transactions on Modeling and Computer Simulation*, Vol. 7, no. 1, pp. 78-103, 1997.
25. J. Rulnick and N. Bambos, "Mobile Power Management for Wireless Communication Networks," *Wireless Networks*, vol. 3, no. 1, pp. 3–14, 1997.
26. T. Sakurai and T. Kuroda, "Low-Power Circuit Design for Multimedia CMOS VLSI," in *Workshop on Synthesis and System Integration of Mixed Technologies*, pp. 3–10, Nov. 1996.
27. K. Sivalingham, M. Srivastava et al., "Low-power Access Protocols Based on Scheduling for Wireless and Mobile ATM Networks," *Int.l Conference on Universal Personal Communications*, pp. 429–433, 1997.
28. M. Srivastava, A. Chandrakasan. R. Brodersen, "Predictive System Shutdown and Other Architectural Techniques for Energy Efficient Programmable Computation," *IEEE Transactions on VLSI Systems*, vol. 4, no. 1, pp. 42–55, March 1996.
29. V. Tiwari, D. Singh, S. Rajgopal, G. Metha, R. Patel and F. Baez, "Reducing Power in High-Performance Microprocessors," *DAC - Proceedings of the Design Automation Conference*, 1998, pp. 732-737.
30. L. Torvalds, "Linux Kernel Implementation," *Proceedings of Open Systems. Looking into the future. AUUG'94*, pp. 9-14, 1994.
31. K. Trivedi, *Probability and Statistics with Reliability, Queueing and Computer Science Applications*, Prentice Hall, 1982.
32. S. Udani and J. Smith, "The Power Broker: Intelligent Power Management for Mobile Computing," *Technical report MS-CIS-96-12*, Dept. of Computer Information Science, University of Pennsylvania, May 1996.
33. M. Zorzi and R. Rao, "Energy-Constrained Error Control for Wireless Channels," *IEEE Personal Communications*, vol. 4, no. 6, pp. 27–33, Dec. 1997.
34. *http://www.intel.com/ial/powermgm/specs.html*, Intel, Microsoft and Toshiba, "Advanced Configuration and Power Interface specification", Dec. 1996.
35. *http://developer.intel.com/IAL/powermgm/apmovr.htm*, Intel, "Advanced Power Management Overview," 1998.
36. *http://www.microsoft.com/hwdev/pcfuture/ ONNOW.HTM*, Microsoft, "OnNow: the evolution of the PC platform," Aug. 1997.
37. *http://www.storage.ibm.com/storage/oem/data/ travvp.htm* Technical specification of hard-drive IBM Travelstar VP 2.5-inch, 1996.
38. *http://now.cs.berkeley.edu/Xfs/AuspexTraces/auspex.html*, Auspex File System Traces, 1993.
39. *http://www.storage.ibm.com/storage/oem/data/travvp.htm*, Hard Drive IBM Travelstar VP 2.5-inch, 1996.

COMPILER GENERATION TECHNIQUES FOR EMBEDDED PROCESSORS AND THEIR APPLICATION TO HW/SW CODESIGN

MASAHARU IMAI*, YOSHINORI TAKEUCHI*, NORIMASA OHTSUKI*, AND NOBUYUKI HIKICHI**

*) *Department of Informatics and Mathematical Science, Graduate School of Engineering Science, Osaka University, Japan*
**) *Software Research Associates, Inc., Japan*

1. Introduction

Due to the advancing semiconductor technology it is becoming possible within ten years to fabricate a highly complex and high performance VLSI that includes more than hundred million transistors on a single silicon chip [1]. Using such a technology, so-called systems-on-a-chip, that includes CPU cores, DSPs, memory blocks (RAM and ROM), application specific hardware modules, FPGA blocks, as well as analog and radio frequency blocks, as shown Figure 1. Systems-on-a-chip will be suitable for embedded applications, such as consumer electronics products that perform sophisticated data and information processing, telecommunication equipment that perform movie picture and audio transmission, control systems for industrial manufacturing, automobile, and avionics.

Embedded system is a one that performs data and information processing, communication, and/or control operations in an application specific end product described above. Then an *embedded processor* is such a one that is used for embedded systems. Requirements to embedded processors are somewhat different from those to *general purpose processors*. Regarding the performance, it can be high enough to perform the application tasks to be given. However, other design quality metrics [2], such as power consumption and hardware cost (module area on chip), would be different in some applications. For example, power consumption should be kept very low in portable application systems driven by battery cells, such as cellular phone and PDA (Personal Digital Assistant). Hardware cost is also a crucial measure especially for consumer products.

A. A. Jerraya and J. Mermet (eds.), System-Level Synthesis, 293–320.

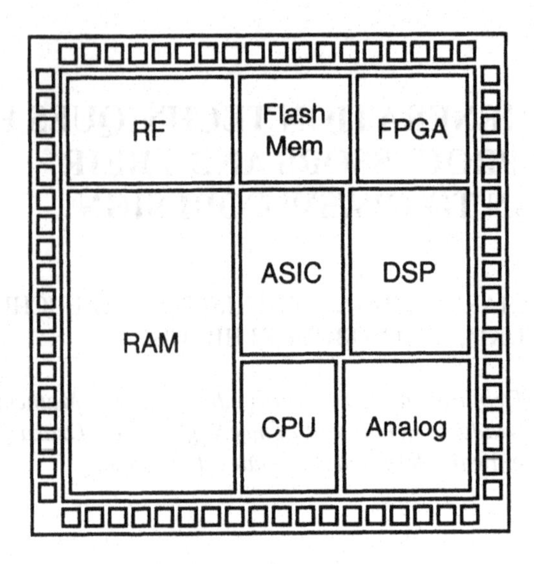

Figure 1 - Example of an Embedded System

HW/SW co-design is one of the most suitable design methods to design embedded systems. Compilation techniques play a very important role in the HW/SW codesign of embedded processors, that is, to estimate the performance of the system and to generate an efficient object code for the processor.

In the rest of this paper, the reason behind the importance of compilation technology to HW/SW co-design is described in Section 2. Then, basic compilation techniques are briefly surveyed in Section 3. In Section 4, code optimization techniques for advanced embedded processor is introduced. In Section 5, retargetable compiler projects are surveyed and the outline of GCC (GNU C Compiler) is introduced. In Section 6, a case study on HW/SW co-design system PEAS-I and –II is introduced with relation to retargetable compilers.

2. Background: Approaches to HW/SW Co-design

HW/SW co-design method is one of the most effective methods to design embedded systems. HW/SW co-design is a system design method that takes the tradeoff of hardware and software components in the system. Approaches to the HW/SW co-design method can be classified in two categories as shown in Figure 2.

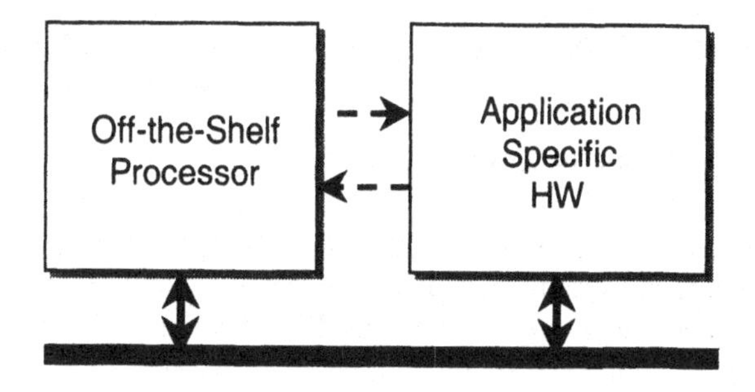

(a) Off-the-Shelf Processor Approach

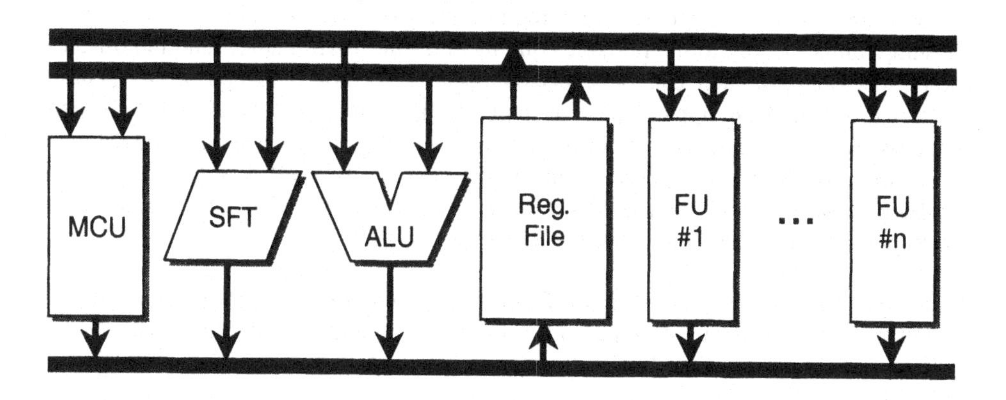

(b) Dedicated Processor Approach

Figure 2 - Two Approaches to HW/SW Codesign

2.1 OFF-THE-SHELF PROCESSOR APPROACH

The first approach is to utilize an off-the-shelf processor as an embedded core. In this approach, the processor core itself could be obtained from IP (Intellectual Property) venders, then some application specific functions would be realized as co-processors or peripherals to be attached to the system bus. This approach has following advantages:

- Predesigned processor core and its compiler can be reused.
- Development time and cost of the system could be reduced by the design reuse.

This approach is taken in HW/SW co-design systems as COSYMA [3], LYCOS [4], and VULCAN [5].

2.2 DEDICATED PROCESSOR APPROACH

The second approach is to design a dedicated processor for a specific application domain. In this approach, a dedicated processor core is supposed to be designed or generated by designers, and application specific functionality can be realized as extended instructions, co-processors, or peripherals. This approach has the following advantages:

- Better design quality in terms of performance, module area, and/or power consumption can be expected, compared to the off-the-shelf processor approach.
- No processor core needs to be purchased from IP venders.

Better design quality is sometimes required in critical applications. For example, power consumption is the key to the portable equipment. Better performance than the off-the-shelf processor approach can be expected because part of the application specific functionality could be implemented as part of the instruction set of the dedicated processor with less execution cycle overhead. While the design cost of the dedicated processor could be more expensive compared to the off-the-shelf processor approach, the design cost/chip could be reduced if the dedicated processor core can be reused for other applications in the same or similar application domain.

This approach is taken in PEAS-I [6], PEAS-II [7], and ASIA [8]. In this approach, a compiler plays much more important role in estimating the performance of the processor under design, as well as to generate an efficient object code for the processor.

3. Basic Compilation Techniques

The compilation process, in general, consists of front end processing and back end processing, as shown in Figure 3. In this section, the front end process is introduced briefly, then the back end process is described more in detail. Interested readers are recommended to read compiler textbooks, such as [9], [10], and [11], for more detail.

3.1 FRONT END PROCESSING

Front end processing in a compiler consists of lexical analysis, syntax analysis, semantic analysis, and intermediate code generation, as shown in Figure 3. Each process in the front end performs the following task.

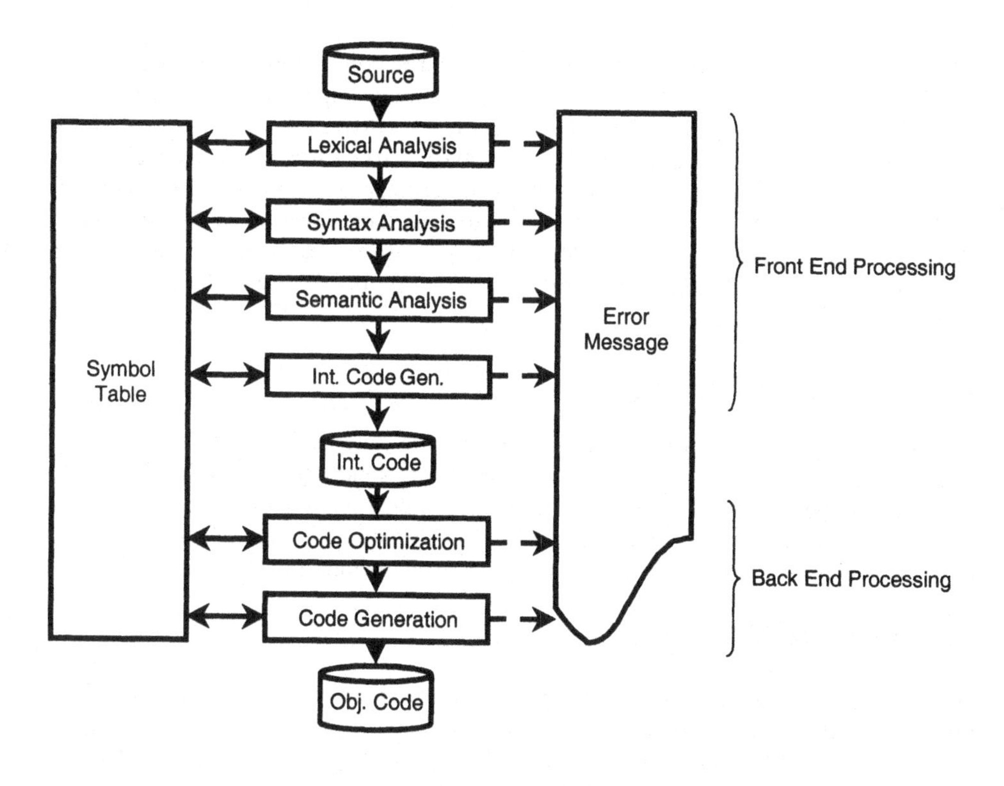

Figure 3 - Typical Compilation Process Flow

3.1.1 *Lexical Analysis*

A lexical analyzer is a filter that translates a given source program into a sequence of *tokens*, or a token stream. A token is a string separated by appropriate characters or meta-characters. In most programming languages, a set of reserved words are defined. These reserved words and user-defined identifiers for variables, functions, procedures, data types, records, as well as constants, are recognized as tokens.

3.1.2 *Syntax Analysis*

A syntax analyzer investigates the token stream and generates a *parse tree* according to a given language grammar. During this process, a symbol table is generated, which maintains the user-defined names associated with their data types. In this process, semantic information of identifiers and constants, such as data types consistency, are not considered.

3.1.3 *Semantic Analysis*

A semantics analyzer analyzes the semantic information of the token stream and translates the parse tree into a *syntax tree*. For example, suppose that a symbol "+" is defined as an addition operator for two integers or two floating point data types, and a source program contains an expression such as "base + 10", where "base" is declared as a floating point data type. In this case, the semantics analyzer detects the inconsistency in the semantic relation among the operator and the data types of its two operands. Then, the semantic analyzer would inform the user of the inconsistency of data types if the source language is strongly typed, or would try to fix the inconsistency or output an error message otherwise.

3.1.4 *Intermediate Code Generation*

The final stage of the front end process is the intermediate code generation. Intermediate code can be generated by traversing the syntax tree of the given source program. In many compilers, an intermediate form called *three-address code* is used to represent the intermediate form that consists of an operator and three operands (source, target, and destination), in case of arithmetic or logical operations.

3.2 BACK END PROCESSING

Back end processing includes *code optimization* and *object code generation*. The purpose of code optimization is to improve the quality of the intermediate and object code for the target machine. On the other hand, object code generation is to generate the final object code that is executable on the target machine.

The optimization of intermediate code is necessary because the quality of that generated by the front end process of a compiler is not usually very efficient. In the classic compilation theory for main frame computers, the code performance would be the primary concern and the code size would be the second concern.

Where the code performance can be measured by the number of execution cycles required to perform the object code on the target processor; and the code size can be measured by the amount of memory to store the object code including constants embedded in the object code.

Code optimization techniques can be divided into two categories: *machine independent* and *machine dependent* techniques.

3.2.1 *Machine Independent Code Optimization*

Machine independent code optimization includes the following techniques developed in the past decades [10]:

- Common Sub-expression Elimination

- Copy Propagation
- Redundant Code Elimination
- Loop Unrolling
- Loop-Invariant Expression Movement
- Induction Variable Elimination
- Strength Reduction
- Unreachable Code Elimination
- Control Flow Optimization
- Arithmetic Optimization
- Operation Combining

Loop unrolling is a transformation that expands a loop body into multiple instances of the loop body and reduces the number of iterations of the loop, which decreases the number of branch operations. Because a larger basic block can be obtained by this transformation, better scheduling results can be expected.

3.2.2 *Machine Dependent Code Optimization*

Machine dependent optimization includes Machine Specific Instruction Mapping, Spill Code Reduction, and Instruction Pipeline Scheduling.

(a) Machine Specific Instruction Mapping

To take advantage of the machine specific instructions, a sequence of intermediate code is replaced by a sequence of codes specific to the target architecture. Some examples of machine specific instructions include following ones:

- Auto Increment and Auto Decrement Indexed Memory Access Instructions
- Stack Instructions (push, pop)
- MAC (Multiply and ACcumulate) Instruction

(b) Spill Code Reduction

In the intermediate code, intermediate computation results are stored in *pseudo resisters*. If the required number of pseudo registers to perform the given program does not exceed the number of actual registers in the target processor, all of these pseudo registers can be assigned to actual registers. Otherwise, pseudo registers that could not be assigned to actual register should be assigned to data memory space, which is usually allocated on a stack area to keep track of local variables. An instruction to access this memory space (or stack area) is called a *spill code*. Because the spill codes degrade the performance of the object code, less spill codes are preferable. On the other hand, the number of required pseudo registers depends on the sequence of intermediate instructions. Then, the compiler tries to reduce the number of the required pseudo registers by changing the order of intermediate instructions.

(c) Instruction Scheduling

When the target processor has an instruction pipeline architecture [12] , the whole performance of the processor would be improved. However, an instruction pipeline architecture could suffer from *pipeline hazards*, which could degrade the performance. The purpose of instruction pipeline scheduling is to eliminate these hazards and enhance the utilization of machine cycles as well as functional units in the processor. Pipeline hazards can be classified into *structural hazard*, *data hazard*, and *control hazard*. Part of structural hazards is avoidable if some additional hardware resources were added to the target architecture. Part of data hazards can be eliminated using register bypass techniques.

One of the easiest ways to avoid data hazards would be to fill NOP (No OPeration) instructions into appropriate places in the code as *delay slots*. However, these delay slots would degrade the performance of the pipeline processor. The purpose of instruction scheduling is to reduce delay slots, as much as possible, by filling the delay slots with other instructions that do not affect the execution results of the code. Because this problem belongs to the class of NP hard problems, several heuristic scheduling algorithms have been proposed for this problem [13]. These algorithms try to find sub-optimal solutions in a reasonable computation time by sacrificing the optimality of the solution.

3.2.3 *Object Code Generation*

The purpose of object code generation is to translate an intermediate code into a machine executable object code for the target processor. However, the final code from a compiler is often described in an assembly language. It is partly because the development of an assembler program for the target processor is much easier compared to the development of a compiler for the processor. Another reason is that application system developers sometimes describe part of their system programs, such as device drivers, in an assembly language.

It is still not unusual for them to describe part or whole of their application programs in assembly language to squeeze even the last drop of performance of the target processor.

Major tasks in object code generation are as follows, where some more code optimization is performed:

- Code Reordering
- Instruction Pattern Matching
- Register Allocation
- Register Assignment

4. Code Generation Techniques for Advanced Embedded Processors

4.1 ADVANCED EMBEDDED PROCESSOR ARCHITECTURES

There are several advanced processor architectures adopted in embedded processors, including *VLIW* (Very Long Instruction Word) architectures and *superscalar* architectures. Because these architectures can execute several operations simultaneously, much higher performance can be expected compared to conventional scalar architecture, if the application program has intrinsically high *instruction level parallelism* (ILP). Some of the recent DSPs (Digital Signal Processors) take advantage of VLIW architecture, and high performance general purpose processors, such as the Pentium series, adopt a superscalar architecture. Common features of these advanced architectures are as follows:

- Instruction pipeline is performed
- ILP is implemented both specially and temporally

4.1.1 *VLIW Architecture*

VLIW and superscalar architectures are able to perform multiple operations simultaneously but the instruction issue mechanism is quite different. In the VLIW processor, operations are scheduled a priori by a compiler and are packed in VLIW instructions, where each VLIW instruction contains multiple operations to be executed simultaneously. Then these operations in the same VLIW instruction are executed simultaneously. Therefore, compilers for VLIW processors are requested to extract as much parallelism as possible from the intermediate code. Some of the ideas to extract more parallelism are surveyed in Section 4.3.

4.1.2 *Superscalar Architecture*

On the other hand, the object code of the superscalar processor is the same as conventional scalar processor. In the superscalar architecture, instructions fetched in an instruction buffer are dynamically reordered according to the data and control dependencies of instructions as well as the status of functional units in the processor. Then, simultaneously executable instructions are supplied to the functional units at each clock cycle.

Therefore, a superscalar processor can be designed as compatible with a corresponding scalar processor at object code level. Object code compatibility is sometimes very important for general purpose processors. Because the instructions (operations) are reordered and grouped just before execution, compilers do not need to pay more attention to the object code sequence. However, the instruction reordering and grouping are performed by a complicated hardware module, which consumes large amount of hardware resources in the processor.

4.1.3 *Comparison of VLIW and Superscalar*

Suitability of the above mentioned processor architectures for embedded processor would be a discussion matter. There are certainly some tradeoffs between these architectures. If object code level compatibility is not a strong requirement for embedded processor, VLIW architecture could be a better choice because of the following reasons:

- More parallelism than superscalar could be extracted.
- Design could be easier because architecture is simpler.
- Less hardware cost is required.

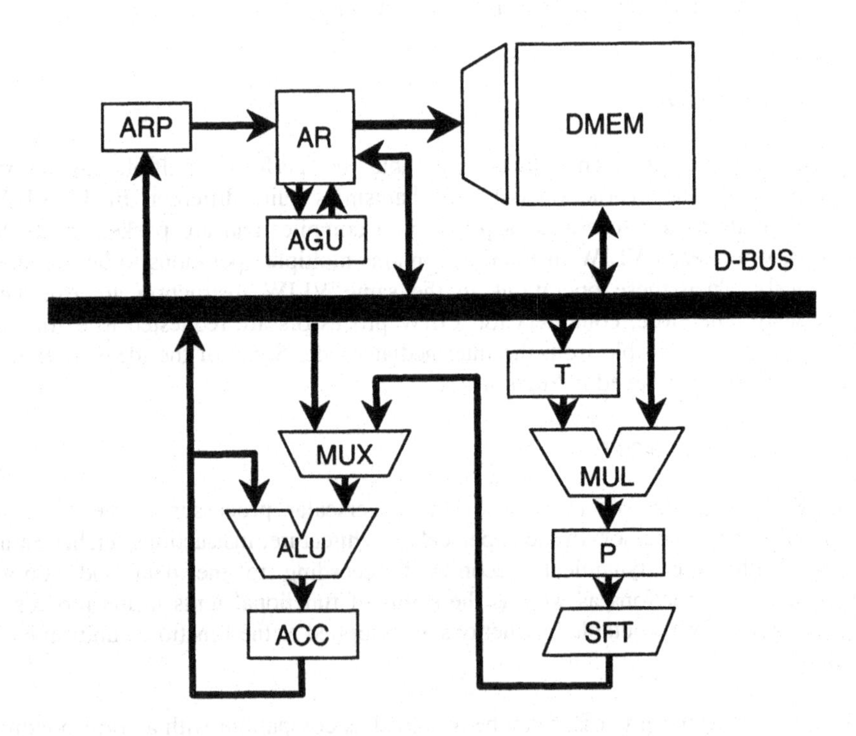

Figure 4 - Block Diagram of TI TMS 320 C25

4.1.4 *DSP Architecture*

There are many DSPs developed and sold so far in the market. The architecture of a DSP could be CISC (Complex Instruction Set Computer) or RISC (Reduced Instruction Set Computer). Even if a DSP has a RISC type architecture, it sometimes has a heterogeneous structure. That is, DSPs often have special compounded instructions such as:

- Data transfer and MAC (Multiply and ACcumulate)
- Multi level loop control

Some of them also have application specific addressing modes such as:

- Modulo addressing
- Bit reverse addressing

The former is very useful to control an FIFO (First-In First-Out) queues; and the latter can be used in FFT (Fast Fourier Transform) computation. Some of them also have special registers for fixed point multiplication and division, floating point calculation, and so on. The block diagram of a typical DSP (TI TMS320 C25) is shown in Figure 4.

4.2 CODE QUALITY MEASURES FOR EMBEDDED PROCESSORS

Following measures allow to evaluate the code quality for embedded processors:

- Performance (Execution Cycles) of code
- Code and data sizes in memory
- Power consumption

In the classic compilation theory for main frame computers, the code performance was usually the primary concern, where the compilers try to generate object code that requires less execution cycles. However, in embedded processors in a systems-on-a-silicon, not only the code performance but also the code and data sizes (in other words memory requirement) are often of interest. It is because the embedded memory (ROM or RAM) modules on a silicon chip often occupies a very large area of the chip.

The third measure of code quality is the power consumption by the computation using the object code. Generally, power consumption depends on the instructions used in the object code or the sequence of these instructions that appears in the object code. Power consumption reduction is sometimes a very crucial issue in such an embedded system as portable equipment driven by battery cells.

4.3 CODE OPTIMIZATION FOR ADVANCED ARCHITECTURES

One of the common features of VLIW, superscalar, and DSP is instruction level parallelism (ILP). In order to take advantage of the ILP, the compiler must extract as much parallelism as possible in the source program and generate object code suitable to these architectures. Several techniques have been proposed for this purpose, as described in the following subsections. Most of these techniques have been developed for super computers, such as vector processors.

4.3.1 *Trace Scheduling*

Fisher proposed the concept of *trace* [14], which is an extension of basic block. In this technique, the intermediate code is profiled to detect execution paths of codes, as shown in Figure 5, that have higher execution counts. The paths that have the highest execution counts are called *traces*. Then instruction scheduling is performed for traces regarding them as basic blocks. Because traces have branches and junction points, additional code sequences have to be inserted at which branches are targeted. This code insertion operation is called *book keeping*. Trace scheduling could increase the parallelism of the resulting code, which would increase the performance, while the code size could increase.

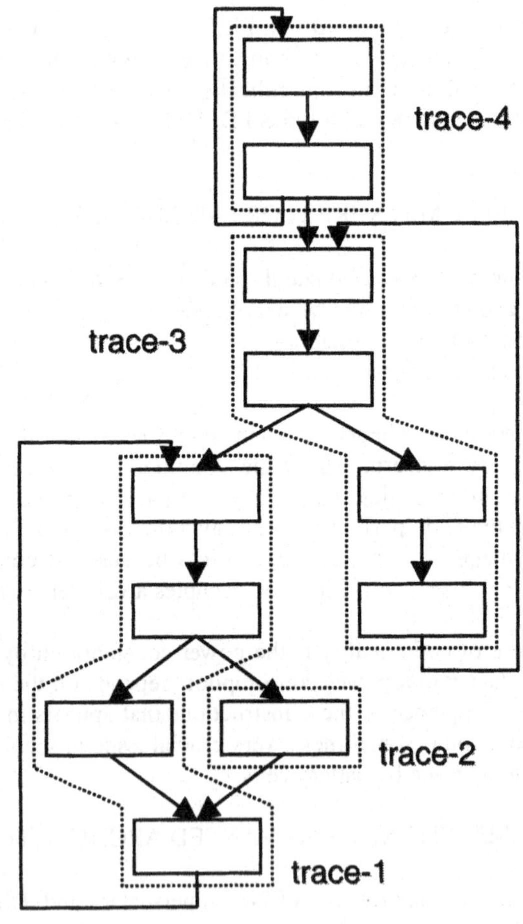

Figure 5 - Example of a Trace

4.3.2 *Superblock Scheduling*

The trace optimization is complicated by the bookkeeping for merge points. The *superblock scheduling* [15] simplifies the optimization by removing merge points from traces. A trace without merge points is called *Superblock*. The control in a superblock may only enter from the top but may leave at one or more exit points. A superblock is formed by removing the tail portion of the trace from the first merge point to the end. The removed portion of the trace becomes new trace. The process is called tail duplicating, and the example is shown in Figure 6. A better code can be expected by superblock scheduling than trace scheduling and conventional scheduling algorithms. The drawback of this method is the increase in the code size due to the code duplication.

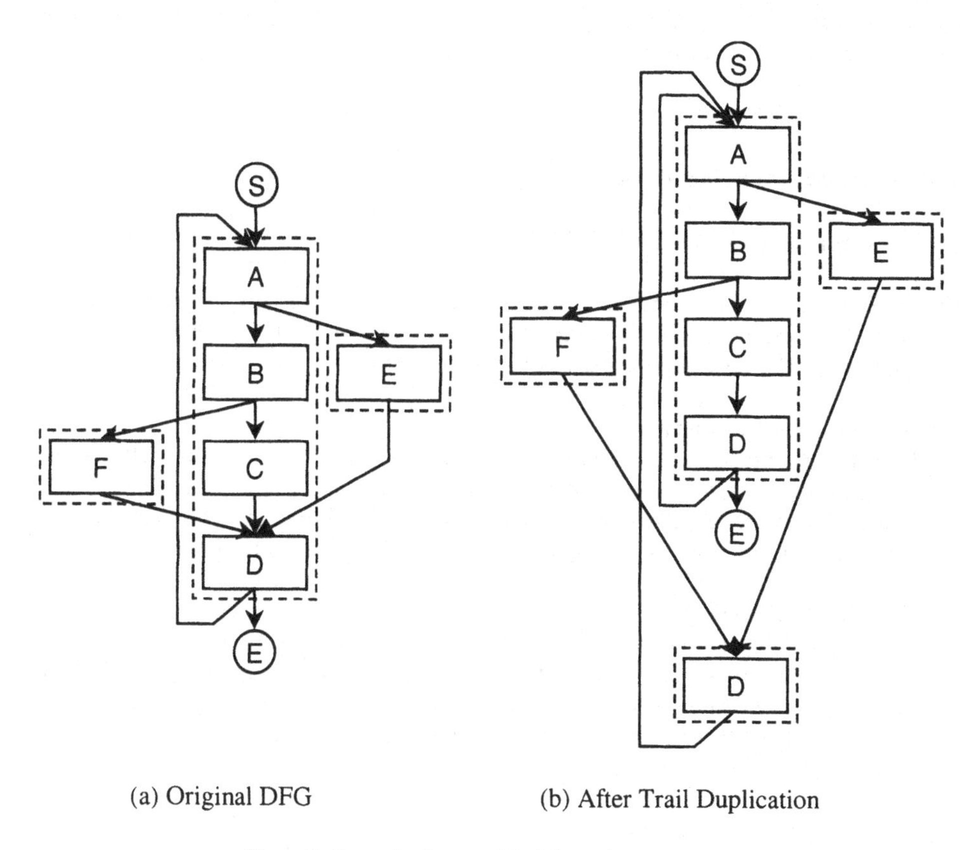

Figure 6 - Example of a superblock formation

4.3.3 *Hyperblock Scheduling*

Predicated execution is a technique to handle conditional branches in application programs. In predicated execution, a compiler appends conditional execution to each instruction as a predicate. Predicate execution refers to the predicate of each instruction. When the predicate has the value "True", the instruction is executed in normally and when the predicate has the value "False", the instruction is treated as a NOP. Predicated execution treats conditional branches without pipeline stalls.

A *hyperblock* is a set of predicated basic blocks in which control may only enter from the top, but may exit from one or more locations. Hyperblock is similar to superblock. Superblock contains only one control flow, but Hyperblock contains multi control flow to deal with predicated execution. *Hyperblock scheduling* [16] is a scheduling algorithm which utilizes predicated execution for scheduling.

5. Retargetable Compiler Projects

5.1 COMPILER RETARGETABLILITY

Compiler retargetability can be classified into three types of implementations: *Automatically Retargetable*, *Compiler User Retargetable*, and *Compiler Developer Retargetable* [17]. These approaches has following features:

(1) Automatically Retargetable:
In this implementation, the "generic" compiler itself contains a set of well defined parameters or switches that enable complete retargeting to new processors. A full range of knowledge about the target architectures should be contained in the compiler a priori. The retargeting time is on the order of seconds to minutes.

(2) Compiler User Retargetable:
In this implementation, the compiler user is supposed to provide the compiler generator with the specification of the target architecture and instruction set. Then, the compiler generator produces a source program of the target compiler, which can be compiled by an ordinary compiler on the compiler development environment. The retargeting time is on the order of hours to days.

(3) Compiler Developer Retargetable:
In this implementation, the compiler developer is supposed to rewrite the compiler system program according to the target processor architecture and instruction set. The compiler developer should have the expertise of the compilation techniques as well as the target processor. This approach is applicable to the widest range of processors, but the retargeting is on the order of weeks to months.

5.2 RETARGETABLE COMPILER PROJECTS

5.2.1 *GNU C Compiler*

The GNU C Compiler (GCC) [18] is one of the earliest retargetable compiler projects. GCC is developed and distributed by the Free Software Foundation originated by Richard Stallman, Jack Davidson, and Christopher Fraser with many contributors. GCC has been ported to a number of computers and retargeted to many more processors including DSPs. In this sense, GCC is a kind of de facto retargetable compiler.
The intermediate code of GCC, which is similar to LISP, is called Register Transfer Language (RTL).

When the GNU C Compiler (GCC) project has started, GCC was designed as a retargetable optimizing C Compiler on UNIX machines or compatible. The target architecture of GCC can be characterized as a 32 bit general register machine with byte addressing mechanism, including quite a number of microprocessors such as a29k, Alpha, ARM, AT&T, DSP1610, Clipper, Convex cN, Elxsi, Fujitsu Gmicro, i386, i370, i860, i960, m68k, m88k, MIL-STD-1750a, MIPS, ns32k, PA-RISC, PDP-11, Pyramid, ROMP, RS/6000, SH, SPARC, SPUR, Tahoe, VAX, we32k.

One of the advantages of GCC is its ability in object code optimization, which employs conventional and extended optimization techniques. Another advantage is its easiness to retarget the compiler.

5.2.2 *SUIF*

SUIF is a research oriented compiler that utilizes Stanford University Intermediate Form (SUIF) supported by the National Compiler Infrastructure project in the USA. One of the objectives of the project is to develop compilers that enable to generate machine instructions with much parallelism. There are many compiler development projects based on SUIF. One of such project examples is the SPAM Compiler Project for DSPs, which is performed as an international joint project by Synopsys, Princeton University, MIT, Aachen University of Technology, UNICAMP, and UC Berkeley. Another example is the Valen-C Compiler Project performed in Kyushu University, Japan.

5.2.3 *FlexWare*

The FlexWare system [19] is a software/firmware development environment for application specific instruction set processors and general purpose processor. Target processor architecture of FlexWare includes ASIPs (DSPs) and general purpose processor. FlexWare consists of the following subsystems:

- INSULIN: an instruction set level simulator which provides a cycle VHDL based simulation environment for a given instruction set.
- CodeSyn: a retargetable code generator, which takes one or more algorithms expressed in a high-level language and maps them onto a user defined instruction set.

5.2.4 *CHESS*

CHESS [20] is a retargetable code generation environment developed by IMEC. CHESS addresses fixed-point DSPs in a range of commercial as well as application specific processors. Where the following characteristics of the target architecture are to be described:

- Memory Structure,
- Register Structure,
- Code Type, and
- Instruction Format.

6. Application to HW/SW Co-design

This chapter describes a compiler application to HW/SW codesign. We have been developing PEAS (Practical Environment of ASIP Development) series for embedded processor design. PEAS systems are workbenches of HW/SW Co-design for embedded processors. In PEAS systems, compilers play important roles for application analysis, generated processor evaluation and so on. In this chapter, the HW/SW Co-design methodology is explained, the PEAS series are introduced, and the compiler's roles in the PEAS-II systems are explained.

6.1 HW/SW CODESIGN

HW/SW codesign is a design methodology for system optimization. It features:

- Quantitative estimation of design qualities,
- Architecture optimization taking HW & SW tradeoff into account.

Key technologies of HW/SW co-design are

- Design Quality Estimation,
- HW/SW Partitioning,
- Compiler Generation.

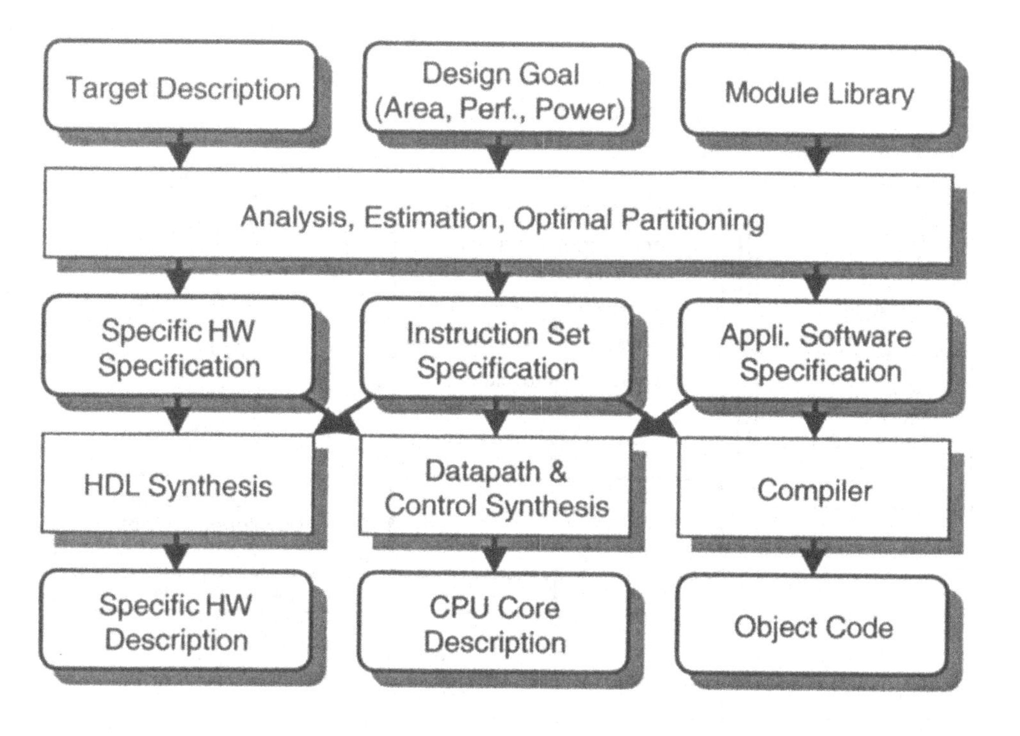

Figure 7 - " Ideal " HW/SW Codesign

An "ideal" HW/SW co-design methodology is illustrated in figure 7, where system design starts with "target (system) description," "design quality goal," and "module library." The target description is analyzed taking Design goal and module library into accounts and HW/SW optimal partitioning can be executed at this point. From here, the process is divided into three parts: Specific HW generation, CPU core generation and Object generation. The specific HW generation part is for special hardware part of the target application. The CPU core generation part is for the application specific processor. The Object generation part is for the specific HW and CPU core control. An "ideal" HW/SW codesign must be executed owing to this flow.

6.2 THE PEAS SYSTEM

PEAS (Practical Environment for ASIP Development) [6], [7] , [21] is our project for HW/SW co-design. We have been developing the PEAS system for about eight years since 1989. PEAS consists of 4 series: PEAS-I, PEAS-II, PEAS-III, and PEAS-V. Each version is specialized for a type of ASIP architecture and target. PEAS-I targets the pipeline (Scalar) processor for small area and low hardware cost. PEAS-II targets the VLIW processor for high performance computing. PEAS-III is a processor designers' workbench. PEAS-V targets the processor with ASIC, namely, system on silicon. In this paper, the PEAS-II system is the main target of compiler applications. This is because PEAS-II is the system for high performance computing like coming multimedia applications.

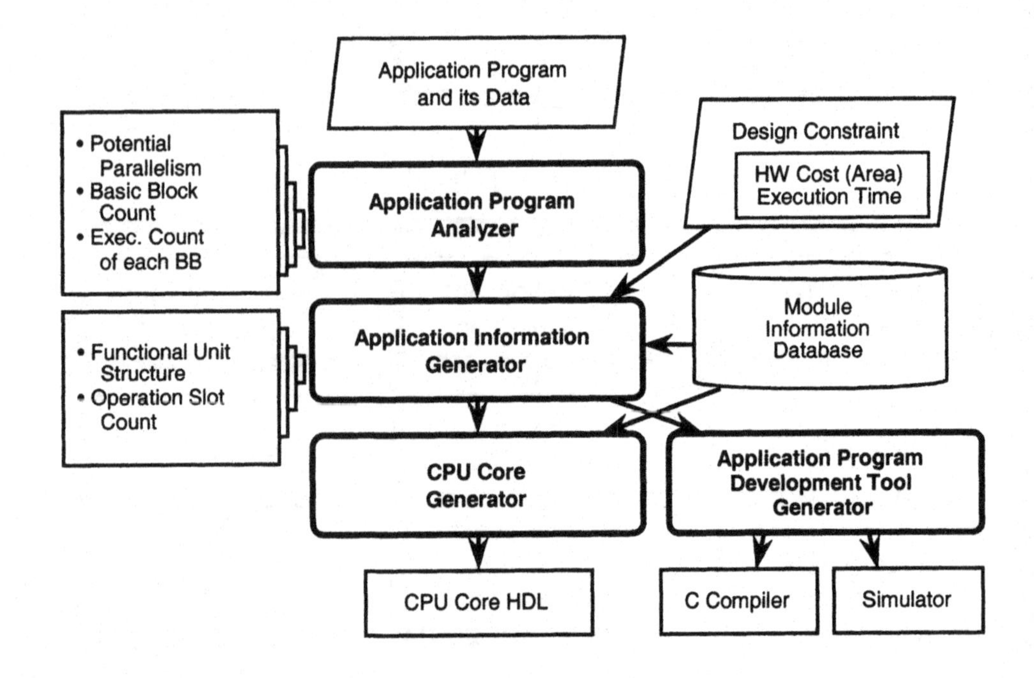

Figure 8 - PEAS-II System

6.3 THE COMPILER IN THE PEAS-II SYSTEM

6.3.1 *Outline of PEAS-II*

The configuration of the PEAS-II system is shown in figure. 8. The PEAS-II system accepts as input (1) an application program written in C language, (2) the data set corresponding to the application program, and (3) the design constraint of hardware cost. PEAS-II generates a VHDL description of a CPU core design which executes the given application program most efficiently, and the development tools for the CPU core such as a C compiler and a simulator.
The PEAS-II system consists of the following subsystems.

(1) Application Program Analyzer (APA):
APA analyzes a given application program and its data statically and dynamically. The analysis results of APA include basic block structure, potential parallelism of each operation, data flow graph, and execution frequency of each basic block. The basic block structure, potential parallelism and the data flow graph of a given application program are obtained by using traditional compiler [9][10][11]. The execution frequency is measured by simulating the program with its corresponding data set.

(2) Architecture Information Generator (AIG):
AIG accepts the analysis results obtained by APA, and decides the construction of functional units and the number of operation slots for the optimal CPU core, which executes the given application program in the least execution cycles under the given constraint on hardware cost. The optimal parameters of the optimal CPU core can be obtained by solving the optimization problem shown in section 6.3.3. The necessary data for the optimization, such as operation cycles and hardware cost of each functional unit, are obtained from the module information database.

(3) CPU Core Generator (CCG):
CCG generates a VHDL description of the CPU core design based on the architecture information obtained by AIG. The CPU core consists of a control module generated by the controller generator and a set of datapath components retrieved from the module information database.

(4) Application Program Development Tool Generator (DTG):
DTG generates application program development tools, which include a compiler and a simulator for the CPU core based on the architecture information obtained by AIG. The compiler generator is realized by adding a parallel scheduling path based on the list scheduling method [13] to achieve the instruction level parallelism of an input program into GCC (GNU C Compiler).

6.3.2 *Generalized VLIW Architecture Model*

The CPU core of the VLIW processor that the PEAS-II system generates has the following features.

- Harvard Architecture
- Load/Store Architecture
- Global Register Set
- Three Stage Pipeline
- Pipelined Multi-cycle Functional Units

The PEAS-II system generates CPU cores shown in figure 9. When a VLIW instruction is fetched from the instruction memory into the instruction register that includes several operation slots, the VLIW instruction is divided into several operations for operation slots. The operation in each operation slot is issued to an appropriate functional unit through the operation routing network. The operation routing network is illustrated in figure 10. The functional unit, to which an operation is issued, starts the execution of each operation simultaneously, but other functional units do not start new executions.

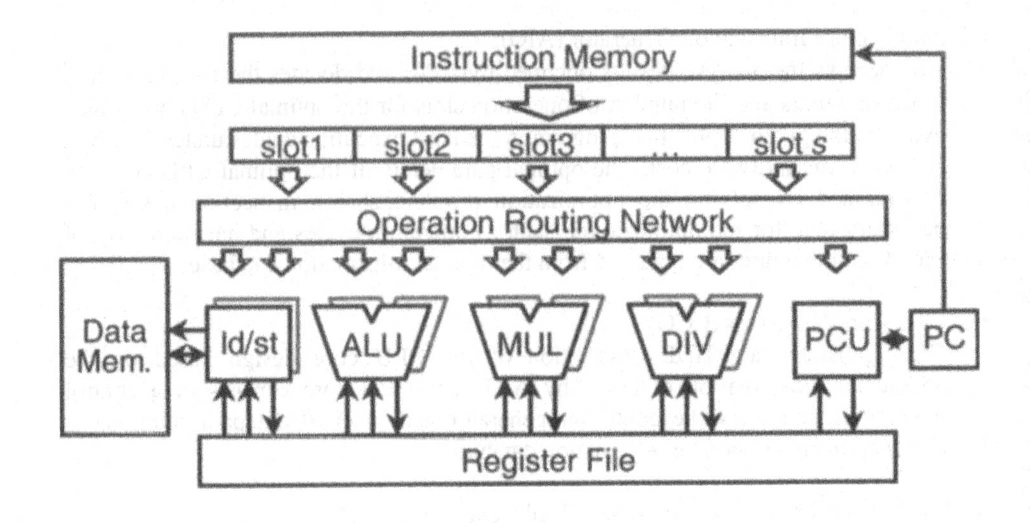

Figure 9 - Target VLIW Architecture

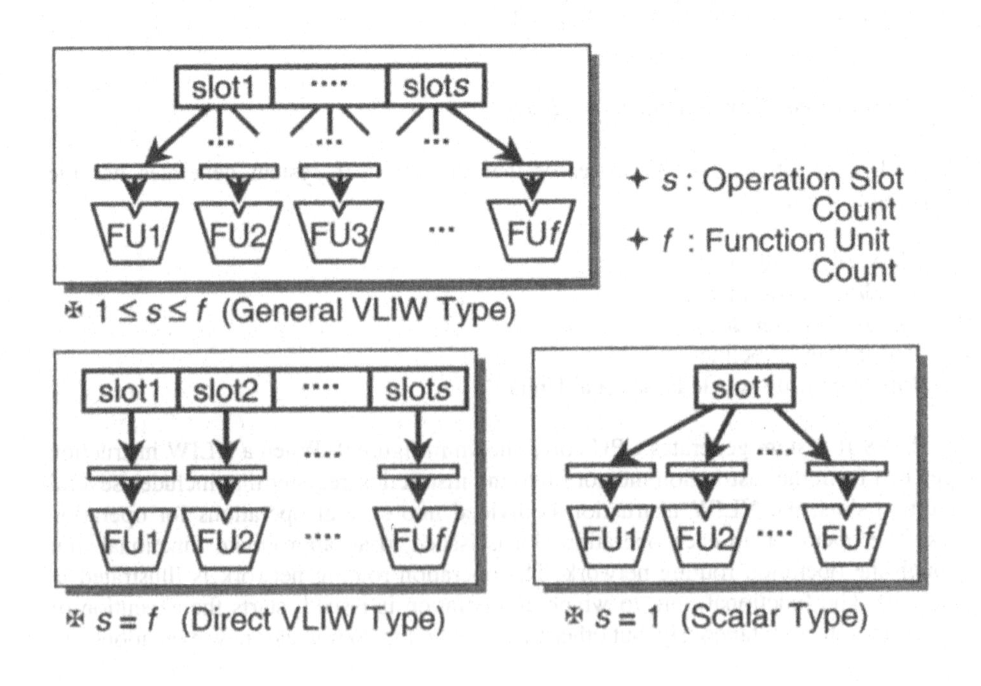

Figure 10 - Operation Routing Network

Table 1- Class and the number of functional units in the VLIW processor of PEAS-II

FU Class		Quantity
PCU	(PC Control Unit)	1
ld/st	(Load/Store Unit)	≥1
ALU	(Arithmetic and Logic Unit)	≥1
MUL	(Multiplier)	≥1
DIV	(Divider)	≥1

Because the generated CPU core does not have any hardware interlock mechanism and it has one branch-delay slot, the compiler must remove all pipeline hazards and fill up the branch-delay slot by the scheduling of instructions.

The PEAS-II system has an ability to tune the number of functional units, operation slots, registers, and the implementation of each functional unit class. The class and the number of functional units which can be used in the VLIW processor of PEAS-II are shown in Table 1. Any number of instances can be chosen for any class of functional units except PCU (Program Counter Control Unit). However, in the current implementation, at least one instance must be assigned for each functional unit class. Although several implementations of functional units are prepared for each functional unit class in the module information database, when multiple functional units are assigned in a functional unit class, they must be the same implementation. It is assumed that the number of registers, the data memory size and the clock frequency are to be given by the designers.

6.3.3 *Experiments*

Several experiments have been performed to confirm the effectiveness and efficiency of the execution cycle minimization algorithm in the PEAS-II system. Following, three sample application programs were used in the experiments.

(1) 3-rd order PARCOR filter
(2)1024 point FFT
(3) ADPCM

Table 2 - FUs in Experiments

Class	Name	Result Delay (cycles)	Issue Delay (cycles)	HW Cost (gates)
pcu	pcu	1	1	1,624
ld/st	ld/st	3	1	1,159
alu	alu	1	1	2,892
mul	mul_2_2	2	2	14,427
	mul_32_8	32	8	7,613
	mul_32_16	32	16	3,944
	mul_32_32	32	32	2,183
div	div_21_21	21	21	17,856
	div_34_17	34	17	6,829
	div_34_34	34	34	3,841

- Design Compiler (Synopsys)
- VSC753d (0.5μm CMOS) Library (VLSI Technology)

Constraint : 50MHz

Functional units used in these experiments are shown in Table 2. In table 2, the result-delay indicates the cycles after which the result can be used. The issue-delay indicates the cycles after which the next operation can be issued. Functional units, the issue-delay of which is shorter than the ready delay are pipelined functional units, and the other functional units, the issue-delay of which is equal to the result-delay are non-pipelined functional units.

The hardware cost has been measured under a 50MHz constraint by using a logic synthesis system, the Design Compiler from SYNOPSYS with a 0.5 um CMOS library VSC753d from VLSI Technology, Inc.
In these experiments, the number of register was fixed to eight and the hardware cost of one bit instruction memory was assumed to be 1 [gate/bit], because a DRAM technology is assumed to implement the registers.

Figures 11, 12 and 13 show the tradeoffs between the hardware cost constraint and the execution cycles of each sample application program. In these figure, the horizontal axis represents the hardware cost constraint in gates. The vertical axis represents the execution cycles of the optimized CPU core generated by PEAS-II system under a given hardware cost constraint. The line type indicates the number of operation slots on an optimized CPU core.

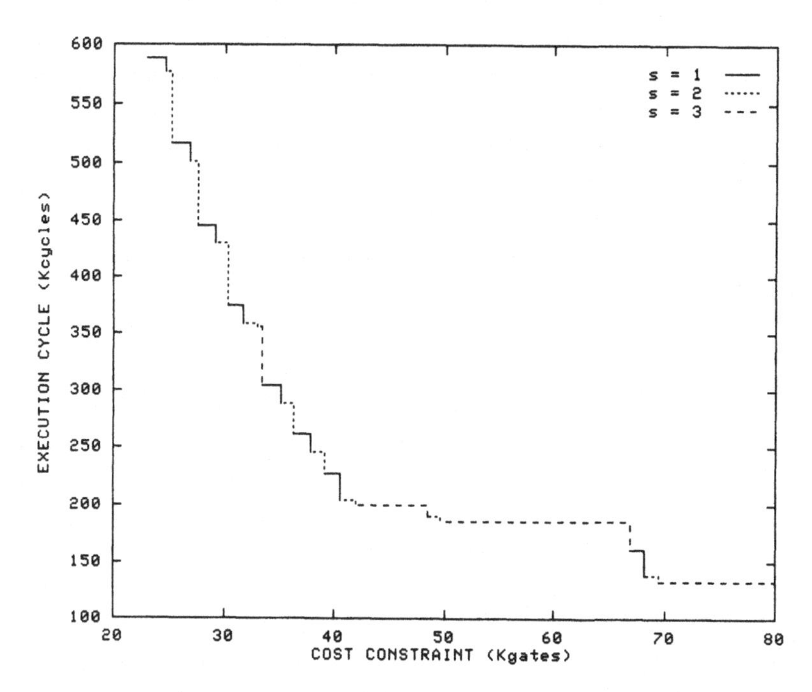

Figure 11 - Tradeoffs between HW Cost and Execution Cycles (PARCOR Filter)

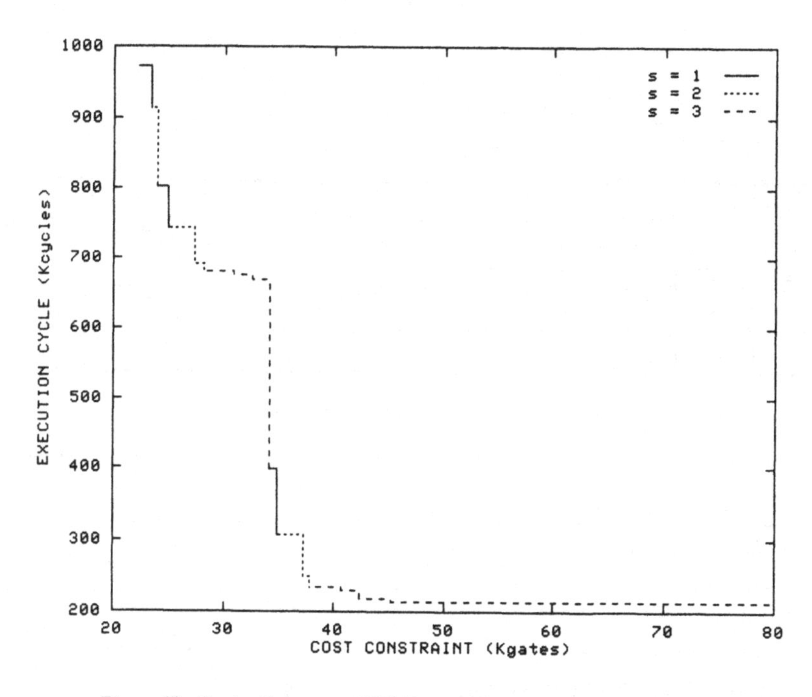

Figure 12 - Tradeoffs between HW Cost and Execution Cycles (FFT)

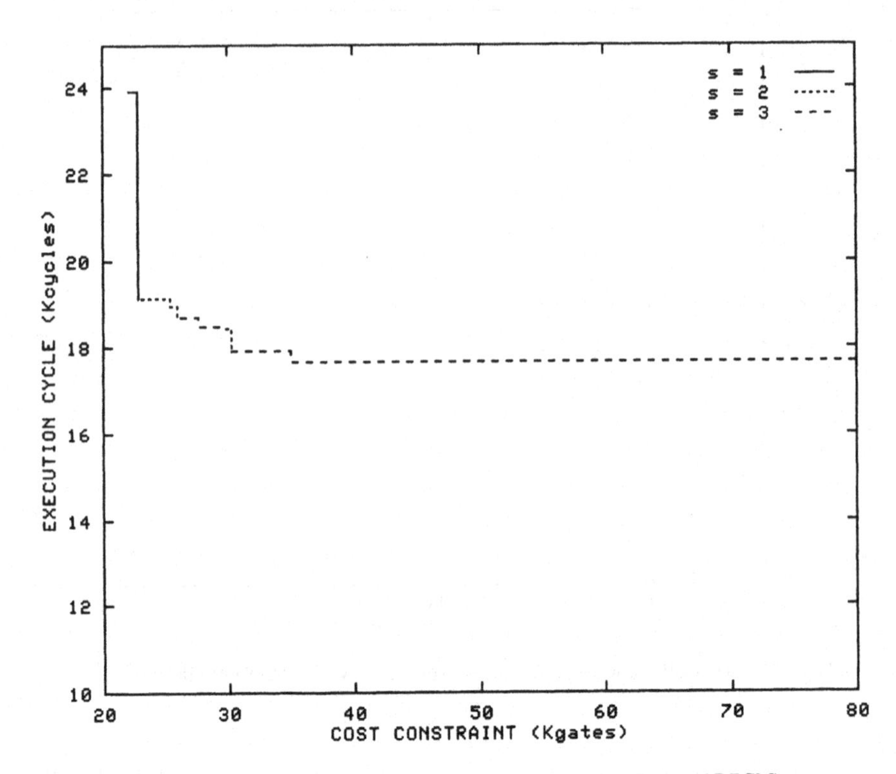

Figure 13 - Tradeoffs between HW Cost and Execution Cycles (ADPCM)

Table 3 - Selected Functional Units

# of Slots	Execution Cycles	pcu	ld/st	alu	mul		div		
					32_16	32_32	21_21	34_17	34_34
1	227,384	1	1	1	—	1	—	—	2
2	245,490	1	2	1	1	—	—	1	—
3	241,425	1	2	1	—	1	—	1	—

– HW Cost Constraint : 40,000 gates

# of Slots	Execution Cycles	pcu	ld/st	alu	mul		div		
					32_16	32_32	21_21	34_17	34_34
1	205,794	1	1	1	1	—	1	—	—
2	189,610	1	2	1	—	1	1	—	—
3	185,545	1	2	1	—	1	1	—	—

– HW Cost Constraint : 50,000 gates

These figures show that a loose hardware cost constraint makes shorter execution cycles. Generally, the looser the hardware cost constraint becomes, the larger the number of operation slot. The tighter the hardware cost constraint becomes, the smaller the number of operation slot. However, there are some exceptions. The architecture with a smaller number of operation slots had some possibilities to achieve better performance compared to those with a larger number of operation slots. In Table 3, although the CPU core on 65,000 gates constraint had three operation slots, the CPU core on 67,000 gates constraint had only one operation slot. The reason is that the CPU core with one operation slot had two dividers at the sacrifice of the hardware cost budget for more operation slots and a faster multiplier. After that, when the hardware cost constraint becomes looser, the number of operation slots becomes larger again. This kind of complicated decisions is difficult for human designers. This is one of the effectiveness of execution cycle minimization algorithm on PEAS-II system.

The computation time for the execution cycle optimization on the PEAS-II system is only a few seconds on Intel PentiumPro 200MHz with Linux ver.2.0.27. This result shows the efficiency of the execution cycle minimization algorithm of PEAS-II.

6.4 FUTURE WORK

Our future work includes extensions of the algorithm so that it could optimize register configuration (global register file, local register files, and combination of them), register count(s), and on-chip data memory size. Extension of the architecture model and the optimization algorithm to special functional units (such as multiply-and-add units) and co-processors are also included in the future work.

7. Conclusion

This chapter presents compiler generation techniques for embedded processors and their application to HW/SW co-design. Today's compiler is not only a translator from high-level language to assembly code but also a code optimizer, and the compiler will be a code generator for many kinds of processors including heterogeneous architectures. In this chapter, compiler optimization techniques, retargetable compiler projects, and an example of compiler application to HW/SW co-design PEAS-II system, are introduced. In the "system-on-a-chip" age, various application specific processors will be included on a chip and the compiler for their processors must be generated. Retargetable compilers will play a very important role in such situations and compiler generation techniques becomes an indispensable technology.

8. References

[1] Semiconductor Industry Association (1998) *The National Technology Roadmap for Semiconductors – Technology Needs*, 3rd Edition.

[2] Gajski, D.D., Vahid, F., Narayan, S., and Gong, J. (1994) *Specification and Design of Embedded Systems*, Prentice Hall.

[3] Henkel, J. and Ernst, R. (1995) A path-based estimation technique for estimating hardware runtime in HW/SW-cosynthesis, *Proc. of ISSS '95*, 116-121.

[4] Madsen, J., Grode, J., Kunudsen, P., Petersen, M., and Haxthausen, A. (1997) Lycos: the lyngby co-synthesis sytem," *Design Automation for Embedded Systems*, **2**, **2**, 195-235.
[5] Gupta, R., Coelho, Jr., C.N., and De Micheli, G. (1992) Synthesis and simulation of digital systems containing interacting hardware and software components, *Proc. of DAC '92*, 225-230..
[6] Alomary, A.Y., Nakata, T., Honma, Y., Sato, J., Hikichi, N., and Imai, M. (1993) PEAS-I: A
[7] Ohtsuki, N., Honma, Y., Takeuchi, Y., Imai, M., et al. (1998) Compiler Generation in PEAS-II: A Hardware/Software Codesign System for ASIP with VLIW Architecture, presented at *Third Int'l Workshop on Code Generation for Embedded Processors*.
[8] Huang, I-J., Holmer, B., and Despain, A.M. (1993) ASIA: Automatic Synthesis of Instruction-set Architectures, *Proc. of SASIMI '93*, 15-22.
[9] Aho, A.V. and Ullman, J.D. (1977) *Principles of Compiler Design*, Addison-Wesley Publishers.
[10] Aho, A.V., Sethi, R., and Ullman, J.D. (1986) *Compilers - Principles. Techniques, and Tools*, Addison-Wesley Publishers.
[11] Appel, A.W. and Ginsburg, M. (1997) *Modern Compiler Implementation in C - Basic Techniques*, Cambridge University Press.
[12] Hennessy, J.L. and Patterson, D.A. (1996) *Computer Architecture - A Quantative Approach*, Second Ed., Morgan Kaufmann Publishers.
[13] Gajski, D.D., Dutt, N.D., Wu, A.C-H., and Lin, S.Y-L. (1992) *High-Level Synthesis*, Kluwer Academic Publishers.
[14] Fisher, J.A. (1981) Trace Scheduling, A Technique for Global Microcode Compaction, *IEEE Transactions on Computers* **C-30**, **7**, 478-490.
[15] Hwu, W.W., Mahlke, S.A., Chen, W.Y., Chang, P.P. et al. (1993) The Superblock, An Effective Technique for VLIW and Superscalar Compilation, *J. of Supercomputing*, **7**, 229-248.
[16] Mahlke, S.A., Lin, D.C., Chen, W.Y., Hank, R.E., and Bringmann, R.A. (1992) Effective Compiler Support for Predicated Execution Using the Hyperblock, *Proc. of MICRO-25*, 45-54.
[17] Liem, C. (1997) *Retargetable Compilers for Embedded Cores*, Kluwer Academic Publishers.
[18] Stallman, R.M. (1998) *Using and Porting GNU CC*, Vers. 2.8.1, Free Software Foundation, Inc.
[19] Paulin, P.G., Liem, C., May, T.C., and Sutarwala, S. (1995) Flexware: A Flexible Firmware Development Environment for Embedded Systems, in in Marwedel, P., G. Goossens (eds.), *Code Generation for Embedded Processors*, Kluwer Academic Publishers, 67-84.
[20] Lanneer, D., Praet, J.V., Kifli, A., Schoofs, K., et al. (1995) CHESS: Retargetable Code Generation for Embedded DSP Porcessors, in Marwedel, P., G. Goossens (eds.), *Code Generation for Embedded Processors*, Kluwer Academic Publishers, 85-102.
[21] Binh, N.N., Imai, M., and Takeuchi, Y. (1998) A Performance Maximization Algorithm to Design ASIPs under the Constraint of Chip Area Including RAM and ROM Sizes, *Proc. of ASP-DAC '98*, 367-372.
[22] Alomary, A.Y., Honma, Y., Binh, N.N., Shiomi, A., Imai, M., and Hikichi, N. (1993) An ASIP Instruction Set Optimization Algorithm with Execution Cycle Constraint, *Proc. of SASIMI '93*, 34-43.
Hardware/Software Co-design System for ASIPs, *Proc. of EURO-DAC '93*, 2-7.
[23] Alomary, A.Y., Nakata, T., Honma, Y., Imai, M., and Hikichi, N. (1993) An ASIP Instruction Set Optimization Algorithm with Functional Module Sharing Constraint, *Proc. of ICCAD '93*, 526-538.
[24] Araujo, G. and Marik, S. (1995) Optimal Code Generation for Embedded Processors, *Proc. of ISSS '95*, pp. 36-41, 1995.
[25] Banerjee, U. (1993) *Loop Transformations for Restructuring Compilers: The foundations*, Kluwer Academic Publishers.
[26] Banerjee, U. (1994) *Loop Transformations for Restructuring Compilers: Loop Parallelization*, Kluwer Academic Publishers.
[27] Banerjee, U. (1997) *Loop Transformations for Restructuring Compilers: Dependence Analysis*, Kluwer Academic Publishers.
[28] Balarin, F., Chiodo, M., Giusto, P., Hsieh, H., et al. (1997) *Hardware-Software Co-Design of Embedded Systems – The POLIS Approach*, Kluwer Academic Publishers.
[29] Bernstein, D. and Gertner, I. (1989) Scheduling Expressions on a Pipelined Processor with a Maximal Delay of One Cycle, *ACM Trans. Programming Language and Systems*, **11**, **1**, 57-66.
[30] Berge, J-M., Levia, O., Rouillard, J. (1997) *Hardware/Software Co-Design and Co-Verification*, Kluwer Academic Publishers.
[31] Binh, N.N., Imai, M., and Shiomi, A. (1996) A New HW/SW Partitioning Algorithm for Synthesizing the Highest Performance Pipelined ASIPs with Multiple Identical FUs, *Proc. of EURO-DAC '96*, 126-131.
[32] Binh, N.N., Imai, M., Shiomi, A., Sato, J., and Hikichi, N. (1994) A Pipeline Scheduling Algorithm for Instruction Set Processor Design Optimization, *Proc. of APCHDL '94*, 59-66.

[33] Binh, N.N., Imai, M., Shiomi, A., and Hikichi, N. (1995) A Hardware/Software Codesign Method for Pipelined Instruction Set Processor using Adaptive Database, *Proc. ASP-DAC '95*, 81-86.
[34] Binh, N.N., Imai, M., Shiomi, A., and Hikichi, N. (1995) A Hardware/Software Partitioning Algorithm for Pipelined Instruction Set Processor, *Proc. of EURO-DAC '95*, 176-181.
[35] Binh, N.N., Imai, M., Shiomi, A., and Hikichi, N. (1995) An Instruction Set Optimization Algorithm for Pipelined ASIPs, *IEICE, Trans. on IEICE*, **E78-A**, **12**, 1707-1714.
[36] Binh, N.N., Imai, M., Shiomi, A., and Hikichi, N. (1996) A Hardware/Software Partitioning Algorithm for Designing Pipelined ASIPs with Least Gate Counts, *Proc. of DAC '96*, 527-532.
[37] Chaitin, G.J., Auslander, M.A., Chandra, A.K., Martin, J.C., Hopkins, E., and Markstein, P.W. (1981) Register Allocation via Graph Coloring, *Computer Languages*, **6**, 47-57.
[38] Chao, L-F., LaPaugh, A.S., and Sha, E. H-M. (1997) Rotation Scheduling: A Loop Pipelining Algorithm, IEEE, *Trans. on CAD*, **16**, **3**, 229-239.
[39] Daveau, J.-M., Marchioro, G.F., Ben-Ismail, T. and Jerraya, A.A. (1997) Protocol Selection and Interface Generation for HW-SW Codesign, IEEE, *Trans. of VLSI Systems*, **5**, **1**, 136-144.
[40] Ellis, J.R. (1985) *Bulldog, A Compiler for VLIW Architecture*, MIT Press.
[41] Ernst, R. (1998) Codesign of Embedded Systems: Status and Trends, *IEEE, Design & Test of Computers*, April-June 1998, 45-54.
[42] Fraser, C. and Hanson, D. (1995) *A retargetable C Compiler: Design and Implementation*, Addison-Wesley.
[43] Gibbons, P.B. and Muchnich, S.S. (1986) Instruction Scheduling for a Pipelined Architecture, *Proc. of SIGPPLAN '86 Symp. on Compiler Construction*, 11-16.
[44] Goodman, J.R. and Hsu, W. C. (1988) Code Scheduling and Register Allocation in Large Basic Block, *Proc. of 1988 International Conference on Supercomputing*, 442-452.
[45] Gupta, R. and De Micheli, G. (1993) Hardware/Software Cosynthesis for Digital Systems, *IEEE, Design & Test of Computers*, 29-41.
[46] Hennecy, J. and Gross, T. (1988) Postpass Code Optimization of Pipelined Constraints, *ACM Trans. Programming Language and Systems*, **5**, **3**, 422-448.
[47] Honma, Y., Imai, M., Shiomi, A., and Hikichi, N. (1994) Register Count Optimization in Application Specific Integrated Processors, *Proc. of APCHDL '94*, 263-270.
[48] Huang, I-J. and Despain, A.M. (1994) Synthesis of Instruction Sets for Pipelined Microprocessors, *Proc. of DAC '94*, 5-11, 1994.
[49] Huang, I.-J. and Despain, A.M. (1995) Synthesis of Application Specific Instruction Sets, *IEEE, Trans. on Computer-Aided Design of Integrated Circuits and Systems*, **14**, **6**, 663-675.
[50] Imai, M., Alomary, A.Y., Sato, J., and Hikichi, N. (1992) An Integer Programming Approach to Instruction Implementation Method Selection Problem, *Proc. of EURO-DAC '92*, 106-111.
[51] Imai, M. (1989) Synthesis of Application Specific CPU Core, *Proc. of the Synthesis and Simulation Meeting and International Interchange (SASIMI '89)*, V-1.
[52] Imai, M., Shiomi, A., Takeuchi, Y., Sato, J., and Honma, Y. (1996) Hardware/Software Codesign in the Deep Submicron Era, *Proc. of IWLAS '97*, 236-248.
[53] Inoue, A., Tomiyama, H., Eko, F. N., Kanbara, H., and Yasuura, H. (1998) A Programming Language for Processor Based Embedded Systems, *Proc. of APCHDL '98*, 89-94.
[54] Imai, M., Takeuchi, Y., Morifuji, T, and Shigehara, E (1998) Flexible Hardware Model: A New Paradigm for Design Reuse, *Proc. of APCHDL'98*, 3.
[55] Jerraya, A.A. (1998) System Level Synthesis (SLS), *TIMA Laboratory Annual Report 1997*, 77-90.
[56] Krishnamurthy, S.M. (1990) A Brief Survey of Papers of Scheduling for Pipelined Processors, *ACM, SIGPLAN notices*, **25**, **7**, 318-328.
[57] Liem, C., May, T., and Paulin, P. (1994) Instruction-Set Matching and Selection for DSP and ASIP Code Generation, *Proc. of ED&TC '94*, 31-37.
[58] Liem, C. and Paulin, P. (1997) Compilation Techniques and Tools for Embedded Processor Architectures," in Staunstrup, J., Wolf, W. (eds.), *Hardware/Software Codesign: Principles and Practice*, pp. 149-191, Kluwer Academic Publishers.
[59] Marwedel, P., G. Goossens (eds.), *Code Generation for Embedded Processors*, Kluwer Academic Publishers, 1995.
[60] Marwedel, P. (1993) Tree-Based Mapping of Algorithms to Predefined Structures, *Proc. of ICCAD '93*, 586-593.
[61] Mahlke, S.A., Hank, R.E., Bringmann, R.A., Gyllenhaal, J.C., et al. (1994) Characterizing the Impact of Predicated Execution on Branch Prediction, *Proc. of the 27th International Symposium on Microarchitecture*, 217-227.

[62] Mahlke, S.A., Hank, R.E., McCormick, J.E., August, D.I., et al. (1995) A Comparison of Full and Partial Predicated Execution Support for ILP Processors, *Proc. of the 22nd International Symposium on Computer Architecture*, 138-149.
[63] Rau, B.R. and Glaeser, C.D., Some Scheduling Techniques and an Easily Schedulable Horizontal Architecture for High Performance Scientific Computing, *Proc.14th Ann. Microprogramming Workshop*, 183-197.
[64] Rau, B.R., Kathail, V., Gupta, S.A. (1998) Machine-Description Driven Compilers for VLIW Processors, presented at *Third Int'l Workshop on Code Generation for Embedded Processors*.
[65] Sato, J., Alomary, A.Y., Honma, Y., Nakata, T., Shiomi, A, and Imai, M. (1994) PEAS-I: A Hardware/Software Codesign System for ASIP Development, IEICE, *IEICE Trans. on Fundamentals of Electronics, Communications and Computer Sciences*, **E77-A**, **3**, 483-491.
[66] Sato, J. and Imai, M. (1990) A Study on the Application Specific Microprocessor Design Environment, *Proc. of SASIMI '90*, 88-94.
[67] Sato, J., Hikichi, N., Shiomi, A., and Imai, M. (1994) Effectiveness of a HW/SW Codesign System PEAS-I in the CPU Core Design, *Proc. of APCHDL '94*, 259-262.
[68] Sato, J., Imai, M., Hakata, T., and Hikichi, N. (1991) An Integrated Design Environment for Application Specific Integrated Processor, *Proc. of ICCD '91*, 414-417.
[69] Staunstrup, J., Wolf, W. (eds.) (1997) *Hardware/Software Codesign: Principles and Practice*, Kluwer Academic Publishers.
[70] Stanford SUIF Compiler Group (1994) *The SUIF Parallelizing Compiler Guide*, Version 1.0, Stanford University.
[71] Sudarsannam, A., Marik, S., and Fujita, M. (1998) Implementation of a Developer-Retargetable Compiler for Embedded Processors, presented at *Third Int'l Workshop on Code Generation for Embedded Processors*.
[72] Walter, N.J., Lavery, D.M., and Hwu, W.W. (1993) The Benefit of Predicated Execution for Software Piplining, *Proc. of the 26th Annual Hawaii Int'l Conference on System Sciences*, 497-506.
[73] Warter, N.J., Backhaus, J.W., Haab, G.E., and Subramanian, K. (1992) Enhanced Modulo Scheduling for Loops with Conditional Branches, *Proc. of 25th Annual ACM/IEEE Int'l Symposium on Microarchitecture*, 170-179.
[74] Yoshioka, K., Sato, J., Takeuchi, Y., and Imai, M. (1997) A Performance Optimization Method for an On-Chip Two Level Cache Memory System, *Proc. of SASIMI '97*, 98-104.
[75] Yasuura, H., Tomiyama, H., Inoue, A., and Eko, F. N. (1997) Embedded System Design Using Soft-Core Processor and Valen-C, *Proc. of APCHDL '97*, 121-130.

IP-CENTRIC METHODOLOGY AND DESIGN WITH THE SPECC LANGUAGE

System Level Design of Embedded Systems

DANIEL D. GAJSKI, RAINER DÖMER AND JIANWEN ZHU
Department of Information and Computer Science
University of California, Irvine
Irvine, California, USA

Abstract.

In this paper, we demonstrate the application of the *specify-explore-refine* (SER) paradigm for an IP-centric codesign of embedded systems. We describe the necessary design tasks required to map an abstract executable specification of the system to the architectural implementation model. We also describe the final and intermediate models generated as a result of these design tasks. The executable specification and its refinements should support easy insertion and reuse of IPs.

Although several languages are currently used for system design, none of them completely meets the unique requirements of system modelling with support for IP reuse. This paper discusses the requirements and objectives for system languages and describes a C-based language called SpecC, which precisely covers these requirements in an orthogonal manner.

Finally, we describe the design environment which is based on our codesign methodology.

1. Introduction

New technologies allow designers to generate chips with more than 10 million transistors on a single chip. The main problem at this complexity is designer productivity. Although the chip complexity measured in number of transistors per chip has increased at the rate of 60 percent per year in the past, the productivity measured in number of transistors designed per day by a single designer has increased only at the rate of 20 percent. This growing gap between the complexity and productivity rates may have the catastrophic effect of slowing down semiconductor industry.

A. A. Jerraya and J. Mermet (eds.), System-Level Synthesis, 321–358.

One of the main solutions for solving this problem is increasing the level of abstraction in design of complex chips. The abstraction level increase should be reflected in descriptions, components, tools, and design methodology.

First, modelling or describing designs on the gate or RT level is not sufficient. Moving to executable specifications (behavior) and architectural descriptions (structure) is necessary to improve design productivity.

In order to explore different architectural solutions, we must use higher-level components beyond RTL components, such as registers, counters, ALUs, multipliers, etc. These higher-level components, frequently called IPs, are changing the business and design models. In order to use IPs, we need a methodology that will allow easy insertion of IPs in designs. This new methodology must have well-defined models of design representation, so that IP can be easily inserted or replaced when supplies disappear or IPs get discontinued. In order to achieve easy insertion and replacement, the design models must separate computation from communication, in addition to abstracting those two functions. This way, IP can be inserted by changing only the communication interface to the rest of the design.

Finally, the above IP-centric design methodology must be supported by CAD tools that will allow easy capture of executable specification, architecture exploration with IPs, and RTL hand-off to semiconductor fabs.

In this paper, we present such an IP-centric methodology, starting with an executable specification, define the abstract models used for architectural exploration, synthesis, and hand-off, and describe the necessary tools to support this methodology.

We also describe a C based language, called SpecC, for describing all the models in the methodology, and the SpecC Design Environment which supports all the transformations and explorations indicated in the methodology.

2. Related Work

For system-level synthesis, in particular codesign and coverification, academia, as well as industry, has developed a set of promising approaches and methodologies. Several systems already exist that assist designers with the design of embedded systems. However, none of todays systems covers the whole spectrum of codesign tasks. Instead, most systems focus on a subset of these problems.

2.1. UNIVERSITY PROJECTS

Table 1 lists some system-level projects developed by universities. Although all systems try to cover all aspects of system-level design, each of them really focuses on a subset of the tasks. Also, the target architectures addressed by the tools in many cases are quite specific and do not cover the whole design space.

TABLE 1. System-level Design Projects in Academia.

Project	University	Main Focus
Chinook	U Washington	Simulation, Synthesis
Cosmos	TIMA	Simulation, Synthesis
Cosyma	TU Braunschweig	Exploration, Synthesis
CoWare	IMEC	Interface Synthesis
Lycos	TU Denmark	Synthesis
Polis	UC Berkeley	Modelling, Synthesis
Ptolomy	UC Berkeley	Simulation
Scenic	UC Irvine	Simulation
SpecSyn	UC Irvine	Exploration
Weld	UC Berkeley	Framework

For the specification of embedded systems, standard programming languages are being used, as well as special languages developed to support important concepts in codesign directly. For the latter, two early approaches must be mentioned. Statecharts [7, 15] and SpecCharts [29, 10] use an extended finite state machine model in order to support hierarchy, concurrency and other common concepts. Both have a textual and a graphical representation. SpecCharts is the underlying language being used in the SpecSyn system [11], which is targeted at design space exploration and estimation.

In the Scenic environment [23, 14], the design is modeled with the standard programming language C++. Features not present in the language, like for example concurrency, can be specified by use of classes provided with the Scenic libraries. The SpecC system, as introduced in [39, 6] and described later in Section 4.3, goes one step further. The standard language C is extended with special constructs that support concurrency, hierarchy, exceptions, and timing issues, among others. For simulation, the SpecC language is automatically translated into a C++ program, which can be compiled and executed. This approach makes it possible for the SpecC system to focus on codesign modelling and synthesis while providing simulation, whereas Scenic mainly targets only simulation.

Similar to the specification language, the design representation being used internally in a codesign system is important. Usually every system has its own representation. The Polis system [2], targeted at small reactive embedded systems, uses the codesign finite state machine (CFSM) model [4] to represent the designs. Since this model is formally defined, it is also a suitable starting point for formal verification.

Most codesign systems can be classified as either simulation oriented, or synthesis oriented. A typical representative for simulation oriented systems is the Ptolomy frame work [22, 19]. Ptolomy models a design as a hierarchical network of heterogeneous subsystems and supports simultaneous simulation of multiple models of computation, such as for example synchronous data flow (SDF).

On the other hand, several systems are mainly synthesis oriented. In this category, Cosmos [35, 18] targets at the development of multiprocessor architectures using a set of user-guided transformations on the design. For the Cosyma [9, 16, 30] and the Lycos [26] system, the target architecture is an embedded micro architecture consisting of one processor with a coprocessor, e.g. an ASIC.

Interface and communication synthesis are addressed in particular by the Chinook [5] and CoWare [31] systems. Chinook is targeted at the design of control-dominated, reactive systems, whereas CoWare addresses the design of heterogeneous DSP systems.

As a special framework, the Weld project [3] addresses the use of networking in electronic design. It defines a design environment which enables web-based computer aided design (CAD) and supports interoperability via the internet.

2.2. COMMERCIAL SYSTEMS

A growing number of commercial tools are being offered by the EDA companies. However, they tend to either solve a particular problem as a point tool in the codesign process, e.g. cosimulation, or focus on one particular application domain, e.g. telecommunications.

For modeling and analysis at the specification level, Cadence and Synopsis offers tools (SPW and COSSAP, respectively) to support easy entry and simulation of block diagrams, a popular paradigm used in the communication community.

Another category of simulation tools is targeted at verification for design after backend synthesis. A representative is Seamless CVS from Mentor Graphics, which speeds up cosimulation of hardware and software by suppressing the simulation of information unrelevant to the hardware software interaction. Such information may include instruction fetch, memory access, etc. A similar tool is Eaglei from ViewLogic.

A variety of backend tools exists. The most widely used retargetable compiler is the GNU C compiler. However, since it is designed to be a compiler for general purpose processors, upgrading it into an aggressive, optimizing compiler for an embedded processor with possibly a VLIW datapath and multiple memory banks can be a tremendous effort. Although assembly programming prevails in current practice, new tools are expected to emerge as research in this area matures. The Behavioral Compiler from Synopsis, Monet from Mentor Graphics, and XE of Y-Explorations, are examples of high-level synthesis tools starting from a hardware description language. The Protocol Compiler of Synopsys exploits the regular expression paradigm for the specification of communication protocols and synthesizes interface circuitries between hardware modules.

There is a limited number of commercial tools offered for system-level synthesis. Among the few is the CoWare system, which targets at the hardware software interfacing problem. VHDL+ of ICL, also provides an extension of VHDL, which helps to solve the same problem.

There are a rapidly growing number of vendors for reusable components, or IP products for embedded systems. A traditional software component is the embedded operating system, which usually requires a small memory, and sometimes real time constraints have to be respected. Examples are VxWorks from Wind River, Windows CE from Microsoft, JavaOS from Sun Microsystems, to name just a few. The Inferno operating system from Lucent is especially designed for networking applications. The hardware IP vendors offer cores ranging from the gate and functional unit level, for example Synopsis Designware, to block level, for example Viterbi decoders and processors. They are often provided with a simulation model or a synthesizable model in VHDL or Verilog. While integrating these cores into a system-on-a-chip is not as easy as it appears, new methodologies, such as those proposed in the academia, and new standards, such as those prepared in the VSI alliance, are expected to make the plug-and-play capability possible.

3. System Design Methodology

A methodology is a set of models and transformations, possibly implemented by CAD tools, that refines the abstract, functional or behavioral specification into a detailed implementation description ready for manufacturing. The system methodology [12] starts with an *executable specification* as shown in Figure 1. This specification describes the functionality as well as the performance, power, cost and other constraints of the intended design. It does not make any premature allusions to implementation details. The specification is captured directly in a formal specification language such as SpecC (see Section 4.3), that supports different models in the methodology.

Since designers do not like to learn the syntax and semantics of a new

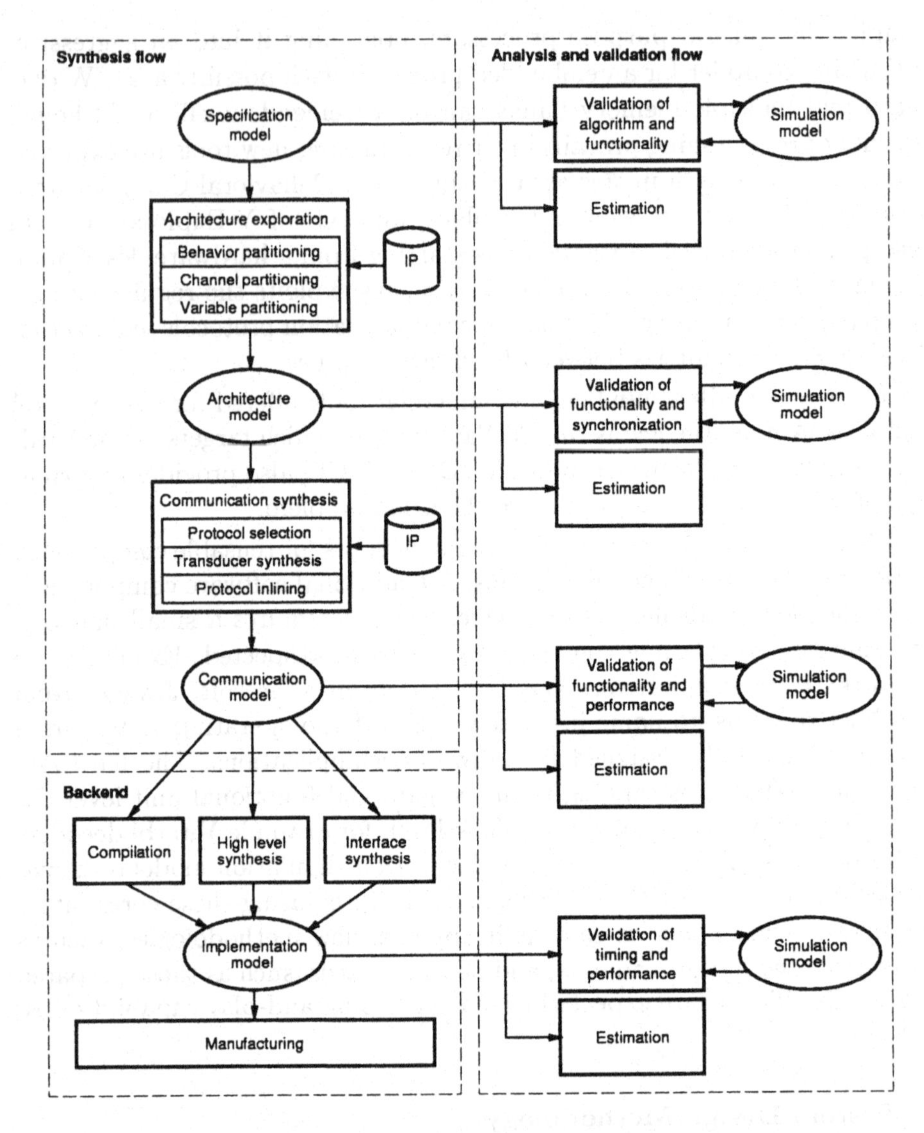

Figure 1. The codesign methodology in the SpecC Design Environment

language, the executable specification can be captured with a graphical editor that generates the specification from well-known graphical forms, such as block diagrams, connectivity tables, communication channels, timing diagrams, bubble charts, hierarchical trees, scheduling charts, and others. Such a graphical editor must also provide support for manual transformations of one model to another in the methodology.

As shown in Figure 1, the synthesis flow of the codesign process consists of a series of well-defined design steps which will eventually map the executable specification to the target architecture. In this methodology, we distinguish two major system level tasks, namely architecture exploration and communication synthesis.

Architecture exploration includes the design steps of allocation and partitioning of behaviors, channels and variables. *Allocation* determines the number and the types of the system components, such as processors, ASICs and buses, which will be used to implement the system behavior. Allocation includes the reuse of intellectual property (IP), when IP components are selected from the component library.

Then, *behavior partitioning* distributes the behaviors (or processes) that comprise the system functionality amongst the allocated processing elements, whereas *variable partitioning* assigns variables to memories and *channel partitioning* assigns communication channels to buses. *Scheduling* is used to determine the order of execution of the behaviors assigned to the processors.

Architecture exploration is an iterative process whose final result is the definition of the system architecture. In each iteration, estimators are used to evaluate the satisfaction of the design constraints. As long as any constraints are not met, component and connectivity reallocation is performed and a new architecture with different components, connectivity, partitions, schedules or protocols is evaluated.

After the architecture model is defined, *communication synthesis* is performed in order to obtain a design model with refined communication. The task of communication synthesis includes the selection of communication protocols, synthesis of interfaces and transducers, and inlining of protocols into synthesizable components. Thus, communication synthesis refines the abstract communications between behaviors into an implementation.

It should be noted that the design decisions in each of the tasks can be made manually by the designer, e. g. by using an interactive graphical user interface, as well as by automatic synthesis tools.

The result of the synthesis flow is handed-off to the backend tools, shown in the lower part of Figure 1. The software part of the hand-off model consists of C code and the hardware part consists of behavioral VHDL or C code. The backend tools include compilers, a high-level synthesis tool and an interface synthesizer. The compilers are used to compile the software C code for the processor on which the code is mapped. The high-level synthesis tool is used to synthesize the functionality mapped to custom hardware. The interface synthesizer is used to implement the functionality of interfaces needed to connect different processors, memories and IPs.

During each design step, the design model is statically analyzed to estimate certain quality metrics such as performance, cost and power consumption. This design model is also used in simulation to verify the correctness of the design at the corresponding step. For example, at the specification stage, the simulation model is used to verify the functional correctness of the intended design. After architecture exploration, the simulation model will verify the synchronization between behaviors on different processing elements (PEs). After communication synthesis, the simulation model is used to verify the performance of the system including computation and communication.

At any stage, if the verification fails, a debugger can be used to locate and fix the errors. Usually, standard software debuggers can be used which provide the ability to set break points anywhere in the source code and allow detailed state inspection at any time.

3.1. IP REQUIREMENTS

The use of Intellectual Property introduces additional requirements on the system design methodology. In order to identify the specification segments that can be implemented by an IP, or to replace one IP by another one, the system specification and its refined models must clearly identify the specific IP segment or the IP functionality must be deduced from the description. On the other hand, if the meaning of a model or one of its parts is difficult to discover, it is also difficult to see whether an IP can be used for its implementation.

This situation is well demonstrated in a much broader problem of design methodologies: simulatable vs. synthesizable languages. We know that almost any language (C, C++, Java, VHDL, Verilog, etc.) can be used for writing simulatable models. However, each design can be described in many different ways, all of them producing correct simulation results. Therefore, an IP function can be described in many different ways inside the system specification without being recognized as an IP description. In such a case, IP insertion is not possible. Simularly, replacing one IP with another with slightly different functionality or descriptions is not possible.

For example, a controller, whose computational model is a finite state machine, can be easily described by a `case` statement in which the cases represent the states. Simularly, an array of coefficients can be described with a `case` statement in which the cases represent the coefficient indices. In order to synthesize the description with these two case statements, we have to realize that the first statement should be implemented as a controller and the second as a look-up ROM. If the designer or a synthesis tool cannot distinguish between these two meanings, there is no possibility that a correct implementation can be obtained from that description although it will produce correct simulation results.

Therefore, in order to synthesize a proper architecture, we need a specification or a model that clearly identifies synthesizable functions including IP functions. In order to allow easy insertion and replacement of IPs, a model must also separate computation from communication, because different IPs have different communication protocols and buses connecting IPs may not match either of the IP protocols. The solution is to encapsulate different IPs and buses within virtual components and channels by introducing wrappers to hide detailed protocols and allow virtual objects to communicate via shared variables and complex data structures. In the methodology presented in Figure 1, the executable specification is written using shared variables for communication between behaviors or processes, while models used for architecture exploration use virtual components and channels for easy insertion and replacement of IPs. The final communication model exposes the protocols and uses again shared variables to describe individual wires and buses used in communication. Thus, the architecture exploration is performed on the model that clearly separates computations (behaviors) from communication (channels) and allows a *plug-and-play* approach for IPs.

However, there is a difference between functions defined in a channel and functions in a behavior. While the functions of a behavior specify its own functionality, the functions of a channel specify the functionality of the caller, in other words, when the system is implemented, they will get *inlined* into the connected behaviors or into transducers between the behaviors. When a channel is inlined, the encapsulated variables are exposed serving as communication media, and the functions become part of the caller. This is shown in Figure 2(a) where the channel `C` connecting behaviors `A` and `B` is inlined, assuming that `A` and `B` will be implemented as custom hardware parts. In such custom parts, the computation and communication will be realized by the same datapath and controlled by one controller.

The situation is different when a behavior is not synthesizable, such as in a processor core with a fixed protocol. This can be modelled using a *wrapper* which is a channel encapsulating a fixed behavior while providing higher-level communication functions that deal with the specific protocol of the internal component. For example, a MPEG decoder component with a wrapper can be used by other behaviors simply by calling the decode function provided by the wrapper. Figure 2(b) shows the inlining of the wrapper in component `A` allowing the communication between `A` and `IP` to use the IP protocol. On the other hand, whenever two channels (or wrappers) encapsulating incompatible protocols need to be connected, as shown in Figure 2(c), an interface component or transducer has to be inserted into which the channel functions will be inlined during communication refinement.

Next, we give a detailed description of each refinement task in the syn-

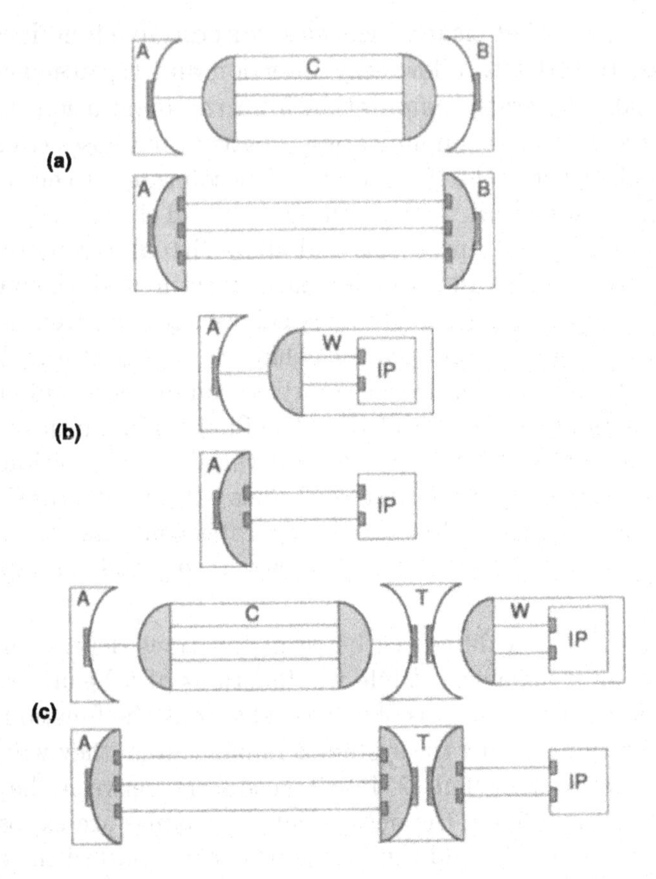

Figure 2. Channel inlining: (a) two synthesizable behaviors connected by a channel, (b) synthesizable behavior connected to an IP, (c) a synthesizable behavior connected to an IP through an incompatible channel.

thesis flow of the codesign process.

3.2. SPECIFICATION

The synthesis flow begins with a specification of the system being designed. An *executable specification* in a formal description language describes the functionality of the system along with performance, cost and other constraints but without premature allusions to implementation details. The specification should be as close to the computational model of the system as possible.

The source code can be executed with the help of a simulator and a set of test vectors, and errors can be located with debugger tools. This step verifies the algorithms and the functionality of the system. Obviously, it is easier and more efficient to verify the correctness of the algorithms at a higher abstraction level than at a lower level which includes the implementation details as well.

In our system, we use the SpecC language, described in detail in Section 4.3, to capture the high-level specification of the system under design. SpecC [39] is a superset of C [37] and provides special language constructs for modelling *concurrency*, *state transitions*, *structural and behavioral hierarchy*, *exception handling*, *timing*, *communication* and *synchronization*. This is in contrast to popular hardware description languages, like VHDL [17] and Verilog [34], which do not include explicit constructs for state transitions, communication, etc., and standard programming languages, like C/C++ [33] and Java [1], that cannot directly model timing, concurrency, structural hierarchy, and state transitions. Thus, SpecC is easily used for specifying FSMD or PSM computational models [11].

In addition, SpecC is synthesizable and aids the designer in developing *"good"* designs by providing the above listed features as language constructs rather than just supporting them in some contrived way. Another important feature of SpecC is its emphasis on *separation* of communication and computation at higher levels of abstraction. This dichotomy is essential to support *plug-and-play* of IPs. SpecC achieves this by using abstract function calls in the port interfaces of behaviors. The function calls are themselves implemented by *communication channels* [39]. The system behavior includes only the computation portion and uses a model similar to remote procedure calls (RPC) for communication. For implementation, the actual communication methods are resolved and inlined during the refinement process.

In the SpecC Design Environment, the SpecC Editor is used to capture the specification model of the system under design. The editor helps in capturing and visualizing the behavioral and structural hierarchy in the specification. It also supports the specification of the state transition tables, component connectivity and scope of variables and channels with a graphical user interface. Only the behavior of leaf nodes is programmed by use of a standard text editor.

We illustrate our codesign methodology with a simple example. The specification model is shown in Figure 3 using the PSM notation. The top level behavior `B0` consists of three sequential behaviors: `B1`, `B2` and `B3`. The system starts execution with behavior `B1`. When `B1` completes, the system transitions to `B2`. Finally, the system transitions to `B3` on behavioral completion of `B2`. Behavior `B2` again is a compound behavior, composed of two concurrent behaviors: `B4` and `B5`. Behavior `B4` is a leaf behavior like `B1` and `B3`. On the other hand, `B5` is hierarchical and consists of two sequential behaviors: `B6` and `B7`.

The leaf behaviors `B6` and `B4` communicate using global variables. First,

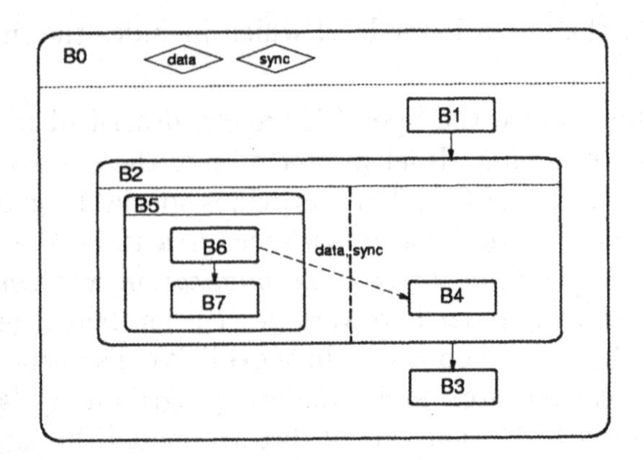

Figure 3. Specification model

B6 synchronizes its execution with B4 by using the sync event, as shown with the dashed arrow in Figure 3. Then, data is exchanged via the (possibly complex) variable data.

It should be emphasized that in the specification, the communication over shared global variables does not imply anything about the way it will be implemented later. For the implementation, this communication scheme could be transformed into a remote procedure call mechanism, or actually a shared memory model. Also, please note that we use the global variable communication to make the example simple. For a larger system, the designer is free to use, for example, communication via channels (as described in Section 4.3) in the specification model as well.

3.3. ARCHITECTURE EXPLORATION

The first major refinement step in the synthesis flow is the task of architecture exploration which includes allocation, partitioning and scheduling.

Allocation is usually done manually by the designer and basically means the selection of components from a library. In general, three types of components have to be selected from the component library: processing elements, called PEs (where a PE can be a standard processor or custom hardware), memories and buses. Of course, the component library can include IP components and already designed parts which can be reused.

The set of selected and interconnected components is called the system target architecture. The task of *partitioning*, then, is to map the system specification onto this architecture. In particular, behaviors are mapped to PEs, variables are mapped to memories, and channels are mapped to buses. In the SpecC system, the *partitioned model*, like the initial specification, is modeled in SpecC.

In order to perform partitioning, accurate information about the design has to be obtained before. This is the task of *estimation.* Estimation tools determine design metrics such as performance (execution time) and memory requirements (code and data size) for each part of the specification with respect to the allocated components. Estimation can be performed either statically by analyzing the specification or dynamically by execution and profiling of the design description. Obviously estimation has to support both software and hardware components. The estimation results usually are stored in a table which lists each obtained design metric for each allocated component.

The table of estimation results can then be used by the designer (or an automated partitioner) to tradeoff hardware vs. software implementation. It is also used to determine whether each partition meets the design constraints and to optimize the partitions with respect to an objective function.

In our methodology, architecture exploration is separated in three steps, namely behavior partitioning, channel partitioning and variable partitioning, which can be executed in any order.

3.3.1. *Behavior partitioning*

First, behaviors are partitioned among the allocated processing elements. This decides which behavior is going to be executed on which PE. Thus, it separates behaviors to be implemented in software from behaviors to be implemented in hardware.

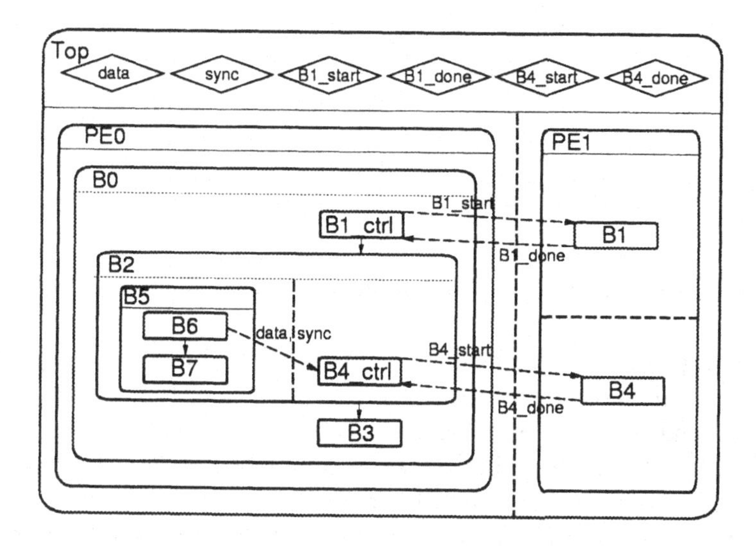

Figure 4. Intermediate model reflecting behavior partitioning

For example, given an allocation of two processing elements PE0 and PE1 (e.g. a processor and an ASIC), the specification model from Figure 3 can be partitioned as shown in Figure 4. Here, the behaviors B0, B2, B3, B5, B6 and B7 are mapped to PE0 (executing in software), and the behaviors B1 and B4 are assigned to PE1 (implemented in hardware). In order to maintain the execution semantics of the specification, two additional behaviors, B1_ctrl and B4_ctrl, are inserted which synchronize the execution with B1 and B4, respectively. Also, for this synchronization, four global variables, B1_start, B1_done, B4_start and B4_done, are introduced, as shown in Figure 4.

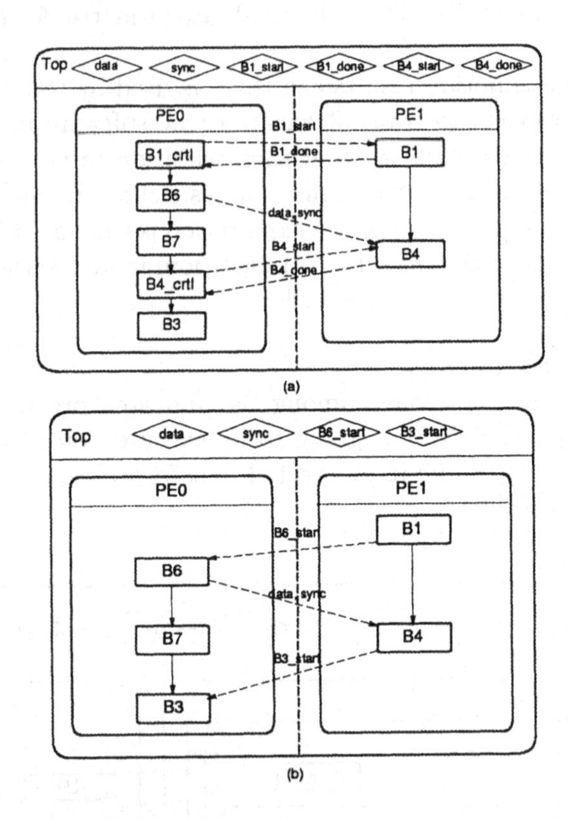

Figure 5. Intermediate model after scheduling: (a) non-optimized, (b) optimized.

The assignment of behaviors to a sequential PE, for example a processor, requires *scheduling* to be performed. As a preparation step, the approximate execution time for each leaf behavior, which was already obtained from estimators for the partitioning phase, is annotated with the behaviors, so that it can be used during scheduling.

The task of *scheduling* determines the order of execution for the behaviors that execute on a processor. The scheduler ensures that the schedule does not violate any dependencies imposed by the specification and tries to optimize objectives specified by the designer. After a schedule is determined, the design model is refined so that it reflects the sequential execution of the behaviors.

In general, scheduling can be either time-constrained or resource-constrained. For *time-constrained scheduling*, the designer supplies a set of timing constraints. Each timing constraint specifies the minimum and maximum time between two behaviors. The scheduler therefore has to compute a schedule, in which no behavior violates any of the timing constraints, and can minimize the number of resources used. On the other hand, for *resource-constrained scheduling*, the designer specifies constraints on the available resources. The scheduler then creates a schedule while optimizing execution time, such that all the subtasks are completed in the shortest time possible given the restrictions on the resource usage. In this methodology, resource-constraint scheduling is used, since the available resources are already determined during allocation.

Scheduling may be done statically or dynamically. In *static scheduling*, each behavior is executed according to a fixed schedule. The scheduler computes the best schedule at design time and the schedule does not change at run time. On the other hand, in *dynamic scheduling*, the execution sequence of the subtasks is determined at run-time. An *embedded operating system* maintains a pool of behaviors ready to be executed. A behavior becomes ready for execution when all its predecessor behaviors have been completed and all inputs are available. With a *non-preemptive* scheduler, a behavior is selected from the ready list as soon as the current behavior finishes, whereas for a scheduler with *preemption*, a running behavior may be interrupted in its computation when another behavior with higher priority becomes ready to execute.

After a schedule is created, the scheduler moves the leaf behaviors into the scheduled order and also adds necessary synchronization signals and constructs to the behaviors. This refined model then reflects the tasks performed for behavior partitioning including scheduling. Since, in the SpecC system, all design models are captured with the same language, the *scheduled model* is also specified in SpecC.

We illustrate the scheduling process with the intermediate model after behavior partitioning, as shown before in Figure 4. Figure 5 shows how scheduling is performed with the example. As shown in Figure 5(a), the behavioral hierarchy inside `PE1` is flattened and its leaf behaviors are sequentialized. For `PE2`, the behavior changes from (potentially) concurrent to sequential execution.

Due to scheduling, some explicit synchronization can become redundant. Figure 5(b) shows the optimized version of the example. Here, the behaviors `B1_ctrl` and `B4_ctrl`), which were introduced in the partitioning stage, are removed, together with their synchronization signals.

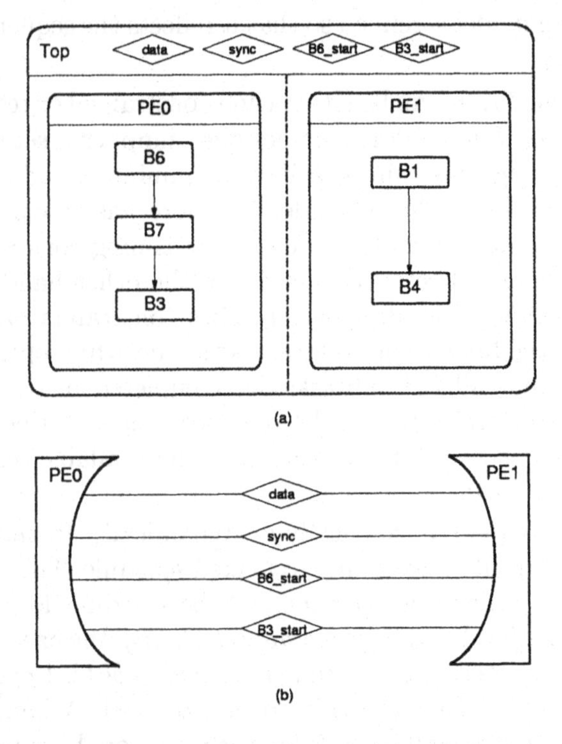

Figure 6. Model after behavior partitioning

After scheduling is done, the task of PE allocation and behavior partitioning is complete. Figure 6 shows the resulting design. In the lower part, it also shows the example from a structural view which emphasizes on the communication structure. This representation helps to explain the insertion of communication channels and memory behaviors which is described next.

3.3.2. *Channel partitioning*

Up to this point, communication between the allocated PEs is still performed via shared variables. In order to refine this abstract communication, these variables are first grouped and encapsulated in virtual channels.

In other words, in order to define the communication structure of the system architecture, channels are allocated into which the variables are partitioned. Later, during communication synthesis, these virtual channels will be refined to system buses.

In our example, channel partitioning is performed as shown in Figure 7. Here, due to the simplicity of the exmple, channel partitioning is easy. Since we have to connect only two PEs, we allocate one channel `CH0` and group all the variables into this channel, as shown in Figure 7(a).

Please note that in Figure 7, the leaf behaviors of `PE0` and `PE1`, which

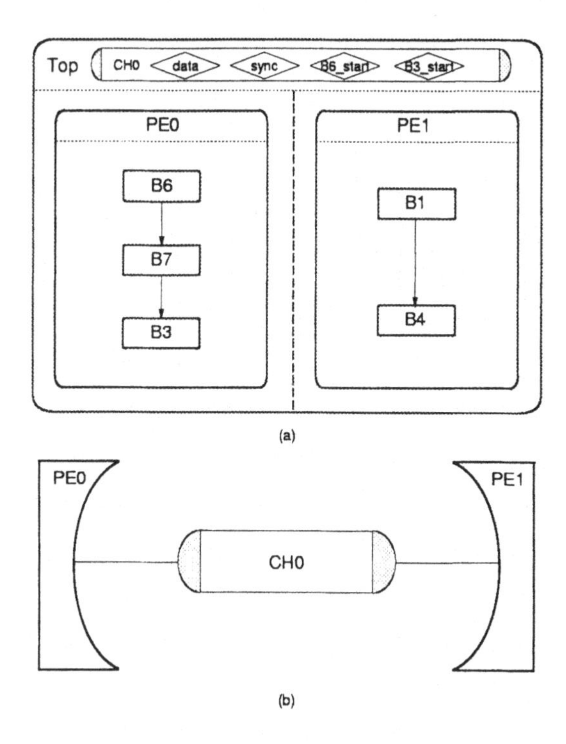

Figure 7. Model after channel partitioning

formerly could access the shared variables directly, are transformed in order to use the protocols supplied by the channel. For example, the behavior B4, formerly containing statements like `x = data`, is now transformed into one which uses statements like `x = CH0.read_data()` instead.

3.3.3. *Variable partitioning*

The last partitioning step is the allocation of memory components and the mapping of variables onto these memories. This is called *variable partitioning.*

Variable partitioning essentially decides whether a variable used for communication is stored in a memory outside the PEs or is directly sent by use of message passing. It also assigns variables to be stored in a memory to one of the allocated memory behaviors.

In our example, a single memory behavior M0 is allocated and inserted in the archtitecture, as shown in Figure 8. The four variables, that were formerly kept locally in the channel CH0, are partitioned into two groups. The possibly complex variable **data** and **sync** are assigned to the memory M0, whereas message passing is used for the synchronization variables **B6_start** and **B3_start**, as illustrated in Figure 8(a).

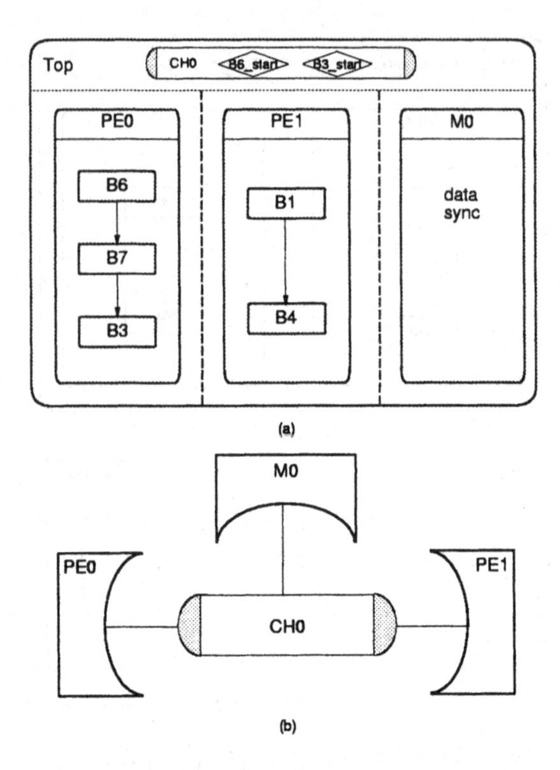

Figure 8. Model after variable partitioning

Please note that the channel CH0 is modified only internally in order to accomodate the communication to the inserted memory. Its interfaces to the PEs and the connected PEs themselves are not affected by this refinement step and, thus, need not be modified.

After variable partitioning, the task of architecture exploration is complete. However, it should be emphasized that in the SpecC environment, the sequence of allocation and partitioning tasks is determined by the designer and usually contains several iterations. The designer repeats these steps based on his experience and the performance metrics obtained with the estimation tools. This designer-driven design space exploration is easily possible in the SpecC Environment, because all parts of the system and all models are captured in the same language. This is in contrast to other environments where, for example, translating C code to VHDL and vice versa must be performed and verified. This design space exploration helps to obtain a "good" system architecture and finally an optimized implementation of the design with good performance and less costs.

3.4. COMMUNICATION SYNTHESIS

The purpose of communication synthesis is to resolve the abstract communication behavior in the virtual architecture through a series of refinements that lead to an implementation consisting of processing elements, buses and memories. During this process, new processing elements may be introduced in the form of transducers which serve to bridge the gap between differing protocols.

In our methodology, communication synthesis consists of three tasks, namely protocol selection, interface synthesis and protocol inlining.

3.4.1. *Protocol selection*

The designer selects the appropriate communication medium for mapping the abstract channels from a library of bus/protocol schemes during the task of protocol selection. Further, the designer has the option of including custom protocols or customizing available protocols to suit the current application. Protocol specifications contained in the library are written in terms of channel primitives of the SpecC language and supply common interface function calls to facilitate reuse. For example, a given VME bus description will supply `send()` and `receive()` as would the PCI specification. In this way, we can easily interchange protocols (as channels) and perform some simulation to obtain performance estimates. Later, the remote procedure calls (RPCs) to the channels will be replaced by local I/O instructions for software, or additional behavior to be synthesized for hardware entities.

The virtual channels in the design model after architecture exploration can now be refined into hierarchical channels which are implemented in terms of selected lower level channels. This process can be either manual or automatic. The cost of manual refinement is still lower than in the traditional way, since the user does not have to bother about issues such as detailed timing, thanks to the abstraction provided by the channel construct. Automatic refinement will generate code which assembles high level messages from low level messages, or vice versa, that can be delivered by the lower level channel.

In our example, this refinement is shown in Figure 9. A single bus channel `BUS`, e. g. a PCI bus, is selected in order to carry out the communication between the three behaviors. The methods of the virtual channel `CH0` are refined to use the methods of the bus protocol that is encapsulated in the channel `BUS`. It should be noted, that the channel hierarchy, as shown in Figure 9(b), directly reflects the layers of the communication between the PEs.

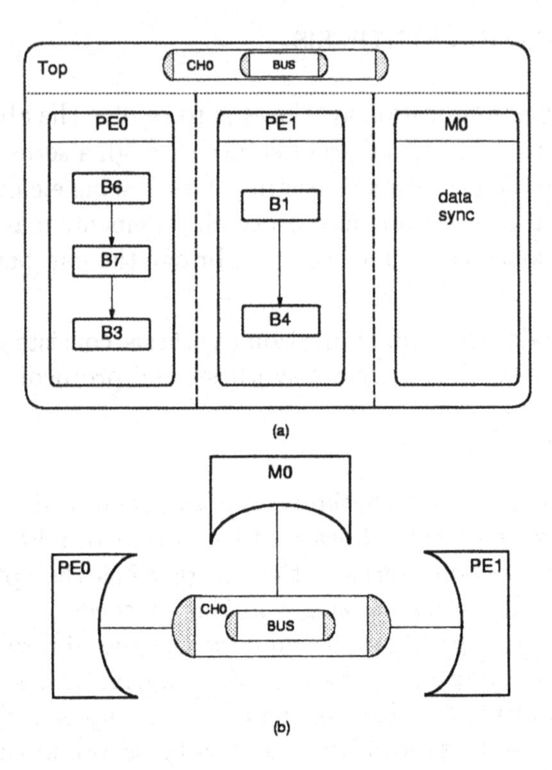

Figure 9. Model after bus allocation

3.4.2. *Protocol inlining*

During the task of protocol inlining, methods that are located in the channels, are moved into the connected behaviors if these behaviors were assigned to a synthesizable component. Thus, the behavior now includes the communication functionality also. Its port interfaces are composed of bit-level signals as compared to the abstract function calls before inlining was done. The "communication behavior" can then be synthesized/compiled with the rest of the component's functional (computational) behavior.

It should be noted that, since all information necessary is available in the design model, protocol inlining is a fully automatic task that requires no designer interaction.

In case, in our example, `PE0`, `PE1` and `M0` are all synthesizable behaviors, the methods of both channels `CH0` and `BUS` can be inlined into the behaviors, as shown in Figure 10. After this protocol inlining, the channel variables `rd`, `wr`, `dt` and `ad` are exposed and serve as interconnection wires between the accordingly created ports of the components.

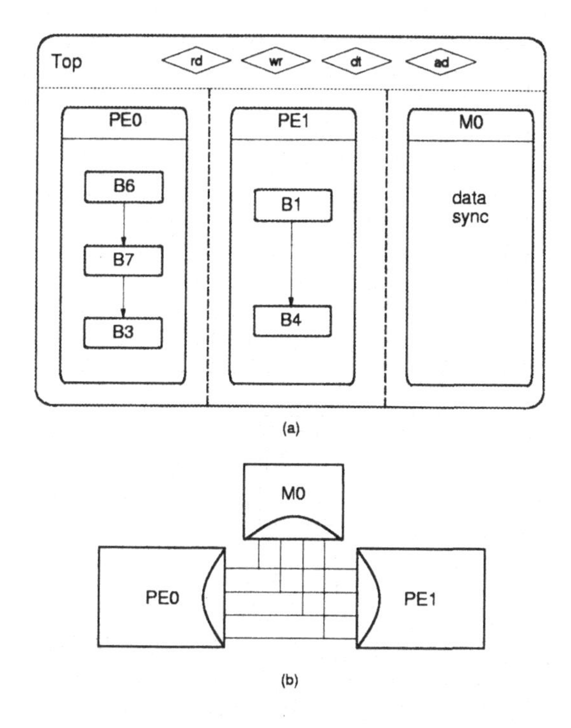

Figure 10. Model after inlining in synthesizable behaviors

3.4.3. *Transducer synthesis*

On the other hand, the designer may decide to use a non-synthesizable IP to implement a behavior in the system architecture. Such an IP can be selected from the component library, which contains both behavior models and *wrappers* which encapsulate the proprietary protocols of communication with the IPs. In the design model, a IP is introduced by creation of a *transducer* which bridges the gap between the IP component and the channels which the original behavior is connected to. Again, such a transducer can be easily created manually thanks to the high level nature of the wrapper and the connected channel.

It should be emphasized that the replacement of synthesizable behaviors with IP components is not limited to the communication synthesis stage. In fact, it is possible at any time during architecture exploration and communication synthesis. The key to this feature is the encapsulation of IP components in wrappers.

In our example, Figure 11 shows the design model where the synthesizable behaviors `PE0`, `PE1` and `M0` are all replaced with non-synthesizable IP components encapsulated in wrappers and connected to the channel `CH0` via the inserted transducers `T0`, `T1` and `T2`.

Finally, Figure 12 shows this new model with the inserted IP compo-

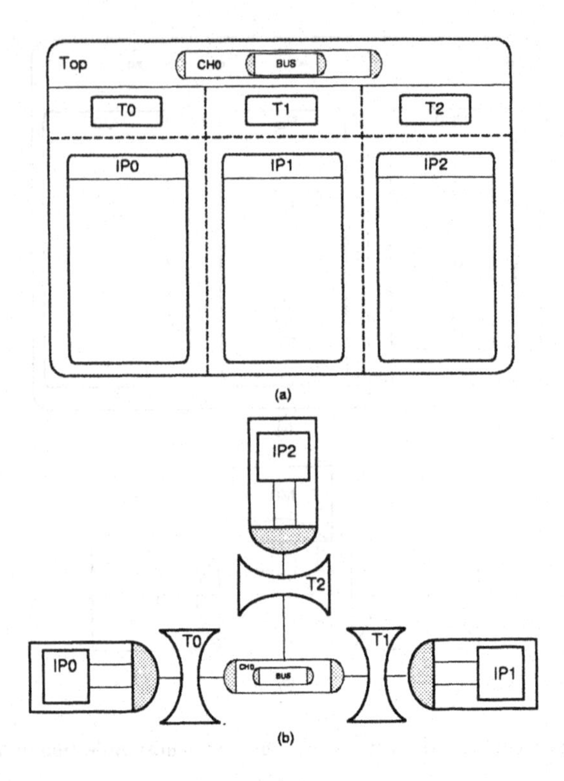

Figure 11. **Alternative model with IPs**

nents after protocol inlining is performed. Here, the methods from all the channels and wrappers, are inlined into the transducers which communicate with the IPs via proprietary buses. Again, the bus variables **rd**, **wr**, **dt** and ad are exposed and serve as interconnection wires between the transducers.

3.5. HAND-OFF

Communication synthesis, as the last step in the synthesis flow, generates the hand-off model for our system. This model is then further refined using traditional back-end tools as shown in Figure 1.

The software portion of the communication (hand-off) model consists of code in C for each of the allocated processors in the target architecture. Retargetable compilers or special compilers for each of the different processors can be used to compile the C code. The hardware portion of the model consists of synthesizable, behavioral models in C or VHDL. The behavioral models can be synthesized using standard high-level synthesis (HLS) tools. The interfaces between hardware and software components are also separated in software (device drivers) and hardware parts (transducers).

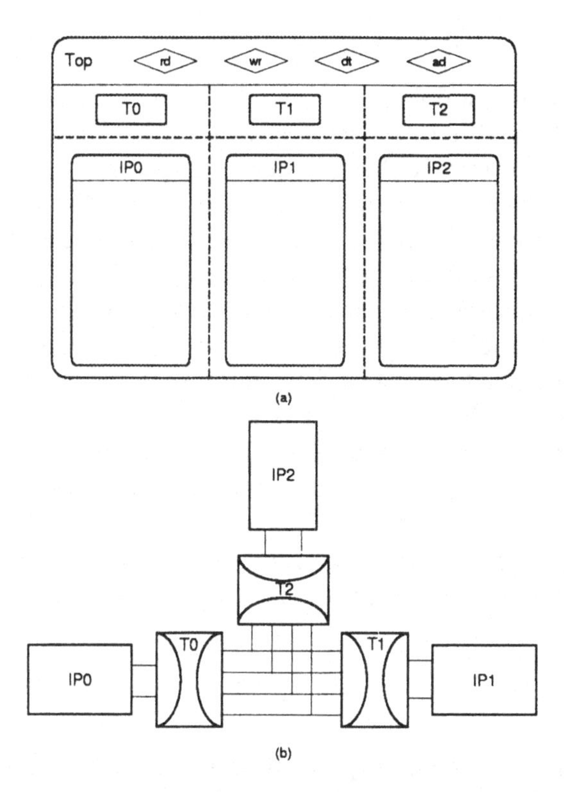

Figure 12. Model with IPs after inlining

Thus, this is just a special case of hardware and software parts and can be handled in the same way.

Finally, this design process generates the *implementation model* consisting of assembly code executing on the different processors and a register transfer level (RTL) or gate-level netlist of the hardware components. Thus, the implementation model is ready for manufacturing.

4. The Language

With this generic methodology in mind, Section 4.1 discusses the requirements and goals for system description languages and Section 4.2 compares traditional languages with these requirements. Since none of the languages supports all concepts a new modelling language called SpecC is proposed and presented in Section 4.3.

4.1. MODELLING LANGUAGE REQUIREMENTS

For the codesign methodology presented above, it is desirable that *one* language is used for all models at all stages. Such a methodology is called *homogeneous* in contrast to heterogeneous approaches [19, 31], where a system is specified in one language and then is transformed into another, or is represented by a mixture of several languages at the same time.

This homogeneous methodology does not suffer from simulator interfacing problems or cumbersome translations between languages with different semantics. Instead one set of tools can be used for all models and synthesis tasks are merely transformations from one program into a more detailed one using the same language. This is also important for *reuse*, because design models in the library can be used in the system without modification (*"plug-and-play"*) and a new design can be used directly as a library component.

System design places unique requirements on the specification and modelling language being used. In particular the language must be

1. executable,
2. modular and
3. complete.

1. Executability of the language is of crucial importance for simulation. The system specification must be validated to assure that exactly the intended functionality is captured. Simulation is also necessary for the intermediate design models whose functionality must be equivalent to the behavior of the model before the refinement.
2. Modularity is required to clearly separate functionality from communication, which is necessary in a model at a high level of abstraction. It also enables the decomposition of a system into a hierarchical network of components. *Behavioral hierarchy* is used to decompose a system's behavior into sequential or concurrent subbehaviors, whereas *structural hierarchy* decomposes a system into a set of interconnected components.
   Modularity is also required to support design reuse and the incorporation of intellectual property. During refinement, modularity helps to keep changes in the system description local so that other parts of the design are not affected. For example, communication refinement should only replace abstract channels with more detailed ones without modifying the components using these channels. The locality of changes makes refinement tools simpler and the generated results more understandable.
3. Completeness is obviously a requirement. A system language must cover all concepts commonly found in embedded systems. In addition

	Verilog	VHDL	Statecharts	SpecCharts	C	Java	SpecC
Behavioral Hierarchy	○	○	●	●	○	○	●
Structural Hierarchy	●	●	○	○	○	○	●
Concurrency	●	●	●	●	○	◑	●
Synchronization	●	●	●	●	○	●	●
Exception Handling	●	○	◑	●	◑	●	●
Timing	●	●	◑	◑	○	○	●
State Transitions	○	○	●	●	○	○	●

○ not supported ◑ partly supported ● fully supported

Figure 13. Language Comparison

to (a) behavioral and (b) structural hierarchy this includes (c) concurrency, (d) synchronization, (e) exception handling and (f) timing, as discussed in detail in [11]. For explicit modelling of Mealy and Moore type finite state machines, (g) state transitions have to be supported. Furthermore, it is desirable that these concepts are organized orthogonally (independent from each other) so that the language can be minimal. In addition to these requirements, the language should be easy to understand and easy to learn.

4.2. TRADITIONAL LANGUAGES

Most traditional languages lack one or more of the requirements discussed in Section 4.1 and therefore cannot be used for system modelling without problems. Figure 13 lists examples of current languages [34, 17, 15, 29, 37, 1, 39] and shows which requirements they support and which are missing.

Because the traditional languages are not sufficient, a new language must be developed, either from scratch or as an extension of an existing language. The SpecC language [6] represents the latter approach as it is built on top of C.

4.3. THE SPECC LANGUAGE

This section introduces the SpecC language and shows how SpecC covers all the requirements discussed before. SpecC is a superset of ANSI-C. C was selected because of its high acceptance in software development and its large library of already existing code.

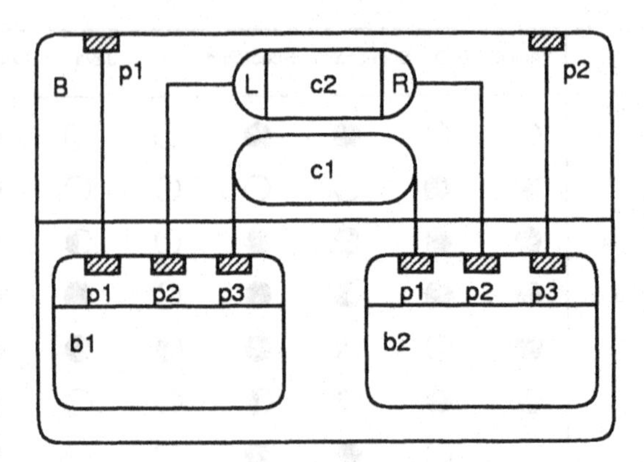

Figure 14. Basic Structure of a SpecC Model

A SpecC program can be executed after compilation with the SpecC compiler which first generates an intermediate C++ model of the program that is then compiled by a standard compiler for execution on the host machine.

Modularity, providing structural and behavioral hierarchy, and the special constructs making SpecC complete are described next.

4.4. STRUCTURAL HIERARCHY

Semantically, the functionality of a system is captured as a a hierarchical network of behaviors interconnected by hierarchical channels. Syntactically, a SpecC program consists of a set of *behavior*, *channel* and *interface* declarations.

A *behavior* is a class consisting of a set of ports, a set of component instantiations, a set of private variables and functions, and a public `main` function. Through its ports, a behavior can be connected to other behaviors or channels in order to communicate. A behavior is called a composite behavior if it contains instantiations of child behaviors. Otherwise it is called a leaf behavior. The functionality of a behavior is specified by its functions starting with the `main` function.

A *channel* is a class that encapsulates communication. It consists of a set of variables and functions, called methods, which define a communication protocol. A channel can be hierarchical, for example subchannels can be used to specify lower level communication.

An *interface* represents a flexible link between behaviors and channels. It consists of declarations of communication methods which will be defined in a channel.

For example, the following SpecC description specifies the system shown in Figure 14:

```
interface L { void Write(int x); };
interface R { int  Read (void);  };

channel C implements L, R
{
int Data; bool Valid;

void Write(int x)
   { Data = x; Valid = true; }
int Read(void)
   { while(! Valid) waitfor(10);
     return(Data); }
};

behavior B1(in int p1, L p2, in int p3)
{
void main(void)
   { /* ... */ p2.Write(p1); }
};

behavior B2(out int p1, R p2, out int p3)
{
void main(void)
   { /* ... */ p3 = p2.Read(); }
};

behavior B(in int p1, out int p2)
{
int c1;
C   c2;
B1  b1(p1, c2, c1);
B2  b2(c1, c2, p2);

void main(void)
   { par { b1.main(); b2.main(); } }
};
```

The example system specifies a behavior `B` consisting of two subbehaviors `b1` and `b2` which execute in parallel and communicate via integer `c1` and channel `c2`. Thus structural hierarchy is specified by the tree of child behavior instantiations and the interconnection of their ports via variables and channels. Behaviors define functionality, and the time of communication, whereas channels define how the communication is performed.

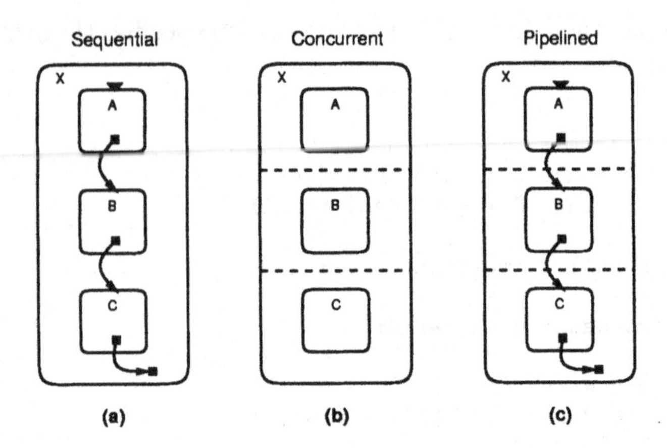

Figure 15. Behavioral Hierarchy

4.5. BEHAVIORAL HIERARCHY

The composition of child behaviors in time is called behavioral hierarchy. Child behaviors can either be executed sequentially or concurrently. Sequential execution can be specified by standard imperative statements or as a finite state machine with explicit state transitions. Concurrent execution is either parallel or pipelined.

For example, we can specify a behavior being the sequential composition of the child behaviors using sequential statements, as shown in Figure 15(a), where `X` finishes when the last behavior `C` finishes. Second, we can use the parallel composition using the `par` construct, as shown in Figure 15(b), where `X` finishes when all its child behaviors `A`, `B` and `C` are finished. Also, pipelined composition is supported using the `pipe` construct, as shown in Figure 15(c), where `X` starts again when the slowest behavior finishes.

Syntactically, behavioral hierarchy is specified in the `main` function of a composite behavior. For example, with `a`, `b`, and `c` being instantiated child behaviors, the sequence of calls

```
a.main(); b.main(); c.main();
```

simply specifies sequential execution of `a`, `b`, `c`. The `par` and `pipe` statements specify concurrent execution. For example,

```
par { a.main(); b.main(); c.main(); }
```

executes `a`, `b`, `c` in parallel, whereas

```
pipe { a.main(); b.main(); c.main(); }
```

specifies execution in a pipelined fashion (**a** in the first iteration, **a** and **b** in the second, ...). The **par** statement completes when its last statement finishes, the **pipe** statement implicitly specifies an endless loop.

SpecC also supports explicit specification of state transitions. For example

```
fsm { a: { if (x > 0)  break;
   if (x <= 0) goto b; }
      b: { if (y > 0)  goto a;
   if (y == 0) goto b; }
      c: { break; }
     }
```

specifies the state transitions of a finite state machine model with three behaviors **a**, **b**, **c**. Implicitly the first label in the **fsm** statement specifies the initial state (**a**). The FSM exits when a **break** statement is executed.

In summary, behavioral hierarchy is captured by the tree of function calls to the behavior **main** methods.

4.6. SYNCHRONIZATION

Concurrent behaviors usually must be synchronized in order to be cooperative. In SpecC, a built-in type *event* serves as the basic unit of synchronization. Events can only be used as arguments to **wait** and **notify** statements (or with exceptions as explained in Section 4.7). A **wait** statement suspends the current behavior from execution until one of the specified events is notified by another behavior. The **notify** statement triggers all specified events so that all behaviors waiting on one of these events can resume their execution.

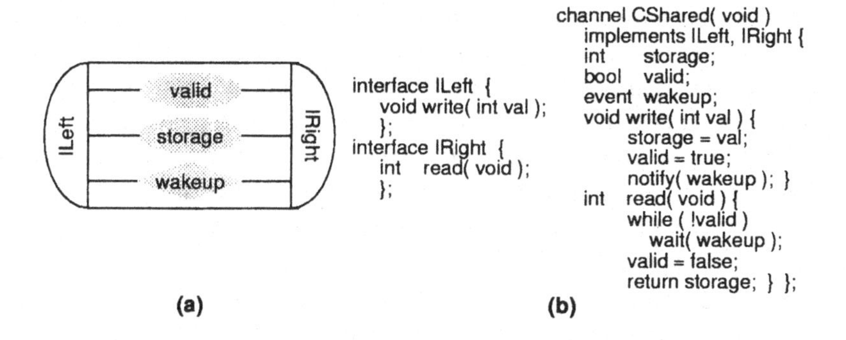

Figure 16. Example for simple Shared Memory Channel

For example, Figure 16 shows a simple shared memory channel **CShared** that, in addition to a **valid** bit, uses the event **wakeup** to allow only syn-

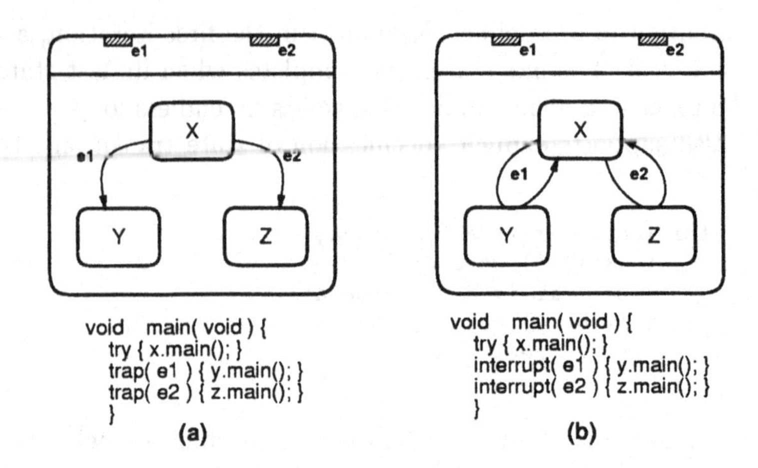

Figure 17. Exception handling: (a) abortion, (b) interrupt.

chronized accesses to its **storage**. With this channel, it is assured that a consumer will always get valid data.

4.7. EXCEPTION HANDLING

SpecC provides support for two types of exceptions, namely *abortion* (or trap) and *interrupt*, as shown in Figure 17.

The **try-trap** construct, illustrated in Figure 17(a), aborts behavior **x** immediately when one of the events **e1**, **e2** occurs. The execution of behavior **x** (and all its child behaviors) is terminated without completing its computation and control is transferred to behavior **y** in case of **e1**, to behavior **z** in case of **e2**. This type of exception usually is used to model the reset of a system.

On the other hand, the **try-interrupt** construct, as shown in Figure 17(b), can be used to model interrupts. Here again, execution of behavior **x** is stopped immediately for events **e1** and **e2**, and behavior **y** or **z**, respectively, is started to service the interrupt. After completion of interrupt handlers **y** and **z** control is transferred back to behavior **x** and execution is resumed right at the point where it was stopped.

For both types of exceptions, in case two or more events happen at the same time, priority is given to the first listed event.

It should be noted that interrupt and abortion type exceptions can be mixed in SpecC. For example, the following code specifies a behavior **B** with a resetable child behavior **b1** and an interrupt handler **b2**.

```
behavior B (in event IRQ, in event RST)
{
B_sub b1, b2;
```

```
void main(void)
   { try { b1.main(); }
     interrupt IRQ { b2.main(); }
     trap      RST { b1.main(); }
   }
};
```

4.8. TIMING

In the design of embedded systems the notion of real time is an important issue. However, in traditional imperative languages such as C, only the ordering among statements is specified, the exact information on when these statements are executed, is irrelevant. While these languages are suitable for specifying functionality, they are insufficient in modeling embedded systems because of the lack of timing information. Hardware description languages such as VHDL overcome this problem by introducing the notion of time: statements are executed at discrete points in time and their execution delay is zero. While VHDL gives an exact definition of timing for each statement, such a treatment often leads to *over-specification*.

One obvious over-specification is the case when VHDL is used to specify functional behavior. The timing of functional behaviors is unknown until they are synthesized. The assumption of zero execution time is too optimistic and there are chances to miss design errors during specification validation. Other cases of over-specification are timing constraints and timing delays, where events have to happen, or, are guaranteed to happen in a *time range*, instead of at a fixed point in time, as restricted by VHDL.

SpecC overcomes this problem by differentiating between two types of timing information, *exact timing* and *timing ranges*. Exact timing is used when the timing is known, for example the execution delay of an already synthesized component. This is specified with a `waitfor` statement which suspends the execution of the current behavior for a specified time. The time is measured in real time units such as nanoseconds. Simulation time is only increased by `waitfor` statements, other statements are always executed in zero time.

Timing ranges are used to specify timing constraints at the specification level. SpecC supports timing information in terms of *timing diagrams* with minimum and maximum time constraints. Timing ranges are specified as 4-tuples $T = \langle l1, l2, min, max \rangle$ with the **range** statement. For example,

```
range(l1; l2; 10; 20);
```

specifies at least 10 but not more than 20 time units spent between labels `l1` and `l2`.

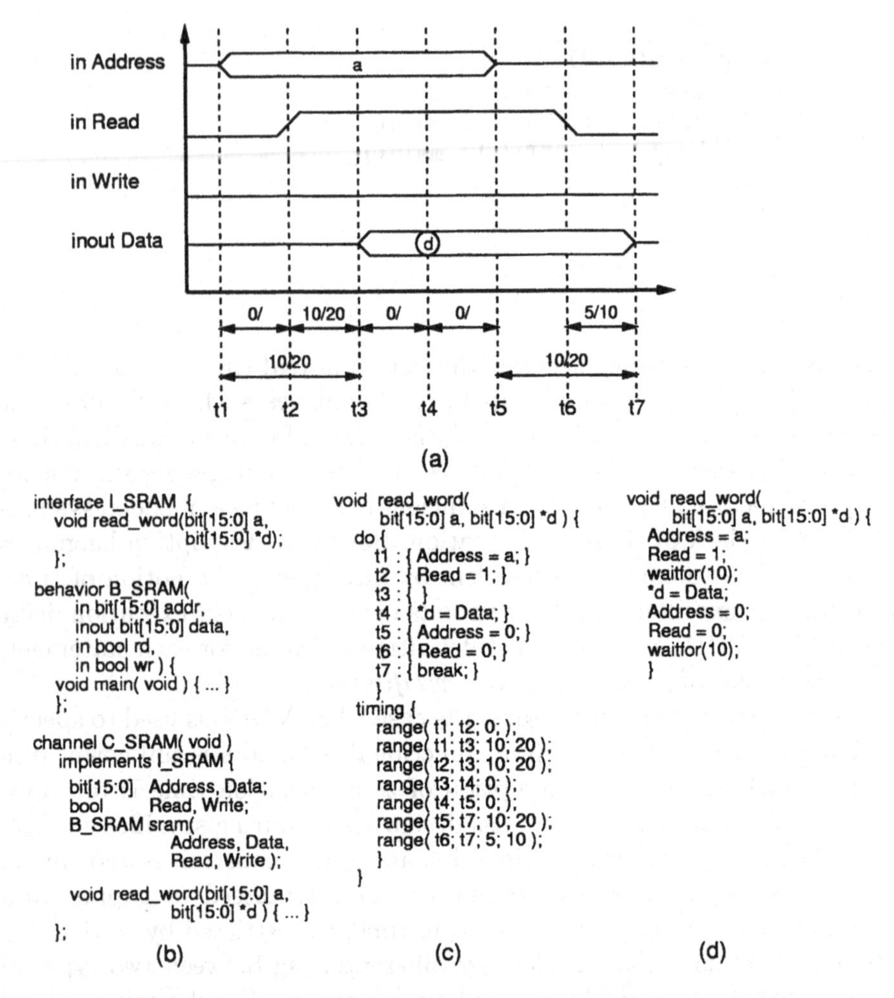

Figure 18. **Timing Example: SRAM Read Protocol: (a) timing diagram, (b) SRAM channel, (c) specification level timing, (d) implementation level timing.**

Consider, for example, the timing diagram of the read protocol for a static RAM, as shown in Figure 18(a). In order to read a word from the SRAM, the address of the data is supplied at the `address` port and the read operation is selected by assigning 1 to the `read` and 0 to the `write` port. The selected word then can be accessed at the `data` port. The diagram in Figure 18(a) explicitly specifies all timing constraints that have to be satisfied during this read access. These constraints are specified as arcs between pairs of events annotated with `x/y`, where `x` specifies the minimum and `y` the maximum time between the value changes of the signals.

Figure 18(b) shows the SpecC source code of a SRAM channel `C_SRAM`, which instantiates the behavior `B_SRAM`, and the signals, which are mapped to the ports of the SRAM. Access to the memory is provided by the `read_word` method, which encapsulates the read protocol explained above (due to space constraints write access is ignored).

Figure 18(c) shows the source code of the `read_word` method at the specification level. The `do-timing` construct used here effectively describes all information contained in the timing diagram. The first part of the construct lists all the events of the diagram, which are specified as a label and its associated piece of code, which describes the changes of signal values. The second part is a list of `range` statements, which specify the timing constraints between the events, as explained above.

This style of timing description is used at the specification level. In order to get an executable model of the protocol, *scheduling* has to be performed for each `do-timing` statement. Figure 18(d) shows the implementation of the `read_word` method after an ASAP scheduling is performed. All timing constraints are replaced by delays, which are specified using the `waitfor` construct.

4.9. ADDITIONAL FEATURES

In addition to the concepts explained in the last sections, the SpecC language supports further constructs that are necessary for system-level design. First, SpecC provides explicit support for *Boolean* (`bool`) and *bitvector* (`bit[:]`) types, in addition to all types provided by ANSI-C.

Also, constructs for binary import of pre-compiled SpecC code and support of persistent annotation for objects in the language are provided. Since these constructs are beyond the scope of this paper, please refer to [6] for further details.

In conclusion, the Sections 4.4 to 4.8 show that the SpecC language satisfies the requirements of executability, modularity and completeness, as discussed in Section 4.1.

It has to be emphasized, that the advantage of SpecC lies in its *orthogonal constructs* which implement *orthogonal concepts*. All SpecC constructs are independent of each other, unlike for example signals in VHDL, which are used for synchronization, communication and timing. The SpecC language covers the complete set of system concepts with a minimal set of constructs. Therefore it is easy to learn and easy to understand.

5. Reuse and IP

This section takes a closer look at how well the SpecC language and the SpecC methodology supports the reuse and integration of intellectual property.

Reuse essentially deals with the check-in ("Design for Reuse") and check-out ("Reuse of Designs") of components in the design library. Because all components in the design library are specified using the same SpecC language, reuse becomes easy. Also, the SpecC language encourages the specification of modular components which are decoupled from each other and therefore can be used independently.

In particular, a SpecC design library consists of behaviors, channels, and interfaces. A new design can be developed from scratch and/or composed from existing parts by selecting components from this library. As described earlier, behaviors represent functional units such as hardware components, and channels encapsulate communication such as bus protocols and bus media. Thus computation and communication are clearly separated. Interfaces connect behaviors and channels, as they declare *what* kind of communication is performed. Channels define *how* the communication is performed by implementing the interface.

A behavior's port of type interface can be mapped to any channel that implements that interface. Thus channels delivering the same type of communication can be exchanged without modification of the connected behaviors, for example a PCI bus can be easily replaced with a VME bus ("plug-and-play"). The same applies to behaviors. A behavior can be replaced with another behavior without affecting the channels as long as both implement the same functionality and have compatible ports.

For integration of intellectual property, three IP configurations are possible with SpecC. First, an IP vendor can offer design specifications which still need to be synthesized. This is called Soft-IP and is useful for standard buses and bus protocols for example. In this case the IP consists of a SpecC interface declaration and a channel definition.

On the other hand, Hard-IP integrates already synthesized components such as cores. Here, in addition to the actual core (layout), the IP vendor delivers a SpecC behavior declaration which only specifies the ports of the component, and an object or library file that can be linked to the executable SpecC code for simulation. Note, in this case the IP vendor keeps the implementation of the core secret.

As a third configuration, a combination of Soft-IP and Hard-IP is possible, where the IP consists of a wrapper (a SpecC channel definition with interface) in addition to the Hard-IP parts. This is exactly the situation as described in Figure 2(b), where the wrapper supplies higher-level functions dealing with the communication to the internal component.

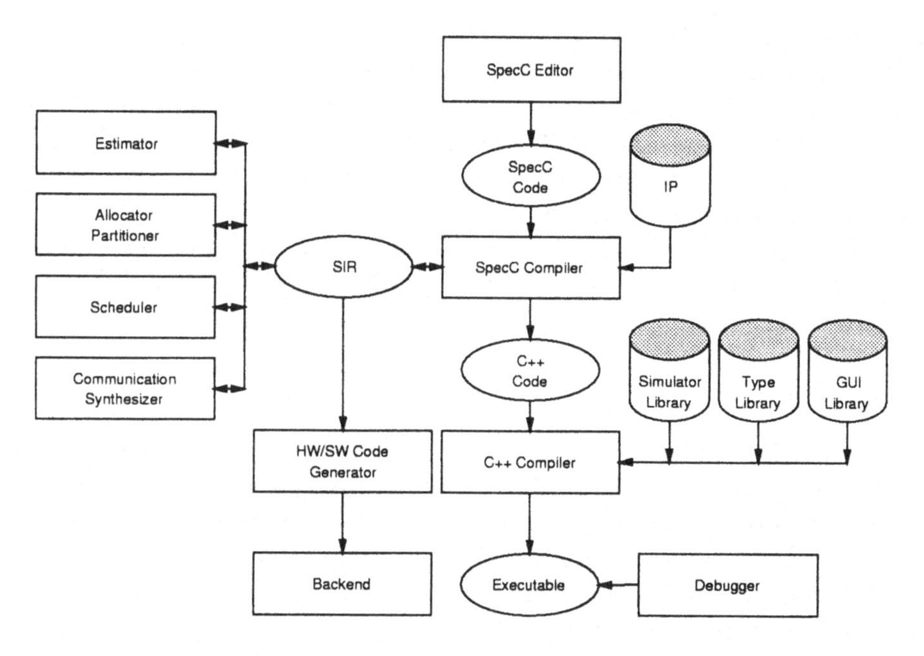

Figure 19. The SpecC Environment (SCE)

6. The System

We have developed the SpecC Environment as shown in Figure 19. The design is specified with the help of the *SpecC Editor* which provides a graphical user interface (GUI). The SpecC Editor is also used for displaying the system models at different design stages and allows the designer to execute transformations on the models in an interactive or automatic manner. Different aspects of the design model are displayed in separate windows. For example, the structural hierarchy of the system under design is displayed in a hierarchy browser, whereas the mapping of ports and variables is shown in a connectivity window. All windows support interactive modification of the design.

In analogy to the methodology described in Section 3, the SpecC synthesis system consists of a set of tools, such as the *Estimator*, the *Allocator* and *Partitioner*, the *Scheduler*, and the *Communication Synthesizer*, which operate on the *SIR*, the SpecC Intermediate Representation. A SIR file can be obtained initially by compiling SpecC source code using the *SpecC Compiler*. It contains the symbol table and the abstract syntax tree of the corresponding SpecC code. It also contains explicit information such as the type of each expression which is implicit in the source.

The SpecC compiler can also automatically generate simulation code in the form of C++, which can then be compiled and linked with a set of predefined libraries in order to generate an executable.

The *Simulation Library* implements a discrete event simulator by maintaining a time wheel which schedules concurrent threads. The *Type Library* provides an implementation of data types such as bitvector and multi valued logic. The *GUI Library* helps to visualize signal waveforms and supports graphical entry of stimuli.

A standard source-level debugger can be used to debug the executable. The *HW/SW Code Generator* exports the implementation level SIR into C or HDL code.

7. Conclusion

With the background of a specify-explore-refine paradigm, an IP-centric methodology for the codesign of embedded systems was presented. The methodology consists of a set of well-defined tasks and design models which allow the easy insertion and reuse of intellectual property.

In particular, the design methodology starts with an executable specification of the system under design and eventually creates an implementation architecture ready for manufacturing. The intermediate tasks of allocation, partitioning, scheduling, and communication synthesis are performed by the designer interactively, either manually or with the help of automatic tools. In other words, for architecture exploration the designer is in the loop.

In order to incorporate IP components and allow "plug-and-play", protocol encapsulation and separation of communication and computation is necessary. A wrapper concept is used to hide details of communication protocols and replace these details with an abstract high-level interface.

For this system design methodology, the language being used is important. Since none of the traditional languages meets all the requirements for system level design, the SpecC language was presented. SpecC precisely satisfies all requirements for codesign languages and explicitly supports structural and behavioral hierarchy, concurrency, state transitions, exception handling, timing and synchronization in an orthogonal way. SpecC encourages reuse and supports integration of IP. Since SpecC is a superset of C, a large library of already existing algorithms can directly be used. SpecC is easy to learn and easy to understand.

Finally, the SpecC Environment was presented. The system is based on the described methodology and the SpecC language.

Acknowledgements

We would like to acknowledge the support of the various granting agencies who have contributed research funding, without which this work would not have been possible. This work was supported in part by grants from: Hitachi, Grant #-H22003; Toshiba, Grant #-TC-20881; SRC, Grant #-97-DJ-146; and Rockwell, Grant #-RSS-24141.

References

1. K. Arnold, J. Gosling; *The Java Programming Language*; Addison-Wesley, 1996.
2. F. Balarin, P. Giusto, A. Jurecska, C. Passerone, E. Sentovich, B. Tabbara, M. Chiodo, H. Hsieh, L. Lavagno, A. Sangiovanni-Vincentelli, K. Suzuki. *Hardware-Software Co-Design of Embedded Systems, The POLIS approach.* Kluwer Academic Publishers, April 1997.
3. F. Chan, M. Spiller, R. Newton. "WELD – An Environment for Web-Based Electronic Design". In *Proceedings of the Design Automation Conference*, San Francisco, 1998.
4. M. Chiodo, P. Giusto, H. Hsieh, A. Jurecska, L. Lavagno, A. Sangiovanni-Vincentelli. "A Formal Specification Model for Hardware/Software Codesign". In *Proceedings of International Workshop on Hardware-Software Codesign*, Oct. 1993.
5. P. Chou, R. Ortega, G. Borriello. "The Chinook Hardware/Software Co-Synthesis System". In *International Symposium on System Synthesis*, Cannes, France, Sept. 1995.
6. R. Dömer, J. Zhu, D. Gajski. *The SpecC Language Reference Manual.* University of California, Irvine, Technical Report ICS-TR-98-13, March 1998.
7. D. Drusinsky and D. Harel. "Using Statecharts for hardware description and synthesis". In *IEEE Transactions on Computer Aided Design*, 1989.
8. R. Ernst, J. Henkel, T. Benner. "Hardware-software cosynthesis for microcontrollers". In *IEEE Design and Test*, Vol. 12, 1993.
9. R. Ernst, et. al. "The COSYMA Environment for Hardware-Software Cosynthesis of Small Embedded Systems". In *Microprocessors and Microsystems*, Vol. 20, No. 3, May 1996.
10. D. Gajski, F. Vahid, and S. Narayan. "SpecCharts: a VHDL front-end for embedded systems". University of California, Irvine, Technical Report ICS-TR-93-31, 1993.
11. D. Gajski, F. Vahid, S. Narayan, J. Gong. *Specification and Design of Embedded Systems.* Prentice Hall, New Jersey, 1994.
12. D. Gajski, J. Zhu, R. Dömer. "Essential Issues in Codesign". In *Hardware/Software Co-Design: Principles and Practice*, edited by J. Staunstrup, W. Wolf. Kluwer Academic Publishers, 1997.
13. R. Gupta, C. Coelho., G. De Micheli. "Synthesis and simulation of digital systems containing interacting hardware and software components". In *Proceedings of the 29th ACM, IEEE Design Automation Conference*, 1992.
14. R. Gupta, S. Liao. "Using a Programming Language for Digital System Design". In *IEEE Design & Test of Computers*, IEEE, 1997.
15. D. Harel; "StateCharts: a Visual Formalism for Complex Systems"; *Science of Programming*, 8, 1987.
16. J. Henkel, R. Ernst. "A Hardware-Software Partitioner Using a Dynamically Determined Granularity". In *Proceedings of the Design Automation Conference*, 1997.
17. IEEE Inc., N.Y. *IEEE Standard VHDL Language Reference Manual*, 1998.
18. T. Ismail, M. Abid, A. Jerraya. "COSMOS: A Codesign Approach for Communicating Systems". In *Proceedings of the International Workshop on Hardware- Software Codesign*. IEEE CS Press, 1994.
19. A. Kalavade, E. Lee. "A Hardware/Software Codesign Methodology for DSP Applications". In *IEEE Design and Test*, Sept. 1993.
20. G. Koch, U. Kebschull, W. Rosenstiel. "A prototyping architecture for hardware/software codesign in the COBRA project". In *Proceedings of the third Inter-*

national Workshop on Hardware/Software Codesign, IEEE Computer Society Press, 1994.
21. D. Ku, G. De Micheli. "HardwareC – A Language for Hardware Design, Version 2.0". Tech. Rep. CSL-TR-90-419, Stanford University, April 1990.
22. E. Lee and D. Messerschmidt. "Static scheduling of synchronous data flow graphs for digital signal processors". In *IEEE Transactions on Computer-Aided Design*, 1987.
23. S. Liao, S. Tjiang, R. Gupta. "An Efficient Implementation of Reactivity for Modeling Hardware in the Scenic Design Environment". In *Proceedings of the 34th Design Automation Conference*, Anaheim, California, USA, 1997.
24. C. Liem, F. Nacabal, C. Valderrama, P. Paulin, A. Jerraya. "System-on-a-chip cosimulation and compilation". In *IEEE Design & Test of Computers*, 1997.
25. C. Liem, P. Paulin. "Compilation Techniques and Tools for Embedded Processor Architectures". In *Hardware/Software Co-Design: Principles and Practice*, edited by J. Staunstrup, W. Wolf. Kluwer Academic Publishers, 1997.
26. J. Madsen, J. Grode, P. Knudsen. "Hardware/Software Partitioning using the LYCOS System". In *Hardware/Software Co-Design: Principles and Practice*, edited by J. Staunstrup, W. Wolf. Kluwer Academic Publishers, 1997.
27. P. Marwedel, G. Goossens. *Code Generation for Embedded Processors.* Kluwer Academic Publishers, 1995.
28. G. De Micheli. *Synthesis and Optimization of Digital Circuits.* McGraw Hill, 1994.
29. S. Narayan, F. Vahid, D. Gajski. "System Specification and Synthesis with the SpecCharts Language". In *Proceedings of the International Conference on Computer Aided Design*, 1991.
30. A. Österling, T. Benner, R. Ernst, D. Herrmann, T. Scholz, W. Ye. "The Cosyma System". In *Hardware/Software Co-Design: Principles and Practice*, edited by J. Staunstrup, W. Wolf. Kluwer Academic Publishers, 1997.
31. K. Rompaey, D. Verkest, I. Bolsens, H. De Man. "CoWare – A design environment for heterogeneous hardware/software systems". In *Proceedings of the European Design Automation Conference*, 1996.
32. W. Rosenstiel. "Prototyping and Emulation". In *Hardware/Software Co-Design: Principles and Practice*, edited by J. Staunstrup, W. Wolf. Kluwer Academic Publishers, 1997.
33. B. Stroustrup. *The C++ Programming Language*, third edition. Addison-Wesley, 1997.
34. D. Thomas, P. Moorby. *The Verilog Hardware Description Language.* Kluwer Academic Publishers, 1991.
35. C. Valderrama, M. Romdhani, J. Daveau, G. Marchioro, A. Changuel, A. Jerraya. "Cosmos: A Transformational Co-design tool for Multiprocessor Architectures". In *Hardware/Software Co-Design: Principles and Practice*, edited by J. Staunstrup, W. Wolf. Kluwer Academic Publishers, 1997.
36. W. Wolf. "Hardware/Software Co-Synthesis Algorithms". In *Hardware/Software Co-Design: Principles and Practice*, edited by J. Staunstrup, W. Wolf. Kluwer Academic Publishers, 1997.
37. X3 Secretariat. *The C Language.* X3J11/90-013, ISO Standard ISO/IEC 9899. Computer and Business Equipment Manufacturers Association, Washington, DC, USA, 1990.
38. T. Yen, W. Wolf. *Hardware-software Co-synthesis of Distributed Embedded Systems.* Kluwer Academic Publishers, 1997.
39. J. Zhu, R. Dömer, D. Gajski. "Syntax and Semantics of the SpecC Language". In *Proceedings of the Synthesis and System Integration of Mixed Technologies*, Osaka, Japan, Dec. 1997.

THE *JAVATIME* APPROACH TO MIXED HARDWARE-SOFTWARE SYSTEM DESIGN

JAMES SHIN YOUNG, JOSH MACDONALD, MICHAEL SHILMAN, ABDALLAH TABBARA, PAUL HILFINGER, AND A. RICHARD NEWTON

Department of Electrical Engineering and Computer Sciences and The Electronics Research Laboratory, University of California at Berkeley

"What has been is what will be,
and what has been done is what will be done,
and there is nothing new under the sun."
...Koheleth, Ecclesiastes 1:9

Abstract

We describe an approach for using Java as a basis for a design and specification language for embedded systems and use our *JavaTime* system to illustrate many aspects of the approach. Java is a pragmatic choice for several reasons. Since it is a member of the C "family" of languages, it is familiar to designers. Unlike C and C++, it has standard support for concurrency. Its treatment of arrays permits better static and dynamic error checking than is conveniently feasible in C and C++. Finally, while Java's expressive power is comparable to C++, it is a much simpler language, that greatly eases the task of introducing additional analysis into compilers.

Successive, formal refinement is an approach we have developed for specification of embedded systems using a general-purpose programming language. Systems are formally modeled as Abstractable Reactive systems, and Java is used as the design input language. A policy-of-use is applied to Java, in the form of language usage restrictions and class-library extensions, to ensure consistency with the formal model. A process of incremental, user-guided program transformation is used to refine a Java program until it is consistent with the policy-of-use. This approach allows systems design to begin with the flexibility of a general-purpose language, followed by gradual refinement into a more restricted form necessary for specification.

A. A. Jerraya and J. Mermet (eds.), System-Level Synthesis, 359–396.

1. Introduction

In this work, we are concerned with the design and verification of complex single-chip real-time systems, where such systems are comprised of large amounts hardware and software and must interact with a complex, time-dependent environment. We refer to such a single-chip design implementation as a System-On-a-Chip (SOC) for an embedded application. As the complexity of SOC designs grows, under the relentless need to maintain or improve product time-to-market, there is ongoing pressure to move to ever-higher levels of abstraction for SOC design, to increase the overall percentage of design re-use, and to construct systems from components drawn form a wide variety of sources. Couple these factors with the constraints applied by a complex and time-dependent operating environment, and embedded SOC design and verification becomes one of the major challenges the electronics industry is facing today.

Present design tools and methodologies do not adequately address many of the challenges posed by modern embedded systems design. In particular, they fail to meet the needs of increasing heterogeneity in implementation technologies, source of design, and of increasing variety in levels of integration . The majority of systems today include behavior implemented as both hardware and software, and many also incorporate newer technologies, such as reconfigurable circuits, that present their own unique set of design and integration challenges as well. As the number and importance of electronic systems continues to grow, techniques must be found that allow reliable, embedded systems to be developed correctly, quickly, and efficiently.

The *JavaTime* system is an experimental test bed we are developing for the design of such heterogeneous, embedded electronic systems. It uses the Java language[22] as a syntax for design representation, and takes *a component-based approach* to system specification, modeling and implementation. *JavaTime* addresses heterogeneity by supporting multiple underlying models of computation, and supports integration by providing a common representation format at a reasonably wide range of levels of abstraction. In particular, we describe an approach below for using Java for describing both the specification and implementation of a heterogeneous embedded system. In a sense, we use the Java language as a kind of *high-level intermediate format*, as shown in Figure 1.1. We embed domain-specific "languages" in the Java syntax, always in legal Java and using all of the primitive Java facilities.

Designs may be created in various ways—textually, as Java source code, or graphically, as block diagrams or hierarchical finite state machines (currently, all three modes are supported in the prototype). Designs are then analyzed, executed, and iterated using the *JavaTime* tools, eventually resulting in a specification for a system in terms of specific hardware implementation styles, software and associated run-time environments, and possibly other implementation media as well.

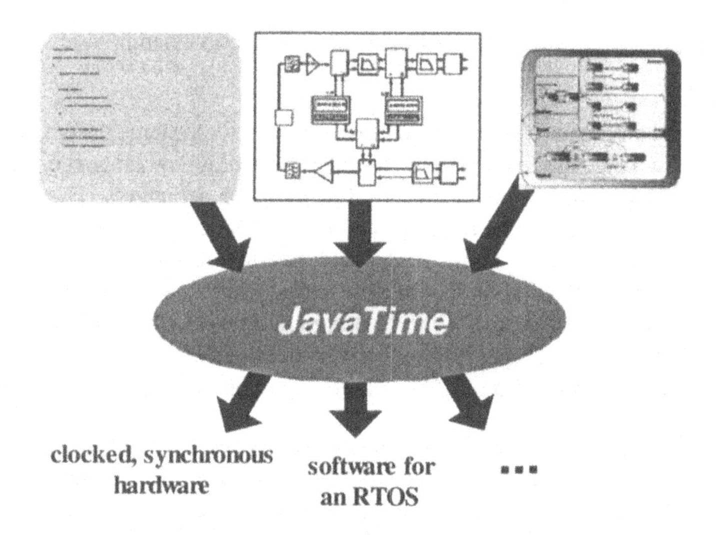

Figure 1.1: Java as a Syntax for Design Representation

The choice of Java in this role is a pragmatic one for several reasons. Since it is a member of the C "family" of languages, it is familiar to designers. Unlike C and C++[24], it has standard (built-in) and simple support for concurrency. While thread packages are available for C and C++, they are not standardized in the language and hence the concurrency semantics are not understood or supported by standard development tools. Java has an abstract memory model and does not allow users to manually manage memory. By forbidding pointer arithmetic, automated analysis and optimization of memory usage is simplified. Finally, while Java's expressive power is comparable to C++, it is a much simpler language, that greatly eases the task of introducing additional analysis into compilers. Using Java also has many practical benefits–its widespread adoption by the science and engineering community promises a large base of support, in the form of compilers, debuggers, development environments, and class libraries. To that end, another of our design goals in *JavaTime* has been to remain strictly compatible with existing Java development environments. All extensions we introduce to Java are introduced via standard classes and do not involve any incompatible extensions to Java syntax.

On the other hand, when compared to hardware description languages, such as Verilog[1] and VHDL[2], we believe Java is more appropriate as a system description language partly because it has no built-in notion of time. That is, execution of a line of Java code does not necessarily imply that time elapses in the system. In contrast, time plays an integral role in the definition and execution of Verilog and VHDL specifications, defined with respect to their simulation models. While time is appropriate for describing and simulating physical hardware, it is not desirable when the intent is to design at the behavioral level. As is explained later, it is possible to interpret a Java program in such a way that the semantics of languages like VHDL and Verilog can be embedded in the system, supporting efficient compilation to existing hardware synthesis back-ends.

All other things being equal, using an existing language for systems design is preferable to creating a new language from scratch. Development of a new language also requires the creation of supporting tools, such as compilers and debuggers. Furthermore, the language also needs to acquire a base of loyal users in order to bring the technology to maturity. If the needs of systems design can be met using an already successful language, the benefits afforded by having preexisting users, tools, and libraries makes it difficult to justify development of a newly design system description language. This is not to say that new languages should never be developed, only that the added capabilities or features provided must be compelling in order to justify the investment.

Almost all embedded systems must operate at speeds dictated by their external environments, and must be reliable and predictable. Therefore, embedded systems should behave deterministically and operate within bounded resources, including time and memory. However, neither Java programs nor digital hardware in general guarantee determinacy or bounded resource usage. To address this fundamental incompatibility, our approach anchors programs on formal models of computation. For example, the Abstractable Reactive (AR) model described below is one model we have developed that has properties suitable for embedded systems specification. Restrictions on the usage of Java are then applied to programs in order to ensure that they are consistent with the AR model, or with any higher-level model built on top of AR as a refinement. This principle of language restriction is one area where our approach differs most from other proposals to use Java for system specification, such as that of Helaihel and Olukotun[3].

However, restricting the use of Java limits its flexibility and expressiveness. Designers may be discouraged if the resulting language is overly restrictive. We have addressed this need by introducing an associated methodology for the development of specifications called *Successive Formal Refinement* (SFR). SFR supports the transformation of a program written in a general-purpose programming language into one that can be embedded within a more restricted model of computation. It consists of a series of static analyses of programs, coupled with an iterative process of incremental, semi-automated program transformation. SFR enables a system to be designed initially in an unrestricted manner using a general-purpose programming language and then be gradually refined into a restricted form for use as a system specification.

Design tools for heterogeneous systems must accommodate two apparently diametrically opposed forces: the drive for increased specialization and the drive for increased integration. On one hand, specialization entails more domain-specific tools and methods. On the other hand, integration encourages development of do-it-all design environments. *JavaTime* is a design philosophy that seeks to find a solution that adequately meets both needs.

In the following section we introduce and number of the fundamental concepts used in *JavaTime*. In particular, we define components, interfaces, events, and the basic model of computation used in components. The notion of time is then introduced, using a reactive model, and our approach is compared with the synchronous reactive model from which it has evolved. Abstraction and modularity are required in any modern design system; *JavaTime*'s current approach to abstraction is presented, especially with respect to abstracting variables in space and in time.

Finally, as mentioned earlier, the *JavaTime* system should not be considered an end in itself, but rather a platform we are developing to experiment with and evaluate a number of approaches to system specification, design, and analysis. The fun is just beginning!

2. Components, Interfaces and Systems

2.1 INTRODUCTION

In *JavaTime*, a *component* is defined as any fully encapsulated behavior. A component is analogous to an object in object-oriented programming in that it has an associated state, behavior, and identity. We use the term component, rather than object, to stress the fact that the implementation medium—logic, memory, software, reconfigurable logic, or some combination—is not a factor in the specification of the component itself. Access to components is provided via an *interface* and the interface is the *only* way to interact with the component[1]. A *system* is defined as one or more, possibly interacting, components and their associated *environment*. The environment specifies all of the external constraints and all possible inputs the collection of components might be asked to respond to and so closes the system.

Later, we introduce time into the design by defining a *reactive component* in *JavaTime* as a component that *continuously reacts* to it's environment at a speed determined by the environment. In many ways, a reactive component in *JavaTime* may be thought of as a fundamental-mode asynchronous process.

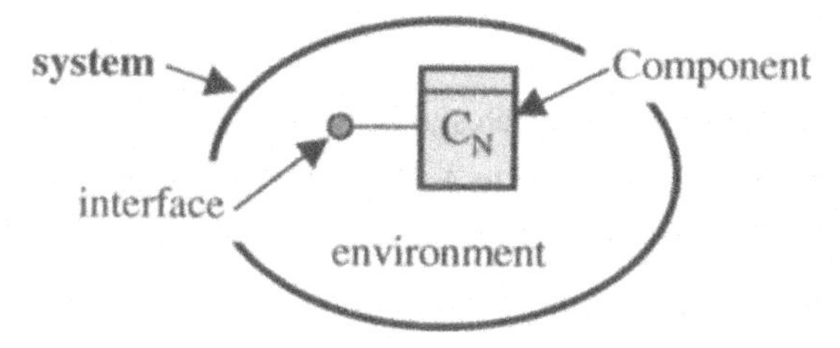

Figure 2.1: Specification versus Implementation

We maintain a semantic distinction between a *description*, that we will refer to here as *specification*, and the *described system*, that we refer to here as the *implementation*.

[1] Components may have multiple interfaces.

Loosely speaking, we use the terms to distinguish between the "what" (specification), that describes the behavior in terms of results, and the "how" (implementation), that defines behavior in terms of a set of procedures or actions. As an aside, most specification languages tend to be "nonprocedural" in nature, to avoid the need to include "implementation details" that are not germane to the task at hand (e.g. specific values for loop indicies, array bounds, a specific order of evaluation that is not actually needed for the specification). These details are "abstracted out" from the specification. In that sense, a conventional imperative programming language is far better suited to implementing a system behavior than specifying it—a point we refer to again later.

2.2 COMPONENT SEMANTICS

The behavior of any *JavaTime* component is modeled as a deterministic finite automaton (DFA), consisting of a set of internal states and a set of state-to-state transitions that occur on input events chosen from an alphabet Σ[4]:

$$C = (Q, \Sigma, \Delta, \delta, \lambda, q_0)$$

where Q is a finite set of internal states, Σ is a finite input alphabet, q_0 in Q is the initial internal state, Δ is the finite output alphabet, δ is a transition function mapping $Q \times \Sigma$ to Q, and λ is the output function mapping $Q \times \Sigma$ to Δ. In classical terms, this is a Mealy machine where the output function depends on both Q and Σ[2]. Q is used to capture the history of the component and so not all variables internal to the specification of a *JavaTime* component need to be included in forming Q. Selecting the minimum possible number of program variables to represent the internal state of the component is an optimization task. In traditional finite-state machine design, variables that are part of "feedback loops" are selected as internal state variables, a point we return to later.

For convenience, we also define the *total state t* of a *JavaTime* component as the concatenation of the current input value σ and the present internal state *q*. The behavior of a *JavaTime* component may then be written:

$$C = (T, \Delta, \delta, \lambda, t_0)$$

Where $T = Q \times \Sigma$ is the set of total states, δ is the transition function mapping *T* to *Q*, and λ is the output function mapping *T* to Δ. When the component is in total state *t* and its' input changes, it produces an output $\lambda(t) \in \Delta$ and transitions to internal state $\delta(t) \in Q$. Note that implied in the model is the fact that σ cannot change "in the middle of" a state transition—state transitions are atomic.

Not that in *JavaTime*, the alphabets accepted and produced by the DFA may contain values that represent a nondetermined value in any implementation of the machine. That

---

[2] This fact is required later for a reactive model.

is, while we can compute a value for the variable in the model, the value does not have a particular, concrete meaning in any possible final implementation of the behavior. This aspect of our approach is presented in more detail later.

2.3 INPUTS AND OUTPUTS

In *JavaTime*, the values associated with component inputs and outputs are considered persistent and change only with an associated change in the environment (change in system input) or a component state transition. That is, in a world where time was a factor (as introduced later), such values would be considered *levels*, rather than pulses, for example. We refer to the change in any value as an *event* on that value. Note that in our definition of event, the variable *must take on a different value* for an event to have occurred.

Definition: An *event* is defined as a change in the value of any variable visible in the system.

So far, we have considered a single input symbol and output symbol as if they were also atomic. Real components have multiple inputs and outputs, a fact that must be comprehended in the model. The simplest approach is to consider the input and output symbols to be comprised of the concatenation of all inputs and outputs respectively. In a classical DFA, one implicitly assumes that all inputs and outputs are updated concurrently and atomically. Such an assumption is supported directly by a synchronous model of computation, as described later, but in many real systems it is not practical or efficient to implement such an approximation. To support a broader range of possible system behaviors and consistent with a *fundamental-mode* asynchronous assumption, in *JavaTime* we only insist that no distinguishable inputs may change their value "simultaneously"—that input events remain distinct, albeit in an order that may not be deterministic.

That is, rather than making synchronization fundamental to the basic *JavaTime* model of computation, we have chosen a fundamental mode asynchronous model as a basis and use abstraction (via the Java syntax) to implement synchronization wherever it is appropriate or desired. A single input to a component might be a complex datatype (e.g. 16-bit integer), perhaps specified in terms of more basic types (e.g. Boolean). Events on the complex datatype are considered atomic and not decomposed to the Boolean level unless such a decomposition is implemented explicitly in the syntax. By aggregating variables into complex types, with a single complex input per component, a fully synchronous system specification can be constructed. While such an approach might seem clumsy at first, to date we have found it practical and the flexibility provided in description useful, as is illustrated in the examples later.

2.3.1 Resolving Values for Variables

As introduced earlier, in *JavaTime* the domains of all variables are defined as finite sets. Most system specifications are deterministic, often built on top of an abstract model that guarantees determinism (e.g. a traditional computer instruction-set architecture (ISA) implementing a procedural language). In the real world it is possible to implement nondeterministic behavior and in some situations it is necessary to specify systems that are nondeterminsitic or that contain nondeterministic components (e.g. latches, native Java threads). Therefore, in *JavaTime* we extend the Java syntax to support nondeterminism in the basic *JavaTime* model.

It is common practice in a discrete-valued, asynchronous world to use fixed-point semantics to give meaning to circuit behavior and that is the approach we adopt in *JavaTime*. Specifically, we use the approach described in detail by Edwards[5] for choosing the least-defined solution for the value of an observable variable. Since we admit the values "undefined" (represented as $\perp$, or bottom) and "unspecified" for all datatypes in *JavaTime*, and so all inputs to any component are always "known", all *JavaTime* components are monotonic. As introduced by Kahn[6], by using a fixed-point approach and insisting that all possible values for variables form a complete partial order(CPO)[7], the solution for such a fixed-point calculation is guaranteed to be unique[3]. Of course, a standard Java Virtual Machine doesn't implement fixed-point semantics for resolving the values and so in *JavaTime* we extend the Java syntax via class libraries and use an approach we call Successive Formal Refinement (described later) to introduce the least-fixed-point approach.

3. Introducing Time

> *"We have diverse curious clocks and other like motions of return, and some perpetual motion... We have also houses of deceits of the senses... These, my son, are the riches of Solomon's House."*
>
> *—Francis Bacon, The New Atlantis*

3.1 INTRODUCTION

One of the advantages of a Java program is that it does not have any built-in notion of time. However, in order to use Java as a syntax for describing real-time embedded systems, a mechanism for representing time and the interaction of a component with an accurate model of the physical world must be introduced into the model of computation and then into the *JavaTime*[4] syntax. In the following sections, our basic notion of time is introduced and is contrasted with the synchronous reactive model. Consistent with our overall philosophy, we introduce a very simple and flexible model of time as a basic model and then use syntactic restrictions to build more abstract, domain-specific temporal models. We treat specific restricted models of time as a "technologies," much like specific gate libraries are restrictions on the way a particular function can be implemented in the space dimension. In that sense, we speak of "temporal mapping" of a

---

[3] See [5] for a detailed explanation and proof.

[4] It is the introduction of time as a part of a Java-based specification that led to the name *JavaTime* for our system.

behavior to a specific temporal model (e.g. clocked synchronous, synchronous reactive, synchronous dataflow) much as a logic synthesis methodology speaks of "technology mapping" to a specific technology library. In fact, ultimately we believe both the spatial and temporal aspects of such mappings should be implemented concurrently.

3.2 Real Time and Embedded Systems

There are many definitions of real-time systems in the literature today. The one thing they all have in common are the notions of response and of time—the time it takes for a system to respond at it's ouput(s) to a change in some associated input(s). The Oxford Dictionary of Computing [8] defines a real-time system as:

> "Any system in which the time at which an output is produced is significant. ... The lag time from input time to output time must be sufficiently small for acceptable timeliness."

Young[9] defines a real time system to be:

> "...any information processing activity of a system which must respond to externally generated input stimuli within a finite and specified period."

The ESPRIT Predictably Dependable Computer Systems Project (PDCS)[10] uses the definition:

> "A real-time system is a system that is required to react to stimuli from the environment (including the passage of physical time) within time intervals dictated by the environment."

Practitioners in the field of real-time systems often distinguish between hard and soft real-time systems. In a hard real-time system, it is absolutely imperative that the system respond before the specified deadline, while in a soft real-time system, response times are important but the system will still function correctly if deadlines are occasionally missed.

In our work, we are after guaranteed system correctness and so we take a conservative approach and assume hard real-time behavior is required in all instances.

Definition: A *real-time system* is defined as any system in which the behavior of the implementation may be affected by the time taken to perform a computation.

Note that this definition considers the time taken to perform a computation—the time taken from inputs to outputs for a component—and does not speak directly about the "time at which" an output is produced (which implies at absolute reference). We consider such absolute time requirements as a special case, where an absolute time reference is used to describe temporal constraints among components or in relation to activity external to the system.

The main features of a correctly implemented real-time system seem to be that it involves concurrency, generally exhibits deterministic behavior, is subject to some strict time requirement by its environment or its own internals, has a reliability requirement, and is implemented as a combination of hard and soft components.

3.3 REACTIVE SYSTEMS AND MRT

Reactive systems are defined as systems that *continuously react* to their environment at a speed determined by the environment. The term *reactive system* has been proposed in a variety of places [11] [12] and was introduced to distinguish such systems from what we refer to as *simulation systems*, where the time taken to produce outputs is not a factor in determining the behavior of the system[5]. Reactive systems are distinguished from simulation systems by virtue of being in constant and continual interaction with their environment and being able to execute sufficiently quickly that they are never overdriven by their environment. Their very behavior is defined by these interactions and by the fact that they do not halt. Simulation systems also interact with their environment as well (e.g. word processors, operating systems, logic simulators), but the time taken for such interaction is not a factor in determining system behavior. In our sense of real-time introduced above, reactive systems represent one of a number of possible approaches to meeting the needs of real-time embedded systems design and representation.

> *"All objects perceptible to our senses, all phenomena of whatever kind and whatever aspect they may assume, are constituted by a rapid succession of instantaneous events... the movement is intermittent and advances by separate flashes of energy which follow each other at such small intervals that these intervals are almost non-existent."*
>
> *... Alexandra David-Neel and Lama Yongden, The Secret Oral Teachings in Tibetan Buddhist Sects, City Lights Books, San Francisco, 1967.*

Central to all reactive approaches is the notion of an *instant* (logical instant), where zero or more events may occur and events that occur at the same instant are considered simultaneous. Apart from these instants, nothing happens in either the system or its environment. However, like the Buddhist scholars of ages past, all that is actually required here is that the instants be "fast enough". As explained in [13], the "entire universe could be destroyed and reassembled between consecutive instants—all we are aware of are the instants themselves."

To this end, in the context of hard real-time described above, we prefer to think of there being a *Minimum Resolvable Time* (MRT)[14], that is defined as a finite amount of time

---

[5] Such simulation systems are sometimes subclassified as *transformational* and *interactive*[16] but the key idea in both cases is that the passage of time does not play a role in determining overall system response.

but represents the highest temporal resolution for the system under analysis at a moment. That is, MRT represents the minimum amount of time that can pass to guarantee that the system behaves in an atomic, reactive manner. There is no way of knowing what actually happens within an MRT step, only what the state of the component or system was at the beginning and then at the end. No particular duration is prescribed for MRT, nor is it necessarily a constant for a component. The temporal window defined by the value of it's MRT can be thought of a s a *time slot* (we refer to the window as a *timeslot*.) It is simply a temporal ordering property for the behavior of a component or a system. Later in the design process, specific values *are* assigned to MRT in a "temporal technology mapping" phase, where a particular "temporal technology"—entirely under our control—is assigned to an implementation of the system.

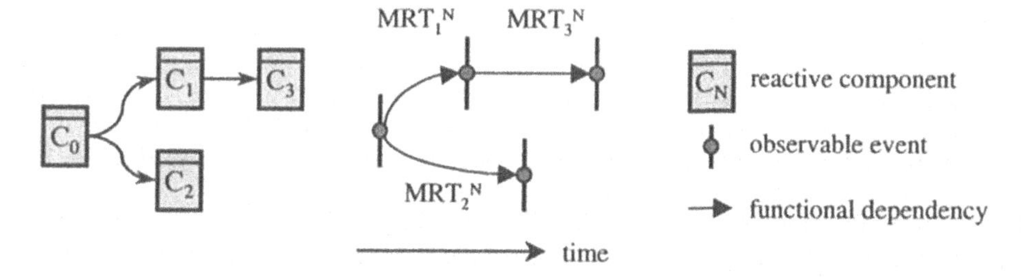

Figure 3.1: A set of components and their interactions, and a schematic illustration of the temporal relationships among the components.

Another important implication here is that values can only be observed at instants, since there is no time "less than" an MRT. That is, output "probes" require the passage of time and must act like a sample-and-hold circuit—across a timeslot—to permit the external observation of a value. In *JavaTime*, all probes are assumed to have reactive semantics.

There are many other ways of representing system component behaviors where time plays a role in determining overall system behavior (e.g. clocked, synchronous systems where the clock runs "fast enough.") Central to the reactive idea is the notion of "continuously reacting," and while such behavior could be "simulated" using other models, the reactive model is appealing especially because it represents behavior in a way that is very close to the behavior of actual hardware. In that sense, a reactive representation of a complex, real-time, asynchronous system can almost always be "compiled" into hardware very efficiently.

Synchronous semantics and associated languages have been proposed recently for reactive systems[15][16]. Synchronous semantics go a step further, by requiring an entire system to compute "infinitely quickly," where reactions can be thought of as being "instantaneous and atomic." Again, the requirement for infinitely fast computation can be replaced by a "fast enough" requirement but the entire system is evaluated to convergence at an instant before time is advanced. One of the most significant advantages of the synchronous approach is that it avoids one of the major sources of nondeterminism in system descriptions: the nondeterministic result of multiple possible orders in a computation caused by asynchronous processes competing with each other for resources.

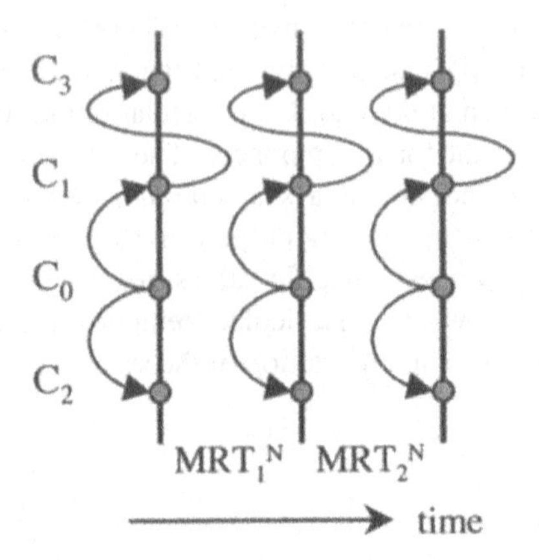

Figure 3.2: The behavior of the components of Figure 3.1 under the synchrony hypothesis

It is interesting to note that David-Neel's *Secret Oral Teachings*[13] suggest that for many centuries Buddhist scholars have believed the entire universe is best thought of as a synchronous reactive system!

In the synchronous reactive systems literature, the notion of simultaneity is captured by the concept of an event. An event is defined as a set of simultaneous occurrences of (possibly valued) signals and a particular run of a program is a sequence of such events called a *history* [16].

JavaTime implements a reactive approach to the representation of time that is not inherently synchronous. However, by aggregating component inputs and outputs into a single, abstract datatype, synchronous behavior can be implemented in a straightforward manner.

3.4 Abstraction

In today's complex SOC world, abstraction is an essential tool for the management of design complexity, design re-use, and for the "safe" integration of components from a variety of source. Components must be able to be abstracted in their interface in a consistent manner that facilitates composition.

Abstraction is a general methodology that enables the isolation of how a compound object is used from how it is implemented. Programs and modern hardware are typically designed to model and interact with complex phenomena and so usually require the designer to build a system with many parts. Abstraction is used to elevate the conceptual level at which a system is designed, to increase modularity, and to enhance the expressive power of language.

At the level of a single component, the abstraction of a *JavaTime* component is the set of ports and attributes seen in the component's interface. For a basic *JavaTime* component, one of those attributes is the behavior of the component, implemented as a Java program. The component's behavior is defined in terms of a fundamental-mode asynchronous Mealy machine, where only a single input is permitted to change in an instant and all internal actions are atomic. That is, it is the responsibility of the implementation environment (*JavaTime* simulator, synthesis tools, compilers, etc.) to ensure that the fundamental mode restrictions are met in the implementation of the component. Note that the order in which the inputs appear is likely to be nondeterministic.

Compound components are comprised of instances of other components that interact via signals. Events from the output of one component are copied to the input of another as they occur. However, for abstraction to be of value, not every event present within the body of the component may see the light of day outside that component. To be precise, in Figure 3.3 below, when an external event occurs on Port A of Component C_T, Component C_N reacts and (possibly) produces an event on it's output. Other components in C_T will continue to react until the behavior within C_T converges to a fixed-point under the CPO-based approach presented earlier. Although events may have been issued to Port B, they are "filtered"[6] in the sense that the values of the signal associated with the port are updated but the event itself is not propagated beyond the component. Once the component has converged (albeit to a possibly "unknown" value), if the value on B has changed, then and only then an event is issued from B into the external environment and the components it affects are processed.

A corresponding schematic *JavaTime* event diagram is shown in Figure 3.4. This illustrates schematically the process outlined above. In this example, components *N* and *M* were evaluated twice before they converged to a consistent state.

In a sense, the temporal resolution within the component is higher than that of the external environment for a particular evaluation of a particular component. This is by no means a limitation and actually provides a convenient mechanism for both propagating actual implementation timing from the bottom up and by providing performance requirements to a component from the top down.

We refer to the combination of using a fundamental-mode (reactive) asynchronous representation of component behavior, coupled with the techniques for abstraction introduced above, the Abstract Reactive model. This is the basic component model and approach to abstraction implemented in the *JavaTime* system.

---

[6] In the current implementation, the port is marked internally "inactive" during this period.

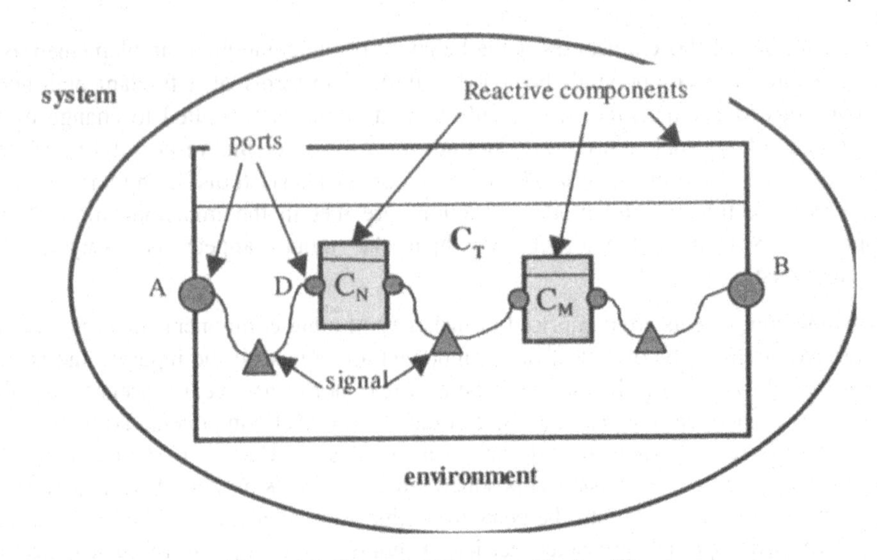

Figure 3.3: Compound Component in *JavaTime*

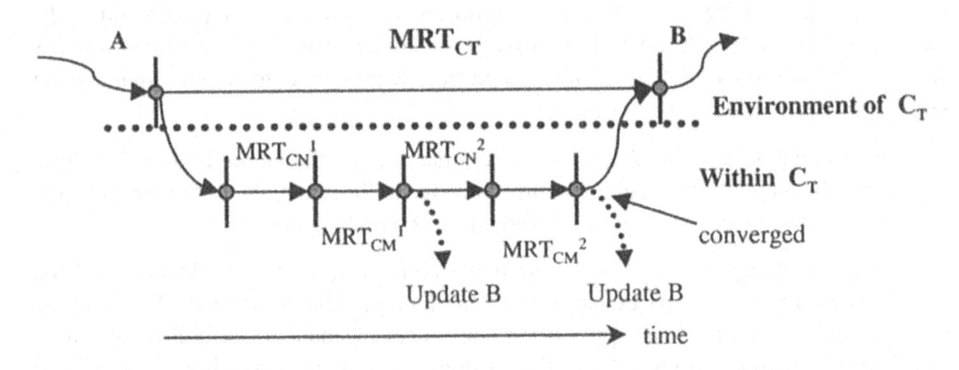

Figure 3.4: A Schematic Event Diagram for Example 3.3 Above

4. Successive, Formal Refinement

The *JavaTime* successive, formal refinement methodology rests on two foundations: a *policy of use* on the source language, and a procedure of *incremental transformation*. In SFR, a policy of use is imposed on the source language, *S*, to make it consistent with the target model, *T*. A policy of use consists of *restrictions* and *extensions*. The restrictions removes portions of *S* incompatible with *T*, while the extensions introduce semantics present in *T* that have no equivalent in *S*. The result is a new language *S'*, as shown in Figure 4.1, whose programs are expressible in *T*.

In this methodology, a user's initial program *P* is contained within language *S* (Figure 4.2(a)) and cannot be mapped directly onto *T*. A process of incremental program transformation is used to refine it into a program in language *S'*, to make it valid with respect to the policy of use.

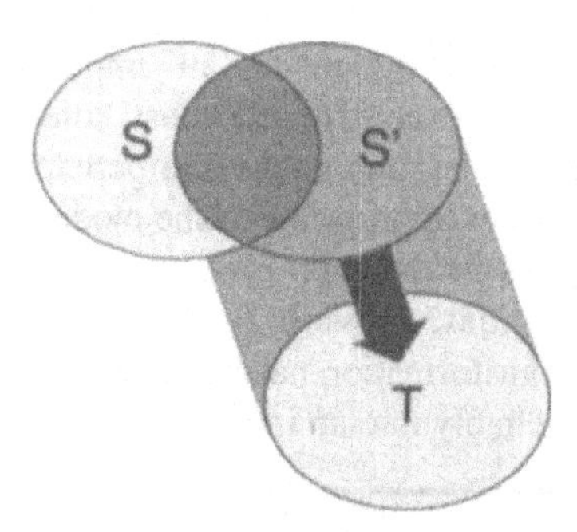

Figure 4.1: A policy of use applied to language S yields S', compatible with model T.

The program is analyzed to verify that the rules in the policy of use are satisfied. If a violation is found, the user is presented with information regarding the nature of the error, and a list of suggested solutions for fixing the problem, including automated program transformations when possible. The user can then modify the program manually or allow the tools to alter it automatically.

This process of analysis and modification is repeated until the program complies with all rules in the policy of use. At this point, the altered version of the program, *P'*, is contained within *S'* (Figure 4.2(b)). Because *S'* is constructed to be compatible with *T*, *P'* corresponds to a system in *T*.

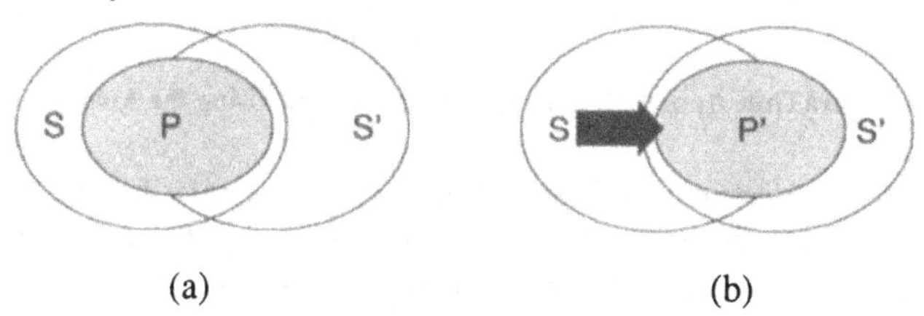

Figure 4.2: (a) Original program is in language S.
(b) Program is embedded in S' through successive, formal refinement.

The SFR process may be applied with respect to a variety of target models, each with its own policy of use. This enables one to use a single general-purpose language for designing systems in an implementation-independent manner, and choose from different models according to the desired implementation style. In addition to the AR model described in these notes, others might include dataflow, discrete-event, and hierarchical finite state machine models, or models defined by the user. This provides a design language and environment that is highly expressive, and yet can be customized to suit the needs of a particular implementation medium.

The overall process we envisage for SFR is illustrated in Figure 4.3. The user writes a Java program, P_J, and "designs" the application using a normal JVM and Java development environment. Once the program has been debugged in the JVM, the user chooses an appropriate target model(s), *T*, of computation (hardware, software, FSM, RTOS, etc.). The user is now in the "implementation" phase. Ideally, the user would like to automatically "compile" the program to an implementation in terms of T. Unfortunately, that is impossible except in the most constrained of cases, since the original Java program is likely to be ambiguous and perhaps nondeterministic in terms of the model T. Using a set of rules appropriate to the model T loaded into the *JavaTime* analysis engine, the user applies incremental program analysis and automatic or semi-automatic program transformations, moving P_J towards another Java program, P_R. However, if the analysis and transformation has been performed correctly, Java program P_R can be mapped easily and reliably into an implementation in terms of T.

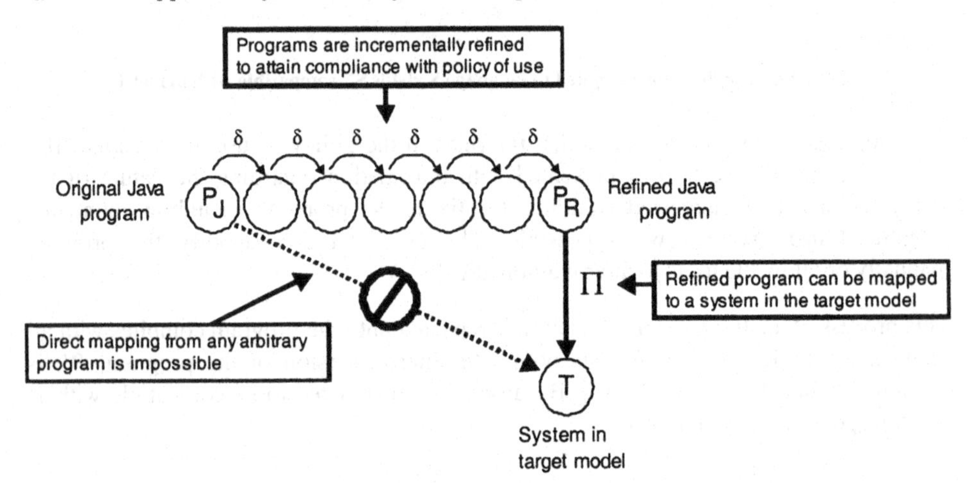

Figure 4.3: JavaTime Approach to Program Refinement Using the Analysis Engine

Systems for program construction through transformations exist for the development of software[17,18], but ours is the first effort we know of to apply such techniques for the purpose of formal modeling of mixed hardware/software systems. Transformation systems for software typically are used to improve program performance while preserving semantics, while in the SFR methodology transformations are used to restrict and alter a program's semantics.

5. Interpreting a Java Program in *JavaTime*

Both software compilers and hardware synthesis tools (also compilers) use a conservative interpretation of the relationship among clauses in the text of the component specification to determine where optimization is possible. For performance optimization, one of the most powerful transformational techniques available to such tools is the use of concurrency. To that end, both hardware and software compilers use dependency analysis to determine the optimal order(s) in which computation may be performed while still satisfying the intent of the programmer under the semantics of the model of computation associated with the specification language. While combinational models and sequential models that operate under strict causal or temporal constraints *may* guarantee a deterministic specification that can result in only a single implementation behavior under the constraints, most concurrent languages and distributed computing systems are based on asynchronous execution schemes, where processes compete for resources, and where this competition is resolved nondeterministically.

The meaning of a text string that conforms to the syntax of a Java program is determined operationally by its interpretation by a Java virtual machine and for an associated set of inputs and expected outputs. That is, the underlying model of computation (the JVM and its associated run-time class libraries) is central to the program's meaning. In *JavaTime*, to begin with we assume that the Java program itself represents simply a *minimal partial order* on the computation. As for Java and for any simulation language, we do not account for any particular execution speed or performance for any particular aspect of the program.

For example, the block:

```
{
    a = 1;
    b = 2;

    c = a;
    d = b;
}
```

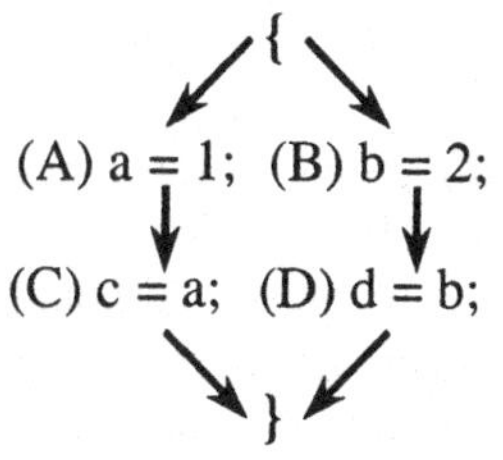

```
              {
           ↙     ↘
(A) a = 1;   (B) b = 2;
     ↓            ↓
(C) c = a;   (D) d = b;
           ↘     ↙
              }
```

can be represented by the flow graph on the right. Here, precedence (causality) requires that the assignment (C) follows assignment (A), but says nothing about the order in which (D) and (C) should occur. Those assignments are nondeterministic under the assumptions of a minimal partial order.

Clearly, good component specifications are written in such a way that any nondeterministic aspects of the specifications are irrelevant in the external behavior of the component, as is the case in statements (C) and (D) or (A) and (B), etc. above. As in the example above, concurrency is often introduced only by the compiler as an optimization and a conservative approach to determinism is used, where the concurrent program faithfully simulates its sequential ancestor. However, when concurrency is an explicit aspect of the system specification and is placed in the hands of the designer, it is very likely to introduce errors. This is especially true when many of the components that execute concurrently are imported from other sources, as must be the case for extensive component re-use. To address this problem, many different approaches have been proposed over the past quarter century to limit the designer's ability to introduce errors, as well as to discover any errors in a formal way should they be introduced.

As stated earlier, in the *JavaTime* framework, we implement a heterogeneous approach by using a very unrestricted, homogeneous underlying model and then use abstraction in the system (via class libraries and syntactic extensions) to implement specific, higher level modeling domains, as needed by the designer.

5.1 THE *JAVATIME* DEVELOPMENT ENVIRONMENT

The *JavaTime* system is a collection of libraries and tools for systems design that uses the Java language as the syntax for design representation and couples the Java environment with multiple models of computation using the successive, formal refinement methodology described below.

While Java is used as a syntax for representing designs, it must be emphasized that the goal of *JavaTime* is not to develop a one-button solution that instantly generates a system from any Java program. Rather, the goal is to implement domain-specific languages in the Java language and couple them with a methodology for refining a Java program into such an embedded language form. In all cases, only legal Java programs are involved in the process.

Fortunately, the Java language provides a simple and useful means for extending its capabilities, through the use of *packages*. This allows features to be added without requiring modifications to the language's grammar. An illustrative example is Java's threads package. Multiple threads are created and managed by invoking methods in the `java.lang.Thread` class. Although syntactically identical to any other method invocation in a program, these methods provide the only means to control thread execution. Of course, in order to adequately support threads, Java development tools must understand the additional semantics implied by the use of the threads package.

In *JavaTime*, we have extended the Java package mechanism include different models of computation. Special meanings may be attached to classes or methods in a package, thereby extending the usefulness of the language without altering its grammar. Just as it is understood that a C program begins execution in the `main` function, or that the `run` method embodies the body of a thread in the Java thread package, additional semantics are implied by the *JavaTime* packages. As long as these added meanings are understood by both the designer and the *JavaTime* tools, this provides a convenient way to help embed different models of computation within Java.

However, the package mechanism by itself is insufficient to embed different models within Java. There must exist a way to ensure that the package is used as intended. Otherwise, if the user is permitted to write any arbitrary Java code, there is no way to ensure that the behavior remains consistent with (implementable in terms of) a particular model of computation. Whe have developed the Successive Formal Refinement (SFR) methodology and have developed tools in *JavaTime* to statically analyze and transform Java source code, in order to ensure that the user written code follows the intended use of the package and so help the designer to produce a conforming specification.

Hence, packages are used to implement the language extensions of a model's policy of use in the SFR methodology, while the *JavaTime* program analysis and transformation engine performs the language restrictions.

5.1.1 A Component-Based Approach

Designs in *JavaTime* are based around the concept of a *component*. A component is an entity with one or more interfaces[7], each of which contains the entire abstraction of a component for a particular use. Each interface contains a number of associated *attributes*, as well as the definition of a set of *ports*. Attributes may include, but are not limited to, the code describing a component's behavior, its control-dataflow graph, one or more graphical representations (symbols), and measured or derived properties. Components are derived from *prototypes*, which specify attributes that are associated with a given category of component, rather than those specific to a particular instance. For example, the behavior specification will be shared by all components of a given type (e.g., 32-bit adder), although different instances may have their own unique attributes (e.g., name, usage statistics, performance requirements). Prototypes and components are implemented as Java classes that store a list of attributes, which are handles to other Java objects.

One essential attribute of any component is its behavior. The behavior is represented as Java source code. Another central attribute of a prototype is its model of computation. In general, the model of computation will correspond to the intended implementation medium. It also determines the model-specific package that the behavior specification must utilize. For example, if the intended target is a real-time operating system (RTOS), the specification must use the accompanying package which allows designers to express RTOS-specific behaviors (e.g., RTOS system calls). Similarly, if one wished to design a clocked, synchronous system, one would use a package that provided a global synchronous clock mechanism.

5.1.2 JavaTime Design Flow

The life of a typical system specification begins as a standard Java program, developed initially using common Java development tools. To enter it into the *JavaTime* system, an intended model of computation is selected by the user, and the Java program is modified so that it uses a Java package that corresponds to the selected model. This usually involves altering the inheritance hierarchy of the program so that classes from the package are used as base classes.

The design is then analyzed, executed, visualized, and transformed using the *JavaTime* tools. Analyses and transformations are performed by the *JavaTime* component analysis and transformation engines. Analysis ensures that the design is in compliance with the policy of use of the model of computation chosen, and transformations can be applied to aid in resolving violations.

---

[7] Interfaces in the *JavaTime* sense, not in the Java programming sense!

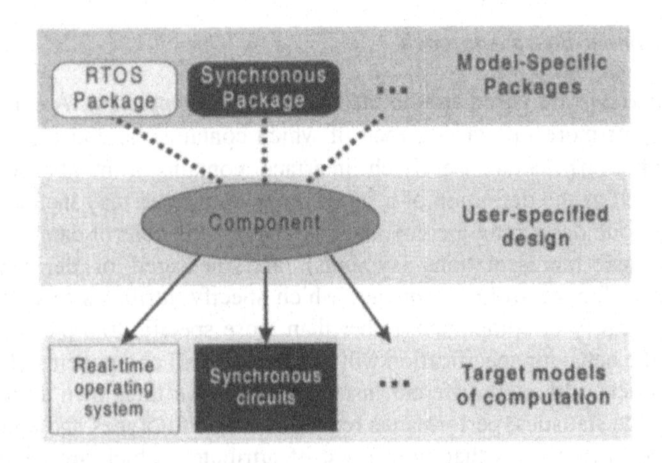

Figure 5.1: The Three Key Elements of a JavaTime Specification

Execution and visualization are implemented on a standard Java virtual machine, and are used to animate the system under development and to provide feedback to the designer regarding its behavioral characteristics. Among the properties that a designer may wish to learn about are the degree of concurrency in the system, dynamic memory usage, or power dissipation.

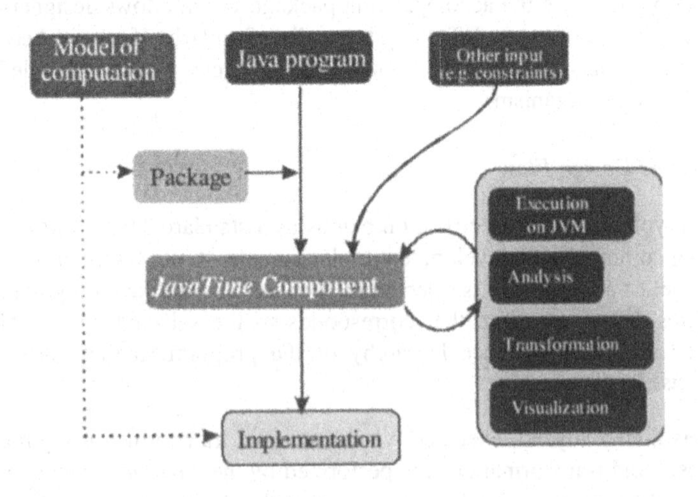

Figure 5.2: *JavaTime* Component Development

The behavior of a component is described with respect to the *JavaTime component model*. In the model, a component behaves by reacting to *events* (changes in values) received on its *ports* from *signals*. Signals play the role of holding the values shared by a number of ports. Components can have any number of ports, and ports may be created or removed dynamically in the system. A port can receive events from its component (output events from the component) or from a set of possible *source* ports (inputs to the component) and it can emit events to a set of *receiver* ports. When a port receives a new event, the component that owns the port is triggered, causing it to *react*. During a reaction, a component may emit events to one or more of its ports, but is not required to do so. As mentioned earlier, a component's previous reaction must terminate (converge) before the next reaction may occur.

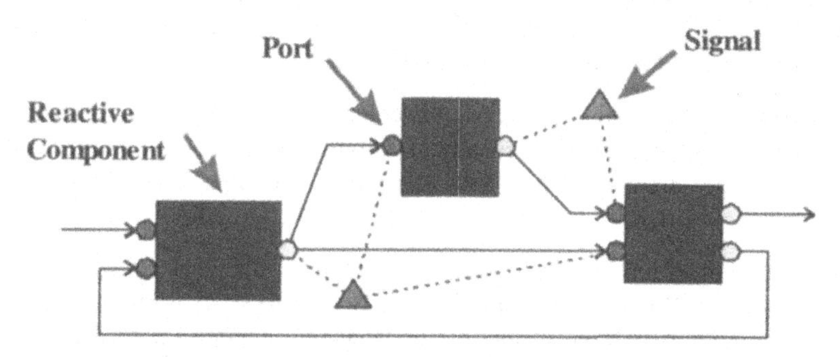

Figure 5.3: The Relationship Between Components, Ports, and Signals

Events are never "lost" by the system; that is, an event that is emitted to a signal will always be received by its intended receivers.

The *JavaTime* component package (JCP) is the concrete realization of the abstract component model in the Java language. The JCP is a Java package, or collection of classes, that implement components, ports, and signals, and allows the model to be executed on a standard Java virtual machine.

5.1.3 Component class

The most important class in the JCP is the `Component` class, from which all components are derived. Components hold a list of ports and a component reacts whenever an event is received on the port. Signals can be handles to any Java object. While ports are not typed in the current implementation, they can be configured to respond only to signals of a particular class and its subclasses. In the *JavaTime* vocabulary, the class describing the component is called the *prototype*, while an object of that class is called a *component*.

Compound components

A component may be described as a composition of references to the behavior of other components and may not contain a separate behavior written directly in Java. Such *compound components*, also called *networks*, provide the mechanism for hierarchical abstraction of designs in the JCP. Networks have ports like any other component, but no behavior of their own. Rather, the signals emitted to a network's ports are simply passed to the components contained within the network. Networks are themselves components, so networks may contain other networks, creating a tree-structured hierarchy.

Copyable Objects

In hardware, latching values and transmitting them over wires "copies" the value in the register. In hardware, copying values is natural and cheap. On the other hand, multiple references to an object with shared state requires wires to be run from each use of the object to the part of the chip where it has been placed. References are, therefore, very expensive. In conventional software systems, the exact opposite is true. References (pointers, handles, OIDs, etc.) are cheap while copying values is expensive. Java provides a mechanism for copying references but not data[8].

While copying references in software is convenient and cheap, it complicates program analysis tremendously and severely constrains a compiler's ability to optimize code. The computer language community has been struggling with this issue for more than a quarter century.

There are two approaches that can be used: (1) allow copying of references, and then use program analysis to ensure that an object is private to some portion of the code or (2) disallow copying of references and define language semantics such that true copies are always made, and then use analysis and run-time techniques like reference counting to try to avoid actually copying objects and just copy references instead. In *JavaTime*, the former approach is taken, in the interest of efficient execution in the Java runtime, as copying objects is an expensive operation.

Ports

In Java, multiple threads communicate through shared objects, which is anathema to hardware since it is highly non-deterministic and difficult to locate through analysis. To resolve this problem, in *JavaTime* we use ports as communication channels that can be used to transmit values between threads, whether the threads are running locally on the same Java virtual machine or in different virtual machines that may reside on different processors connected on a network.

5.1.4 Execution

To execute components described using the JCP, an instance of the component must be created. Then, signals can be emitted to the component's ports, which will cause the component to react and emit signals in response. Note that since the specification behavior describes a partial order of events, there may be many possible behaviors, only one of which will be revealed by any particular execution or simulation.

---

[8] In Java, the 'clone' function copies the contents (values) of objects.

5.2 *JAVATIME* APPLICATION ARCHITECTURE

The *JavaTime application* is a set of software tools that supports the *JavaTime* design flow. These tools allow designs to be edited, executed, and analyzed. One novel feature is that the application itself is composed of *JavaTime* components.

The structure of the application is quite different from that of most programs. Rather than using a centralized, monolithic structure, *JavaTime* comprises a collection of components, interacting over a shared communications bus. This design can provide all the features available in a centralized program, but also provides improved scalability, modularity, and extensibility.

5.2.1 Communications bus

The communications bus allows the various components of the application to communicate with one another. Rather than having to identify and locate other components itself, a component simply attaches itself to various *channels* of the bus. The component can then send a signal to the channel, that is then broadcast to all other components attached to that channel. The bus model is attractive for situations where a component must send signals to multiple, unknown recipients.

As shown in the figure below, the *JavaTime* application is constructed by attaching different application components to the bus. Application features may be added or removed through addition or subtraction of components. The user interface actuates the user's commands by emitting the appropriate sequence of signals to the channels on the bus.

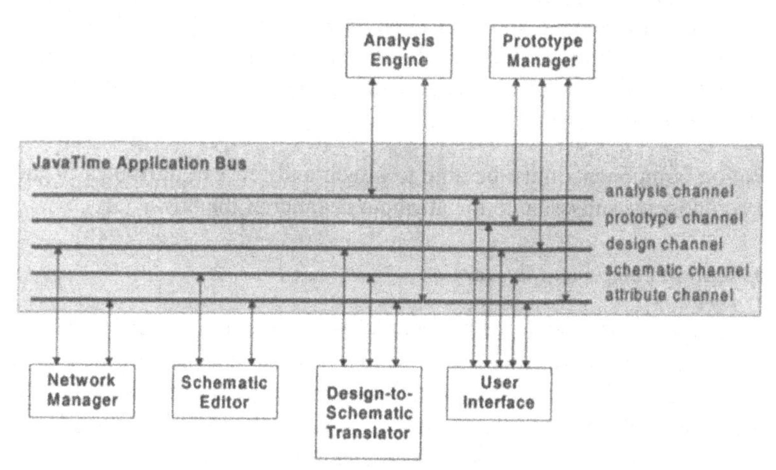

Figure 5.4: The Architecture of the *JavaTime* Development Environment

The implementation of the bus consists of two classes, `AppBus` and `AppAdapter`. `AppBus` maintains the list of channels, that is global to the application, while `AppAdapter` is an utility class for attaching application components to the bus.

5.2.2 *Application components*

The *JavaTime* application consists of a number of components that communicate through the application bus. The major components are the network managers, the prototype managers, the analysis and transformation engine, schematic managers, and the user interface.

At the top level, a design is represented in the application as a *JavaTime* network. Networks are components that serve as containers of other components. Each network has an associated network manager that maintains all the attributes associated with the network's child components. Network managers are the means by which other application components may view and modify the contents of a network.

Prototype managers are responsible for maintaining the attributes associated with prototypes. Prototype attributes are those that are common to all components of a given type. Such attributes include the control-dataflow graph, visual representation, or user documentation.

The program analysis and transformation engine is integrated into the application as a component. The engine acts as a service, accepting and carrying out requests for analyses and transformations from other components in the application.

The schematic editor component enables graphical editing of networks, and allows components to be easily added and connected together. The user interface component is responsible for presenting an integrated interface for the user. It acts as a central coordinator that sends the necessary commands to other components in the application to execute the user's directives.

All application components must be able to attach and retrieve attributes of the design. Hence all components subscribe to the attribute channel of the bus.

5.2.3 *Application component interaction*

Application components interact with one another by broadcasting and receiving signals over the bus. A component attaches itself to one or more channels of the bus that carry signals that the component is interested in. The signals that are sent over the bus may be commands, status updates, or service requests. In general, the usage varies from channel to channel. Consider two examples of how channels are used in the application: the attribute channel and the design channel.

The attribute channel is used to request attributes associated with prototypes and components. Attributes are a general mechanism for annotating component properties, and are used throughout the application. The protocol that components using the attribute channel must implement is straightforward: signals sent over the attribute channel are of class `AttributeAction`, and contain the information to be communicated to other components.

A component requests an attribute associated with a prototype or component by creating a new `AttributeAction` object. The new object is initialized by passing the appropriate parameters to its constructor. The parameters are: the action code (in this case, the action is "get attribute"), the handle to the prototype or component whose attribute is desired, the name of the desired attribute, and a port on which to return the desired attribute. This `AttributeAction` object is then emitted to the attribute channel on the application bus, and is received by all other application components attached to the channel. If one of the listening applications contains the desired attribute of the specified prototype or component, it emits the attribute to the given return port. This sequence of events is depicted in Figure 5.5. If no component contains the requested attribute, then no signal is returned. If multiple components contain the attribute, then the requestor will receive multiple responses, one from each responding component. In practice, the *JavaTime* application is constructed such that at most one application component will respond to an attribute request.

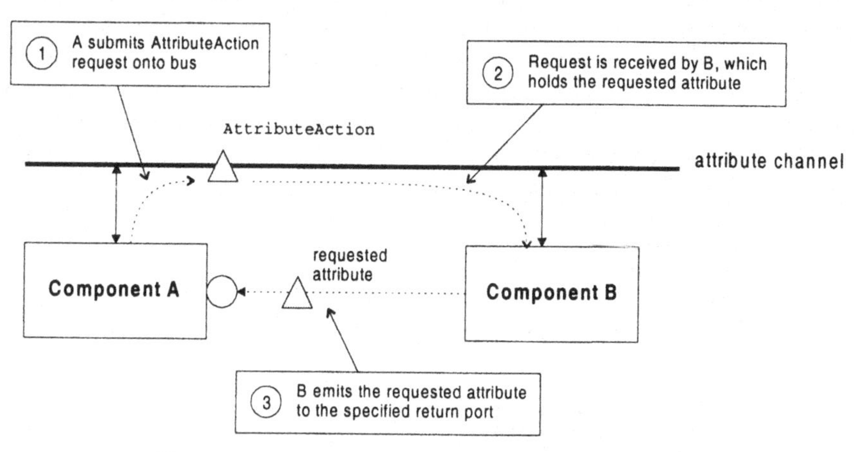

Figure 5.5: Sequence of events in attribute request protocol.

The other type of attribute action is for setting attributes. This initializes or modifies the attributes of a prototype or component. Again, an `AttributeAction` is sent to the attribute channel of the bus, initialized with the appropriate information. In this case, the action code is different ("set attribute"), a handle to the attribute is provided, while the return port is not provided. When the `AttributeAction` signal is emitted to the bus, components attached to the channel that are responsible for managing the specified prototype or component's attributes (either a prototype manager or network manager) will update their own internal data structures to reflect the addition or change of attribute. If no component on the channel is interested in that prototype or component, then the signal is ignored, and a subsequent request to get that attribute will yield no respondents. The component emitting the "set attribute" signal has no means by which to ascertain whether any components responded to its request other than to emit consecutive "set" and "get" actions.

The broadcast communications model of the bus enables servicing of attribute requests to be anonymous. That is, the requesting component need not know the location of the component(s) that fulfills its requests. The bus can also be used for more specialized, point-to-point communications between application components. For example, the application bus is used by the network manager and schematic editor components to communicate changes to the design made via graphical editing.

In the *JavaTime* application, a design is stored in a network manager as a collection of components. The design may be modified by the schematic editor, that uses its own internal represntation of the design, as a collection of graphical nodes and edges. The challenge is to maintain consistency between the two representations of the design, without building a centralized data structure, that would compromise the *JavaTime* system's scalability.

To solve this problem, the components use the application bus to communicate any changes made to their data representation to other components, that then update their own representations to reflect the changes. Network managers emit `DesignAction` signals, that declare any modifications to the network. Changes that are declared by `DesignAction` signals include adding, removing, connecting, and disconnecting components. Network managers both emit and receive `DesignAction` signals. It emits them when it makes a change itself, and if it receives one from elsewhere, it will update its internal data to reflect the change.

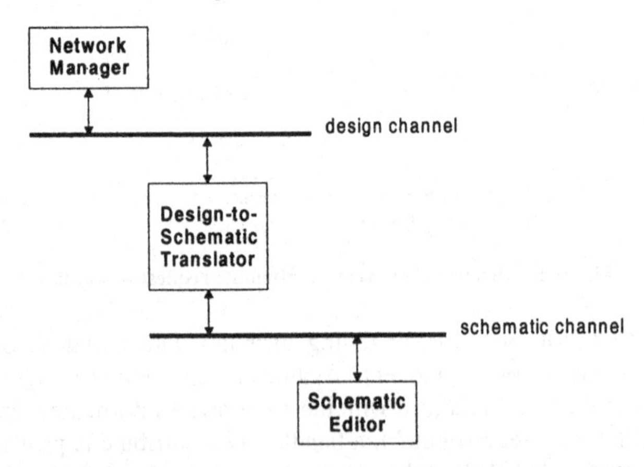

Figure 5.6: Consistency between a design and its visual representation is maintained by the Design-to-Schematic Translator.

Similarly, a schematic editor will emit a `SchematicAction` signal to reflect changes made in the schematic representation of a design. In this case, changes declared include adding/removing, repositioning, and selecting nodes and edges.

`DesignAction` signals are emitted onto the design channel of the application bus, while `SchematicAction` signals are emitted onto the schematic channel. A component called the design-to-schematic translator bridges the two, by transforming `DesignAction` signals to the equivalent `SchematicAction` signals, and vice versa, as shown in Figure 5.6. For example, if a node is deleted from the schematic, a `SchematicAction` will be emitted to the bus. The translator receives this signal, looks up the component associated with that node (stored as an attribute of the node), and then generates a `DesignAction` that will cause the component to be removed from the network. Of course, some actions, such as repositioning of nodes on the screen, need not be translated into `DesignAction` signals.

As illustrated above, the communications bus plays a very important role in the architecture of the *JavaTime* application. It enables application components to implement one-to-one, one-to-many, or many-to-many interactions, while relieving components from the need to have a global view of the entire application.

5.3 THE *JAVATIME* ANALYSIS ENGINE

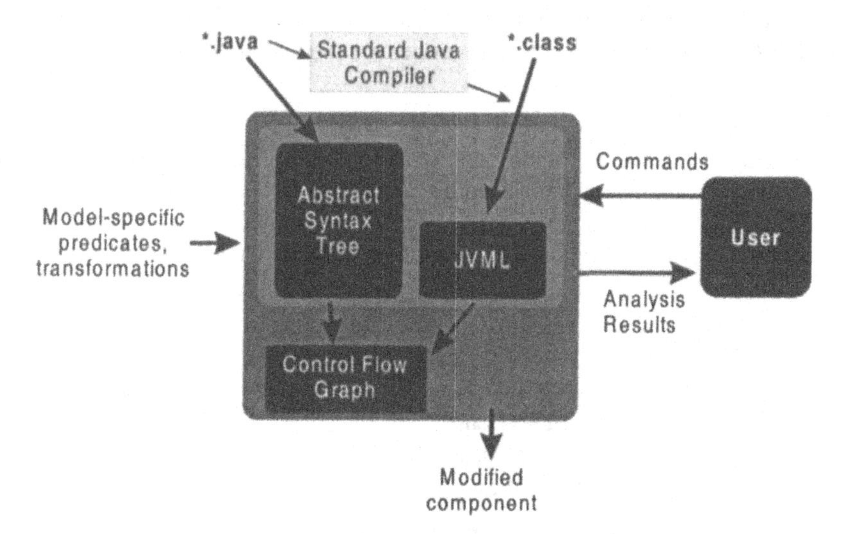

Figure 5.5: The Architecture of the JavaTime Analysis Engine

The *JavaTime* analysis engine is a toolkit for analyzing, manipulating, and generating Java programs. The focus has been on simplifying many common tasks, including parsing, static semantic analysis, name resolution, bytecode generation, classfile reading, and classfile writing. The motivation behind the toolkit comes from the following observation: many conceptually simple tasks are made difficult by the large investment in time and effort required to adapt or construct new tools for the above, common tasks.

The toolkit is organized around three complementary data structures, the abstract syntax tree, the namespace, and the classfile. These data structures are available as a Java class hierarchy and API. An abstract syntax tree represents parsed and resolved Java source code. The namespace serves to lookup and locate class files, compilation units, packages, as well as the contents of these containers (fields, methods, constructors, other classes and packages, etc). The classfile data structure represents a disassembled Java classfile with resolved constant pool entries and instructions. In addition to these central data structures, control flow graphs, dominator trees, and SSA-form may currently be computed, and more fundamental algorithms may be added as they are needed.

The ultimate goal of this project is to explore a different paradigm in programming, one in which modifications or extensions to the language's rules, translation, and implementation have equal status with programming in the language itself. This style of reflexive/meta-programming has not been thoroughly explored in a modern, imperative, widely used language; the experiment is to see whether such a paradigm can succeed and its impact on the software development process, programmer productivity, software evolution, and software testing.

The initial testing ground for the *JavaTime* analysis engine is the *JavaTime* component-based design environment. The toolkit has been integrated with the application and used it for several analyses and transformations. The intention is to use the analysis engine to support the *JavaTime* successive, formal refinement methodology.

5.3.1 Abstract Syntax Tree

The abstract syntax tree is a program representation constructed from concrete syntax in which many irrelevant details have been discarded and ambiguities have been resolved. The abstract syntax tree is only concerned with the hierarchical relation between elements of the language's grammar. For example, the concrete syntax:

```
if (x < 0) { x = 0; System.err.println ("bonk"); }
```

becomes the abstract syntax:

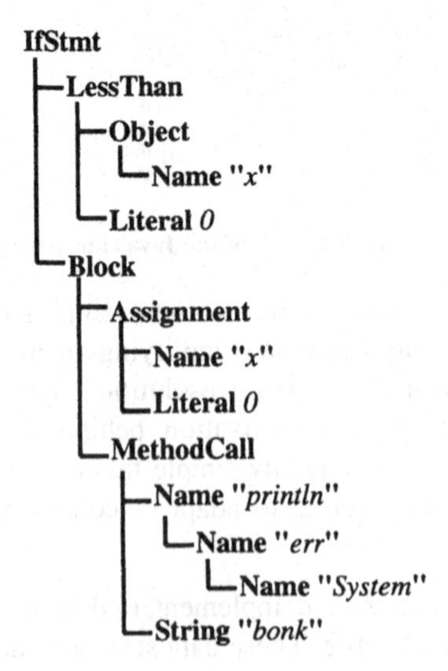

The toolkit provides a parser to translate concrete syntax into an abstract syntax tree as above. Each node in the abstract syntax tree is an object whose class is named after the right-hand-side of grammar production that created it. In the above less than expression, the Java grammar reads:

LessThanExpression ::= *Expression* < *Expression*

and so the toolkit contains a class:

```
class LessThanNode : ExprNode {
/* ... */
   public ExprNode opnd1() { ... }
   public ExprNode opnd2() { ... }
/* ... */
}
```

where the children have been named "`opnd1`" and "`opnd2`". The abstract syntax tree is not yet in a very usable form, for many details in the above tree need to be filled in. For example:

- What does the name "`x`" refer to?
- What does the name "`System.err.println`" refer to?

and in addition we would like to know whether the input program is valid, the types of each expression, and so on. For this purpose, the toolkit provides a routine to compute all of this information. The result might yield:

- x is a local variable declaration
- `System` is the unqualified name for `java.lang.System`
- `err` is a static member `java.lang.System` of type `java.io.PrintStream`
- `println` is an overloaded method belonging to the class `java.io.PrintStream`. This particular method call uses the method with signature `public void print (String)`

These results are filled in as references to various elements of the namespace. For example, each name has a `def()` method that returns the `NamedObject` (*member of the namespace*) that defines it.

Each abstract syntax node class is a subclass of `astagjava.TreeNode`. When the grammar dictates a list of zero or more productions, an object of the class `astagjava.TreeListNode` stores these children. The number of abstract syntax nodes can be overwhelming at first, as there are over 100 node types, but typically only several of these are of interest at any particular time. It is best to keep in mind that the abstract syntax tree structure closely resembles the grammar, so a familiarity with the language should be enough to get started working with the abstract syntax tree.

5.3.2 Namespace

Names in a Java program refer to classes, interfaces, packages, variables, methods, constructors, and so on. Once a program's names have been resolved, each name is resolved to a subclass of `astagjava.NamedObject`. Certain named objects are also containers, and each named object belongs to a container. These containers have lookup methods to locate elements belonging to them. Some lookups are recursive, such as a lookup for the name "`java.lang.String`". All names are reachable from the root package, a container of type `astagjava.ClassHierarchy.Package`.

Elements of the namespace are either Types (Primitive, Reference, Class Type, Interface Type), Containers (Compilation Unit, Class File, Package, Class Declaration, Interface Declaration), or Typed Objects (Method, Field, Local Variable, Parameter).

5.3.3 The Class File

The Java class file is a compiled and simplified form of a Java class. In addition to having Java source code translated into JVML bytecode instructions, most of the information that had to be computed for the abstract syntax tree above is given explicitly in the class file. As a result, many tasks are simpler when performed directly on class files. In addition, since you may not always have direct access to Java source, working with class files is safer when dealing with whole programs.

The toolkit has support for reading, writing, and modifying class files. This includes a disassembled form of the bytecode for each method allowing easy bytecode insertion. The work of analyzing the types of each local variable and types and height of the stack is performed by the toolkit.

Elements of a class file are subclasses of the type `astagjava.ClassFileElt` and instructions (elements of the bytecode array) are subclasses of the class astagjava.Instruction. There are thirteen types of instruction, categorized by their operands. The editing and generation of a method body follows a simple text editor paradigm.

5.3.4 Control Flow and SSA Form

The bytecode instructions found in a class file are very close to a control-flow graph, but this type of information is more difficult to extract from an abstract syntax tree. The toolkit provides methods to generate a control flow graph for each method from the abstract syntax tree. It can then compute a dominator tree and translate the program into a static-single-assignment form. Static-single-assignment is a form in which no variable is assigned to more than once, and is useful in advanced analyses.

5.3.5 Examples

<u>*A Simple Predicate*</u>

Suppose we would like to impose a rule saying that assignment in conditional contexts are illegal, to prevent the classic error:

```
if (x = 1) { ... } else { ... }
```

This can be implemented in several ways. The control-flow analysis marks each variable reference with flags indicating whether the variable is read or written, and whether the variable appears in a control expression. However, this information is not difficult to compute. The following set of methods do not rely on the pre-computed context information:

```
void assignInControl (TreeNode node, boolean in_control) {
    if (node instanceof IfStmtNode) {
            IfStmtNode ifn = (IfStmtNode) node;
            assignInControl (ifn.condition(), true);
            assignInControl (ifn.thenpart(), false);
            assignInControl (ifn.elsepart(), false);
    }
    else if (in_control && node instanceof AssignNode)
                System.err.println ("assignment in if conditional");
        else {
                for (int i = 0; i < node.children().length; i += 1)
                assignInControl (node.children()[i], in_control);
            }
}
```

As described in the future work section, we are working on improving the interface for this type of predicate.

Instrumenting Arithmetic Operations

The toolkit was used to alter class files to count arithmetic operations. For each instruction in the following categories:

- add, subtract, logical or, logical xor, and logical and
- divide, multiply, remainder and for each of the types:
  int, long, float, double

a counter was added to the class. Each arithmetic instruction then had the appropriate 6 instruction sequence:

- aload_0 # Load this (1)
- dup # Dup this (2)
- getfield <the-counter> # Load previous value with (2)
- iconst_1 # Push the constant 1
- iadd # Add to previous value
- putfield <the-counter> # Store with (1)

This instrumentation was performed on *JavaTime* components. The component's "go" method was also modified to output another port with the value of the counters. This allowed a probe to be attached to the counter's output port and the running totals to be viewed during execution.

Presburger Constraint Analysis

We have been applying the toolkit to dead code elimination and loop termination testing with Presburger formulae. This analysis involves generating constraints on all the integer variables and attempting to solve them using a Presburger solver.

The procedure employs Presburger formulae for a constraint style program analysis, inspired by the set and term constraint system employed in the Berkeley Analysis Engine [19]. Presburger integer constraints are used to locate dead code fragments in a style similar to conditional constant propagation after conversion to static single assignment (SSA) form [20,21]. An algorithm for proving strong loop termination is under development that generates constraints on each method in SSA form. These constraints approximate each loop's variation with constant multiplications, where possible, and the resulting Presburger constraints are solved to determine dead code branches and loops that do not strongly terminate.

5.3.6 Future Work

The next goal for the analysis toolkit is to make these analyses simpler to write and incorporate into an application. For example, the above sample function, assignInControl, is not written in the most convenient form. A better form might be:

```
void assignInControl (TreeNode n, boolean c) {
  for (int i = 0; i < n.children().length; i += 1)
        assignInControl (n.children()[i], c);
}

void assignInControl (AssignNode n, boolean c) {
  if (c)
     System.err.println ("...");
}

void assignInControl (StmtNode n, boolean c) {
  for (int i = 0; i < n.children().length; i += 1)
        assignInControl (n.children()[i], true);
}
```

Of course every Java programmer knows that this isn't how overloading works.

We are exploring different techniques that allow code to be written without modifying the toolkit's class hierarchy, yet still allowing some form of dynamic dispatch from the programmer's point of view. The difficulty with this type of toolkit, from an object-oriented point of view is that the programmer has difficulty extending the library's functionality with subclassing, since the toolkit is the one constructing the data structure and doesn't know about subclasses the programmer has created. The approaches taken in CLOS and ML for this type of pattern-matching dispatch seem a good starting point.

The toolkit presented here has proven effective in eliminating much of the overhead involved in writing a small, simple analyses and transformations. The toolkit has also been employed for a fairly serious constraint based analysis. In all of these cases, experience with the toolkit has driven future enhancements and the interface design. As more experience is gained, we continue to improve and develop the toolkit and its interface.

6. Examples

A number of design examples are included below to illustrate the use of the *JavaTime* environment. They include a JPEG encoder/decoder, a digital camera controller, and an MPEG-2 encoder. These real-world examples aid evaluation of *JavaTime*'s suitability for realistic designs.

6.1.1 JPEG Codec

A JPEG image encoder/decoder system has been described using *JavaTime*, and demonstrates *JavaTime*'s component-based design approach. The design was created by porting a preexisting Java JPEG program to the *JavaTime* component model. The original version is structured as a basic sequential program, using no concurrency. The *JavaTime* version is described as a hierarchical design, with components based on different models of computation. Processing of the red, green, and blue color elements of the image is performed by three identical, concurrent image processing components.

Each image processing component is described as a network of components that perform the various computations required by JPEG. Separate components implment the discrete cosine transform (DCT), inverse DCT, quantization, and dequantization functions. These components perform their functions on 8x8 image blocks, and thus additional components are required that divide the input image (of arbitrary dimensions) into a stream of 8x8 blocks, and construct the output image from the stream of processed blocks. They are specified in a reactive manner, generating one output signal for each input signal received. The description of the quantizer block is shown below.

```
package jpeg;
import java.util.*;
import jcp.*;

public class Quantize extends Component
{
      // Work on 8X8 blocks, so 8 is a magic number
      private final static int N = 8;
      private int[][] quantum = new int[N][N];
      private int outputData[][] = new int[N][N];
      private int quality = 10;
      private Port in_p, out_p;
      private Port quality_p;

      // Constructor. Fields are initialized here.
      public Quantize() {
            setName("Quantize");

            // Ports
            in_p = addPort(true,"in");
            out_p = addPort(false,"out");
            quality_p = addPort(true,"quality");
```

```
            initMatrix(quality);
      }

      // This method is executed whenever a new signal is
      // received on an active port.
      public void go(Port port) {
            if (port == in_p) {
                  IntMatrix s = (IntMatrix)in_p.signal();
                  int[][] matrix = s.get_matrix();
                  emit(new IntMatrix(quantize(matrix)),out_p);
            } else if (port == quality_p) {
                  int qual = ((Integer)quality_p.signal()).intValue();
                  initMatrix(qual);
            }
      }

      private int[][] quantize(int inputData[][]) {
            for (int i = 0; i < N; i++) {
                  for (int j = 0; j < N; j++) {
                        double result = (inputData[i][j] / quantum[i][j]);
                        outputData[i][j] = (int)(Math.round(result));
                  }
            }
            return outputData;
      }

      private void initMatrix(int quality) {
            for (int i = 0; i < N; i++) {
                  for (int j = 0; j < N; j++) {
                        quantum[i][j] = (1 + ((1 + i + j) * quality));
                  }
            }
      }
}
```

Figure 6.1: Behavior specification of quantizer component.

While the basic reactive model is used to describe many of the blocks, the blocks that perform the serialization/deserialization process are described using a model of computation called the *clocked* domain. In this domain, components' behaviors are triggered by ticks of a shared clock, that provides a synchronization mechanism between components. In clocked components, the specification of the behavior must call a method called `waitClock`, that suspends execution of the program until the next clock "tick".

In addition, a controller component is needed, whose main task is to coordinate the serialization/deserialization process with signals received from the inputs. The result of this design is that while an image is processed internally as a series of 8x8 blocks, from the outside the JPEG block looks like a component in the reactive model. That is, an input image is received, and a output image is emitted. The serialization/deserialization process is not visible by external observers.

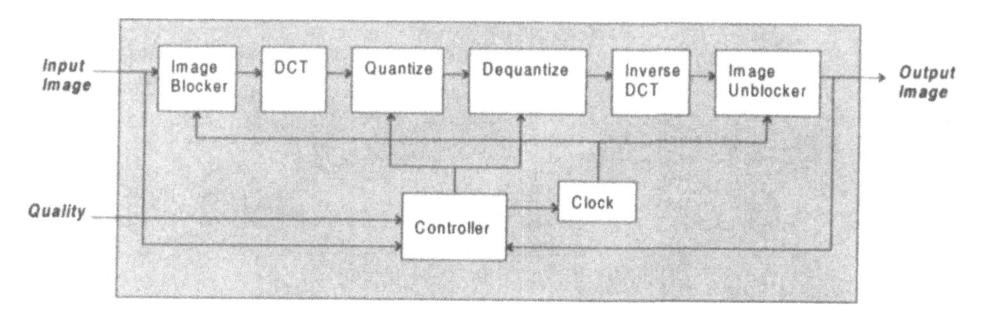

Figure 6.2: Architecture of JPEG example.

Execution of the *JavaTime* version on the Java Virtual Machine is significantly slower than the original program. This is primarily due to two factors: first, the *JavaTime* version must allocate memory in order to represent the signals being communicated between components. Second, the *JavaTime* version is concurrent, while the original is not. Concurrency is implemented in the Java Virtual Machine using threads, that induces a context switch penalty. However, while the *JavaTime* version utilizes threads in its execution, no mention of Java's threads package is made in the specification. Rather, threads are utilized by classes in the *JavaTime* component package, a fact that the designer need not be aware of.

6.1.2 Digital Camera Controller

A controller for a digital camera was also created using *JavaTime*. The controller generates the necessary control signals to display and record images when the user presses the buttons of the user interface.

The controller is specified using the *JavaTime* hierarchical finite state machine package, that provides functionality similar to Harel's Statecharts visual formalism. A Statecharts-style graphical representation of the camera controller is shown in Figure 6.3.

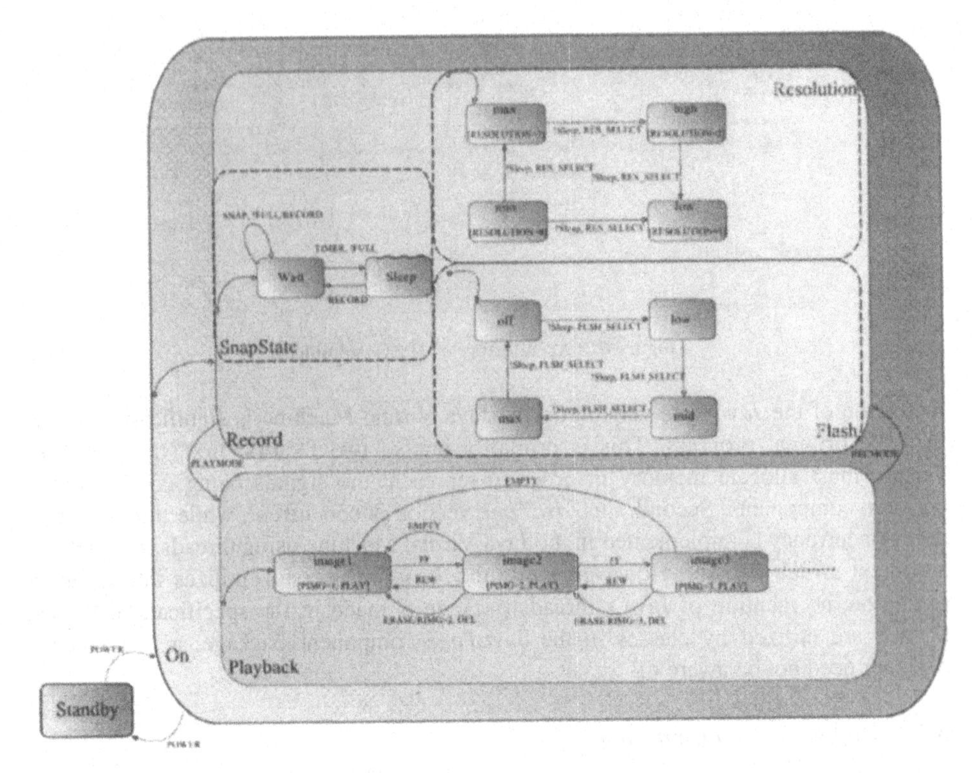

Figure 6.3: Hierarchical FSM Representation of Digital Camera Controller in *JavaTime*

6.1.3 MPEG-2 Encoder

As an example of a larger system, an entire MPEG-2 encoder was designed using *JavaTime*, created from and exisiting C implementation of the specification. The C source code was first ported to "Java-equivalent" C++, meaning that no C++ features were used that have no direct equivalent in Java (e.g., multiple inheritance, pointer arithmetic, operator overloading). The C++ was then converted to standard Java, and then finally modified to use the *JavaTime* component package, that entailed that the functionality of the system be divided into several components, and method invocations converted into signal emissions over ports. No efforts were made to restructure the program to optimize concurrency or memory usage.

The original C program was 7605 lines in length, while the version utilizing the *JavaTime* component package was 4923 lines long. Execution of the *JavaTime* version is slower than the C version, primarily due to the relative immaturity of Java compared to C.

7. Summary

The *JavaTime* system is an experimental test bed we are developing for the design of heterogeneous, embedded electronic systems. It uses the Java language as a syntax for design representation—in many ways, as a high-level "intermediate format"—and takes *a component-based approach* to system specification, modeling and implementation. *JavaTime* addresses heterogeneity by supporting multiple underlying models of computation, and supports integration by providing a common representation format at a reasonably wide range of levels of abstraction. In particular, we described an approach for using Java for describing both the specification and implementation of a heterogeneous embedded system.

We believe that by embedding models of computation in a "source level" language, the system can perform analyses at a much higher level of abstraction and can use modern program analysis and transformation techniques. Our choice of Java is also a pragmatic one, and we expect to be able to continue to leverage the rich set of program analysis, debugging, and compilation tools and techniques being developed for Java. We have also developed a *JavaTime*-specific analysis framework for the transitive application of predicates to a Java AST, followed by a set of analysis or local transformational rules, as described above.

8. References

1. *Verilog HDL Language Reference Manual.* (1995) Draft, IEEE 1364.
2. ANSI/IEEE Standard 1076-1993, VHDL Language Reference Manual. (1994) IEEE Press, New York.
3. Helaihel, R. and Olukotun, K. (1997) Java as a Specification Language for Hardware-Software Systems, *Proc. ICCAD '97.*
4. Hopcroft J. and Ullman, J. (1979) *Introduction to Automata Theory, Languages, and Computation*, Addison-Wesley.
5. Edwards, S. (1997) *The Specification and Execution of Heterogeneous Synchronous Reactive Systems*, Ph.D. Thesis, University of California, Berkeley. Available as UCB/ERL M97/31. http://ptolemy.eecs.berkeley.edu/papers/97/sedwardsThesis/
6. Kahn G. (1974) The semantics of a simple language for parallel programming, *Information Processing 74: Proceedings of IFIP Congress 74*, pages 471–475.
7. Davey, B. A. and Priestley, H. A. (1990) *Introduction to Lattices and Order*, Cambridge University Press.
8. Illingworth, V. (1991) *Dictionary of Computing* (Oxford Paperback Reference), Oxford Univ. Press, 3rd edition.
9. Young S. (1982) *Real Time Languages: Design and Development*, Ellis Horwood.
10. Randell, B. et. al. (1995) *Predictably Dependable Computing Systems*, Springer-Verlag.
11. Pnueli, A. (1977) The Temporal Semantics of Concurrent Programs, *Proceedings of the 18th Symposium on the Foundations of Computer Science*, pages 46-57.
12. Harel, D. and Pnueli, A. (1985) On the Development of Reactive Systems, Logic and Models of Concurrent Systems, K. R. Apt editor, *NATO Advanced Study Institute on Logics and Models of Verification and Specification of Concurrent Systems*, Springer-Verlag, pages 477-498.
13. David-Neel, A. and Lama Yongden. (1967) *The Secret Oral Teachings in Tibetan Buddhist Sects*, City

Lights Books, San Francisco.

14. Newton, A. R. (1979) Techniques for the simulation of large-scale integrated circuits, *IEEE Trans. on Circuits and Systems*, vol. 26, no. 9, pp. 741-749.
15. Benveniste A. and Berry, G. (1991) The synchronous approach to reactive real-time systems, *Proceedings of the IEEE*, 79(9):1270–1282.
16. Halbwachs, N. (1993) *Synchronous Programming of Reactive Systems*, Kluwer.
17. Bauer, F., Moller, B., Partsch, H., and Pepper, P. (1989) Formal Program Construction by Transformations–Computer-Aided, Intuition-Guided Programming, *IEEE Trans. On Software Engineering*, vol. 15, no. 2, pp. 165-180.
18. Smith, D. (1990) KIDS: A Semiautomatic Program Development System, *IEEE Trans. Software Engineering*, vol. 16, no. 9, pp. 1024-1043.
19. Fahndrich, M. and Aiken, A. (1998) *Bane: Analysis Programmer Interface*, University of California, Department of Computer Science, Internal Memorandum.
20. Cytron, R. et. al. (1991) Efficiently computing static single assignment form and the control dependence graph, *ACM Trans. On Prog. Lang. and Syst.*, 13, 4 pp.451-490.
21. Wegman, M. and Zadek, F. (1991) Constant propagation with conditional branches, *ACM Trans. On Prog. Lang. and Syst.*, 13, 2, .pp. 181-210.
22. Arnold, K. and Gosling, J. (1996) *The Java Programming Language*, Addison-Wesley, Reading, MA.
23. Gosling, J., Joy, B., and Steele, G. (1996) *The Java Language Specification*, Addison-Wesley, Reading, MA.
24. Stroustrup, B. (1991) *The C++ Programming Language*, Addison-Wesley, Reading, MA.
25. Stoyenko, A. D. (1992) The Evolution and State-of-the-Art of Real-Time Languages, *J. Systems Software*, 18:61-84.
26. Lamport, L. (1978) Time, Clocks, and the Ordering of Events in a Distributed System, *Communications of the ACM*, 21(7):558-565.
27. Berry, G., and Gonthier, G. (1992) The ESTEREL synchronous programming language: design, semantics, implementation, *Science of Computer Programming*, vol. 19, pp. 87-152.

MODELS AND METHODS FOR HW/SW INTELLECTUAL PROPERTY INTERFACING

ROSS B. ORTEGA
Department of Computer Science & Engineering
University of Washington
Seattle, Washington USA
ortega@cs.washington.edu

LUCIANO LAVAGNO
Department of Electronics
Politecnico di Torino, Italy
lavagno@polito.it

AND

GAETANO BORRIELLO
Department of Computer Science & Engineering
University of Washington
Seattle, Washington USA
gaetano@cs.washington.edu

Abstract. This paper focuses on the problem of enabling system companies to quickly integrate IPs from different sources, and adapt them to different manufacturing technologies. An evolutionary approach from current methodologies is possible with appropriate and extensive CAD support. We cover the main aspects of interfacing Intellectual Property, both in hardware and software form, in an embedded system design context. In particular, we review the main approaches to specification, synthesis and validation of interfaces that have appeared in the literature. From the specification viewpoint, we illustrate the main protocol and timing constraint specification models. From the synthesis and optimization viewpoint, we review software and hardware generation, as well as time-driven interface scheduling techniques. From the verification viewpoint, we discuss various strategies for hardware/software co-simulation, with special attention to the interface layer. Finally, we consider the growing importance of formal verification in this domain, and outline the main techniques for interface protocol verification.

A. A. Jerraya and J. Mermet (eds.), System-Level Synthesis, 397–432.

1. Introduction

The scale of integration of electronic circuits keeps increasing exponentially, even faster than industry specialists often predict. Thanks to the increased computing capabilities and the decreasing cost and power consumption, electronic circuits are becoming more and more pervasive in every aspect of modern life. This consumerization of electronics implies a push towards shortening the design cycle and increasing the variety of design derivatives, in order to cope both with the demands of an extremely competitive market and with regional variations due to, *e.g.*, culture and standards. On the other hand, productivity of designers, in terms of gates or lines of code per hour, is not increasing at the same pace, despite the efforts of CAD tool and methodology developers to raise the level of abstraction. This means that, short of a major breakthrough in high-level synthesis or some other productivity enhancement, the only way to exploit the availability of cheap silicon to meet consumer demands is by *re-use*.

Re-use is already a well-known concept at the electronic system level, where boards are populated with standard, off-the-shelf components as well as ASICs. It is also the rule in software, where a major part of a program or system generally originates from object or source code libraries. The new aspect, that is re-shaping the electronic industry today by changing the way in which systems are designed and manufactured, is that now re-use must occur at the *chip* level. This implies a complete change in the way re-usable System-On-Chip (SOC) components, generally called *Intellectual Property* (IP), are defined, specified, designed, marketed, sold, integrated, manufactured and tested.

So far, semiconductor companies used to identify the component market needs, design chips at the logic and the layout level, manufacture and test them, and document their capabilities, interfacing mechanisms, recommended use and so on via data books. System companies, in turn, used to analyze the end product market requirements, choose and integrate those chips by using custom or programmable glue logic on a printed circuit board. The interface was relatively clear: a functional, written specification that could be checked from both sides by testing.

Now this scenario is changing, with the introduction of a third partner due to a possible splitting of the role of semiconductor companies.

1. IP companies identify re-usable components and design them, but the design stops at the register transfer, logic or layout level.
2. System companies integrate these IPs, from different sources, based on their understanding of the end-user needs.
3. Semiconductor companies manufacture and test the chips that are composed of several IPs glued together.

The complete system, including chips, sensors, actuators, interfaces and so on can be assembled by system companies.

This new scenario for the electronic industry brings in a whole new set of problems to be addressed and solved by the Electronic Design Automation industry in the (very) near future. These problems are, for example:

- Ensuring that the RTL, logic or layout (also known as soft, firm or hard) IP performs according to the specification, both at the level of exchange between IP and system companies, and at the level of exchange between system and semiconductor companies.
- Protecting intellectual property at all levels, thus guaranteeing each party a fair share in the profits. IP cannot be sold by the item any longer, since its transmission is in electronic, infinitely reproducible form (RTL, logic or layout). Hence its use in production must be measured and protected by other means.
- Enabling system companies to quickly integrate IPs from different sources, and adapt them to different manufacturing technologies.

In this paper we focus mostly on the last problem, by showing how, at least in this case, an evolutionary approach from the board level is possible, by providing appropriate and extensive CAD support.

IP can come in a variety of forms, from traditional programmable processors (*e.g.*, microprocessors and Digital Signal Processors), to dedicated devices (*e.g.*, MPEG decoders, timer units, disk controllers and cellular phone radios). It may provide ready-made interfaces between widely used protocol standards (*e.g.*, adapting a given microprocessor to a given bus). IP can also be software (*e.g.*, GSM protocol stacks or FFT routines). This means that interfaces must be designed by the system integrator between a variety of domains ranging from digital to analog to software.

Of course, like in any other CAD problem, the ideal solution is a correct-by-construction approach in which formal specifications of the IP interface protocols are used as a basis to

- automatically verify compatibility (answering questions such as "can these IPs talk to each other?" or "what bandwidth and latency can I achieve, and at what cost?"), and
- synthesize an implementation meeting given performance and cost constraints.

In this paper we summarize, analyze and compare different *partial* solutions to the interface design problem. The rest of the paper is organized as follows. After a background section devoted to the main communication mechanisms used by formal models in the literature, we describe, in Section 2, various approaches to hardware-to-hardware interface synthesis. We then consider, in Section 3 various techniques for interfacing software and hardware components. Software architecture and high-level protocols are also an important part of IP integration. In Sections 4 and 5 we describe aspects of this problem that are most closely related to embedded system design, thus ignoring important issues such as client-server architectures, object broker architectures and so on. Finally in Section 6 we discuss interface validation issues, in terms of both hardware-software co-simulation, and of formal verification.

1.1. BASIC COMMUNICATION PRIMITIVES

In this section, we define some of the communication primitives that have been described in the literature on formal models, since they are useful in order to understand the rest of the paper.

Unsynchronized In an unsynchronized communication, a producer of information and a consumer of the information are not coordinated. There is some connection between them (*e.g.*, a buffer) but there is no guarantee that the consumer reads "valid" information produced by the producer, and no guarantee that the producer will not overwrite previously produced data before the consumer reads the data.

Read-modify-write Commonly used for accessing shared data structures in software, this strategy locks a data structure between a read and write from a process, preventing any other accesses. In other words, the actions of reading, modifying, and writing are atomic (indivisible, and thus uninterruptible).

Unbounded FIFO buffered This is a point-to-point communication strategy, where a producer generates (writes) a sequence of data tokens and a consumer consumes (reads) these tokens, but only after they have been generated (*i.e.*, only if they are valid).

Bounded FIFO buffered This mechanism is similar to the unbounded FIFO, but the producer either blocks or overwrites some data when the FIFO is full.

Rendezvous In the simplest form of rendezvous [34] a single writing process and a single reading process must simultaneously be at the point in their control flow where the write and the read occur. It is a convenient communication mechanism, because it has the semantics of a single assignment, in which the writer provides the right-hand side, and the reader provides the left-hand side. Multiple rendezvous is also possible, but it is much more complex to implement, and hence it is not generally supported in practice.

The essential features of the communication primitives described above are presented in Table 1. These are distinguished by the number of transmitters and receivers (*e.g.*, broadcast versus point-to-point communication),

	Transmitters	Receivers	Buffer Size	Blocking Reads	Blocking Writes	Single Reads
Unsynchronized	many	many	one	no	no	no
Read-Modify-Write	many	many	one	maybe	maybe	no
Unbounded FIFO	one	one	unbounded	yes	no	yes
Bounded FIFO	one	one	bounded	maybe	maybe	yes
Petri net place	many	many	unbounded	no	no	yes
Single Rendezvous	one	one	one	yes	yes	yes
Multiple Rendezvous	many	many	one	no	no	yes

TABLE 1. A comparison of concurrency and communication schemes.

the size of the communication buffer, whether the transmitting or receiving process may continue after an unsuccessful communication attempt (blocking reads and writes), and whether the result of each write can be read at most once (single reads). A "maybe" entry means that both the "yes" and "no" answers have been proposed in the literature.

2. Hardware Interface Synthesis

In this section we review the main approaches to interface synthesis in which both sides are hardware IPs. Most approaches provide protocol specification mechanisms that are easier to use and more powerful than general-purpose Hardware Description Languages. The designer specifies the protocols that are followed by both partners. The synthesis algorithm produces either a high-level specification (*e.g.*, a Finite State Machine to be synthesized further using standard logic synthesis) or directly a circuit implementing the interface. The cost function that is minimized is generally a mix of control circuit cost and performance (latency/throughput).

We will first review asynchronous interfacing techniques, then synthesis from timing diagrams, and then end this section with a discussion of synchronous interface synthesis.

2.1. ASYNCHRONOUS INTERFACES

Interfacing is a classical domain of application of asynchronous techniques, since they provide modularity and robustness characteristics that are quite desirable in this field, and do not make too many assumptions about the partners' clock speeds and skews. While the advantages of asynchronicity seem to be mostly important for board- and backplane-level integration, we include it in this review because reducing problems with clock-skew control may still be important even for Systems-On-Chip.

The main informal specification model for interfaces is still the timing diagram, since it conveys the essential aspects of a low-level communication protocol in a clear and intuitive manner. Traditional timing diagrams used in data books, however, are not formal enough to be used as a specification mechanism for interface synthesis. Researchers in recent years hence have developed a number of formalizations, and of corresponding synthesis algorithms, as described in this section and the next one.

A Signal Transition Graph (STG) [51, 14] is, formally, an interpreted Petri Net (PN), in which transitions are labeled with falling and rising transitions of interface signals. A PN is a bipartite directed graph, in which transition nodes represent activity of the modeled system, place nodes represent its distributed state, and edges represent causal relationships between places and transitions. Places can hold any non-negative number of tokens, and often PNs modeling interfaces are constrained to have at most one token per place (*safe* PNs). A transition may fire when all its predecessor places are marked. If it fires, it removes one token from all its predecessors and adds one token to all its successor places.

One of the main advantages of STGs as interface protocol specification mechanisms is the availability of a huge amount of research on PNs, that can be used in order to analyze their properties and synthesize a logic circuit from them. Figure 1 shows a simple timing diagram, the corresponding Signal Transition Graph, and the State Graph (derived from the STG by playing the token game and exploring all the possible markings). The latter can be used to derive an implementation of the output signals of the modeled interface, as described below.

2.1.1. *Asynchronous interface synthesis*

Techniques for synthesis of STGs can be divided into two main classes.

1. Syntax-directed methods establish a correspondence between patterns in the specification and sub-circuits. They have the advantage of a low computational complexity, but may produce sub-optimal results [58].
2. Boolean minimization-based methods, on the other hand, derive a Boolean function for each interface output signal from the State Graph, and use various techniques and delay models in order to implement that function without hazards [15, 42, 1, 39].

Sun *et al.* [56] and Lin *et al.* [44] proposed techniques to derive asynchronous interface specifications from a description of the protocols followed by the two partners. The basic idea (as initially developed by Borriello [6]) is to match the data transfer parts of the protocol (since data transfer is

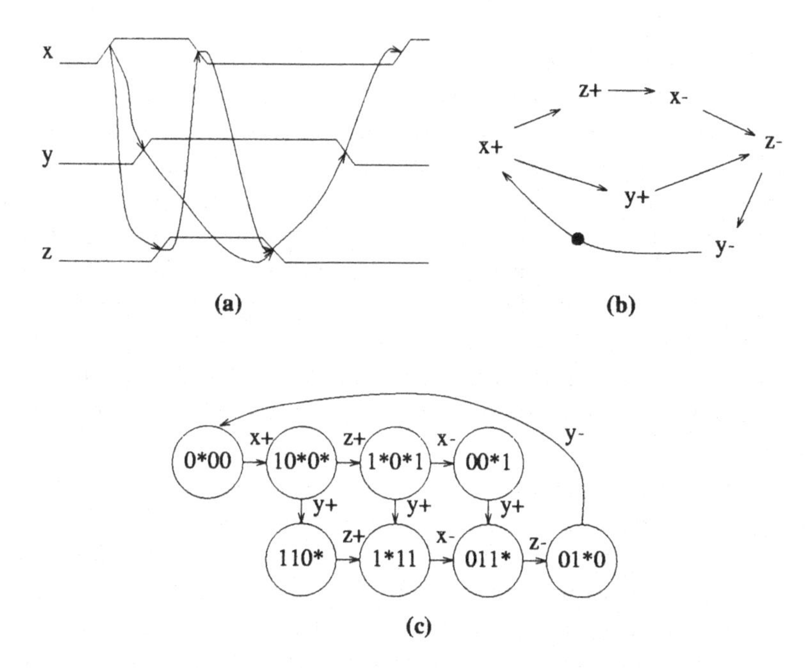

Figure 1. Example of a Signal Transition Graph

generally the main goal of an interface, sometimes in addition to synchronizing the partners) in order to derive a complete specification from the two halves embodied in each side's protocol.

In Sun's approach, the protocols are modeled using STGs, and synchronization edges are inserted between "active" edges (as marked by the protocol designer) to keep the two STGs interlocked into a single handshake.

In Lin's approach, the protocols are modeled in a CSP-like language [34], using channel communication entities. A channel includes a two-way handshake, plus an optional data connection. The language is directly translated into an STG with channel reads and writes labeling the transitions. The STGs are refined to implement reads and writes with low-level signal transitions. Finally, the two STGs representing the two sides of the interface are composed by matching transitions with the same label.

2.2. SYNTHESIS FROM TIMING DIAGRAMS

As mentioned previously, timing diagrams are a common way to describe interfaces. Data books contain pages and pages of waveforms with annotated timing constraints. An advantage of timing diagrams is that they present synchronous and asynchronous signals uniformly. It is simple to identify signal events and associate them with timing constraints. Another advantage of timing diagrams is that they focus solely on the interface between two components without regard to their implementation. As long as the appropriate waveforms are generated, communication can occur. Such an input language is not biased towards a hardware implementation giving the interface synthesis tools freedom to implement the most appropriate interface logic. Unfortunately, timing diagrams can not represent all the information necessary to synthesize an optimized interface. For example, it is impossible to specify that a signal is actually a boolean function of other signals. There is no convention for capturing conditional or looping behaviors. There is also no way of decomposing a timing diagram into subdiagrams. As a software analogy, it is like being forced to program without subroutines.

Formalized timing diagrams [8] [7] address these problems by extending the timing diagram notation. Signals may be specified with a formula that generates the appropriate waveform. This formula may include latching conditions to specify sequential relationships. Diagrams can be decomposed into segments and regular expressions can be applied to the segments. Loops and conditionals can also be supported using this technique. It is also possible to link events in different diagrams to express constraints between different operations. Being based on timing diagrams, both asynchronous and synchronous interfaces can be specified with formalized timing diagrams. This formalism of the interface requirements makes formalized timing diagrams an interesting specification format for interfaces in IP-based design. An IP block consisting of a synthesizable function along with a formalized timing digram description of its interface would enable the creation on an IP library. Such a library could support a design environment where system architects pick and choose various library elements to create SOC circuits.

2.3. SYNCHRONOUS INTERFACES

Synthesizable HDLs and logic diagrams are still the most popular specification mechanism for synchronous interfaces in practice. Both, however, are poorly suited for this task, since they do not exploit any particular characteristics of interfacing. Models with explicit concurrency, sequencing and choice, such as synchronous PNs [3], synchronous languages [29, 30], or Regular Grammars (RGs) [53, 50] can be used more efficiently in this case.

In particular, RGs have been used in several recent approaches to interface design. A Regular Grammar (or, equivalently, a Regular Expression) denotes a regular language, and hence can be recognized/generated by a Finite State Machine. An RG is

– either a terminal symbol, representing an assignment of values to the interface signals and possibly a transfer of data, with a duration of exactly one clock cycle.

– or a non-terminal symbol, defined by a combination of RGs by sequencing, alternative and bounded or unbounded repetition.

The basic idea is that protocol specifications, such as the structure of an ATM cell, are easier to model with an RG than with an HDL.

For example, the ATM frame for one cell (from [53]) at the top level is defined by the following regular grammar. Parentheses are used for grouping, "::" separates a non-terminal from its definition, ";" denotes sequencing, "," denotes concurrency (within the same clock cycle), "#" denotes alternative, ^{n} denotes repetition n times and * denotes unbounded repetition. *cs* is a single-bit input control wire (cell start), *di* is an 8-bit input data bus, *vpi* is a 12-bit output data bus, and *vci* is a 16-bit output data bus.

– *ATM :: ((cs = 0) # (cs = 1; atm_cell))**
  (an ATM stream is an infinite sequence of cells; each cell starts when *cs* is at 1 for at least one clock cycle)
– *atm_cell :: header; payload*
  (an ATM cell is composed of a header followed by a payload)
– *header :: vpi[11:4] ⇐ di; (vpi[3:0] ⇐ di[7:4], vci[15:12] ⇐ di[3:0]); ...*
  (an ATM header is composed of the 8 MSBs of the Virtual Path Identifier, followed by the 4 LSBs of the VPI together with the 4 MSBs of the Virtual Circuit Identifier, and so on)
– *payload :: (buf[adr] ⇐ data, adr ⇐ adr + 1)*^{48}
  (an ATM payload is composed of 48 bytes, that must be stored locally in buffer *buf* at the location pointed by *adr*)
– ...

Note how this declarative specification is much more readable than an equivalent FSM described in synthesizable VHDL or Verilog.

2.3.1. *The DALI protocol compiler*

Seawright *et al.* [53] describe an efficient technique to synthesize an interface protocol, based on a direct translation of RG fragments to synchronous circuit fragments. The RG is used to specify the control structure, while terminal symbols are directly represented in VHDL.

The RG is *directly translated* into hardware, rather than explicitly generating an FSM and performing standard state minimization and state encoding on it. The motivation is that the "natural" state encoding, obtained from an identification of RG constructs with hardware patterns, performs better experimentally (both in terms of area and delay) than the standard FSM-based synthesis flow.

Recent results [54] also show that syntax-directed *partitioning*, by clustering along the structural elements of the RG, can yield better synthesis results than synthesis of the full RG.

2.3.2. *Regular grammar-based interface synthesis*
Passerone *et al.* [50] describe techniques to derive a single FSM from two RGs describing the two partners' protocols. This is different from Seawright's work because in the former case *a single RG* is used as a compact form to describe the *interface*, while in Passerone's case *two RGs* are used to specify the two protocols, and an interface FSM is synthesized in between. This is similar to [44], but Passerone's approach minimizes a cost function (mostly data latency) over all possible non-deadlocking, causal compositions of the two RGs.

The idea is again to match assignments of values to signals with the same name and direction on the two sides of the protocol. However, in this case composition is synchronous, and hence there are some additional issues to take into account:

1. the composition must not enter a sequence of uncontrollable events (responses from the receiver) eventually leading to a deadlock,
2. the composition must be causal (data must be received only after it has been sent),
3. the composition must transfer data as quickly as possible.

The resulting FSM corresponds to a *scheduling* of the non-deterministic concurrent composition of the two FSMs derived from the two RGs, with the objective of satisfying these three main constraints.

2.3.3. *Communication scheduling*
Filo *et al.* [22] describe a technique for scheduling *communication graphs* with the objective of minimizing synchronization (and hence timing overhead). Communication is between processes, each modeled with an acyclic hierarchical constraint graph. Each node in the graph can represent:

- an internal operation (with an associated deterministic delay),
- a rendezvous send or receive operation (non-deterministic delay),
- one or more invocations of a sub-graph (*e.g.*, to model a data-dependent loop).

The authors derive a *message dependency graph* from the composition of several process graphs communicating over channels, by abstracting away all internal operations and removing hierarchy. This graph must be acyclic in order to be schedulable without deadlocks.

The objective is to make as many communications as possible *non-blocking*, meaning that both sender and receiver are ready at the same time, without a need to wait for one another. Communications can be made non-blocking, intuitively, if there are no unbounded loops between them and preceding communications. If so, these predecessors can be used as "anchors" to statically define the initiation times of *controllable* communications. Each controllable communication can thus be assigned an As Soon As Possible and As Late As Possible initiation time, based on the delays of concurrent operations. Intersection of these intervals between communicating partners allows one to both statically schedule communications, and to *share channels* if their utilization times are non-overlapping.

2.3.4. *The SIERA system*

The SIERA system [55] focuses on the problem of system-level synthesis for printed circuit board designs consisting of multiple processors, ASICs, and FPGAs. It is fully integrated with the ALOHA interface synthesis tool [56] mentioned previously. A key aspect of SIERA is its use of parameterizable library components for hardware module generation. The library includes memory subsystems, embedded computer modules, bus-interface modules, *etc.* The library elements can be defined in a hierarchical manner so that increasingly complex functional blocks can be created and re-used in future designs. ALOHA was modified for SIERA to support synchronous as well as asynchronous interface synthesis. SIERA is an example of the type of integration needed between interface synthesis and system composition tools so that a design environment can support IP-based design.

3. Hardware/Software Interface Synthesis

With the increasing levels of integration possible in core-based technologies, system on a chip (SOC) designs frequently include general purpose microprocessor cores in addition to functional IP cores (*e.g.*, an FFT or MPEG processor). For example, the ARM processor [25] from Advanced RISC Machines is a synthesizable core commonly used in SOC designs. When using processors with other IP blocks, a hardware/software interface must be synthesized. The main goal in hardware/software interface synthesis is to generate a communication link between the processor and the IP blocks while respecting timing constraints. In addition to a hardware interface component synthesized using some of the techniques mentioned earlier, a software interface must also be constructed. The device-driver running on the processor must be aware of the synthesized hardware/software interface to allow the processor to communicate with the other IP blocks.

In order to support IP-based design, the interface to the IP functional blocks must be described in sufficient detail to allow interface synthesis. The Virtual Socket Interface Alliance (VSI) [59] is creating an open standard for

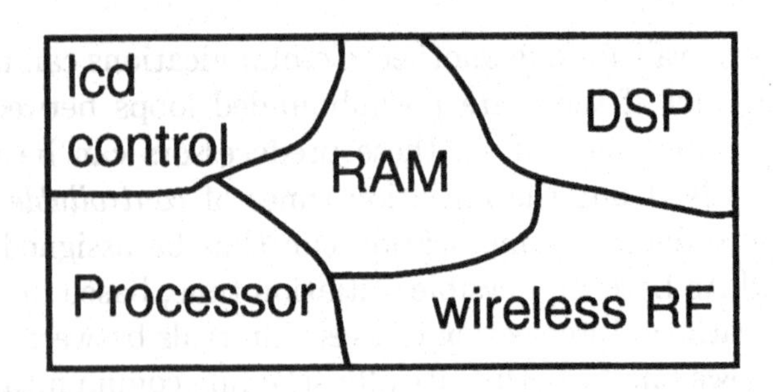

Figure 2. System on a chip (SOC) composed of multiple cores.

IP descriptions so that tools can easily integrate IP blocks from multiple sources into a design. Such a standard allows designers to use "virtual" components in a component-based design methodology. The VSI initiative is looking at two major issues for the encapsulation of IP. The first issue, physical sockets, addresses the physical interface of the IP component and includes information about power requirements, clocks, *etc.* The second issue, functional sockets, addresses the runtime behavior of the IP block and includes information about the interface protocol.

3.1. LIBRARY SUPPORT FOR LOW-LEVEL INTERFACE SYNTHESIS

The information required to automate interface synthesis between a processor and peripheral IP functional blocks consists of both hardware and software descriptions. The processor's interface description must include information about the address and data ports, read/write lines, interrupt lines, and peripheral components that are integrated with the processor's core. These peripherals include timers, DMA controllers, UARTs (serial ports), and other communication modules. Additional information is required to describe the details normally found in a data book such as a range of clock frequencies, standard connections to memory, and which physical ports must be tied to logic one or zero to achieve the appropriate mode of operation. The IP block must also contain similar data so that it can be interfaced to a processor. This information corresponds to the VSI physical socket description.

In addition to the physical description, timing and protocol information is necessary to allow communication from the processor to the IP functional block. From a software perspective this information can be viewed as device-driver templates that will be customized once the low-level hardware interface has been synthesized. From a hardware perspective, this information can be viewed as timing diagrams that will be realized by glue logic. This type of protocol information corresponds to VSI's functional socket description. It should be noted that the protocols specified at this level are very simple (*e.g.* blocking and non-blocking communication). In Section 4 more complex protocols will be addressed.

3.2. MEMORY-MAPPED I/O

Embedded processors have two main methods of interfacing with surrounding functional blocks, direct I/O and indirect I/O. Direct I/O manipulates the interface directly without interfacing or glue logic. Indirect I/O means the processor communicates with the IP blocks via additional logic. This may be necessary for two reasons. First, if the processor does not have sufficient I/O resources, then it requires multiplexing logic. For instance, if an IP block has a 64-bit data port and the processor has a 32-bit data port, then the processor must divide all reads and writes to the 64-bit port into two accesses. When reading from the IP block, glue logic is needed to capture the 64-bit value which can then be read in 32-bit chunks. Second, the processor is restricted to instruction-cycle timing and may require an I/O sequencer (finite state machine) to guarantee intricate low-level signaling constraints such as latching constraints and fast reaction times. Once the hardware interface has been synthesized, the device-drivers must be updated to reflect the introduction of any glue logic.

The most common method for a processor to communicate with other functional blocks is through a memory-mapped interface called memory-mapped I/O (MMIO). The device-driver running on the processor communicates with other devices through load and store instructions. There are two common techniques for designing a memory-mapped interface. Gupta and DeMicheli in [28] proposed an architecture where the processor writes data to a shared memory. An Application Specific Integrated Circuit (ASIC) monitors the address bus and reads the proper memory locations when appropriate (see Figure 3). Communication from the ASIC to processor occurs in a similar fashion. The advantage of this shared-memory technique is that it requires a very simple hardware interface for all functional blocks and can easily be extended to architectures consisting of multiple processors. However, the shared-memory technique has the drawback of requiring the data to be copied into the memory location and then copied back out to the functional block.

Another common MMIO technique is for the processor to write to "phantom" memory addresses. A range of these addresses are designated as I/O addresses (I/O space in the memory map) and are not mapped to a physical memory. When the processor puts an I/O address on its address bus, comparator logic allows the appropriate functional block to either read the processor's data bus or write a value to the data bus. The main ad-

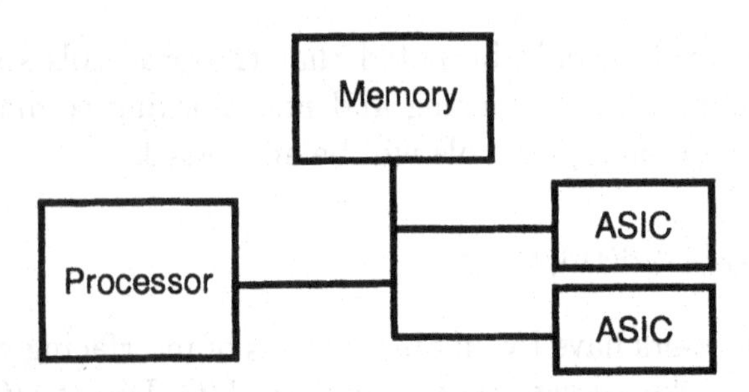

Figure 3. A system architecture where all communication occurs by reading and writing locations in the shared memory.

vantage of the phantom memory approach is that data to and from the processor does not have to be copied to memory and then transferred to its destination.

Automating the synthesis of a memory-mapped interface requires that the functional blocks have a guard-based interface to prevent bus contention from the various IP blocks sharing the processor's data bus. That is, there must be a port or set of ports on the functional block that disables the remaining ports from reading or writing the data bus. The guarded interface to a typical memory chip consists of an address port, data port, read/write port, and a chip enable port (see Figure 4). The chip enable is the guarding port for the memory. IP cores that are designed for a memory-mapped interface contain a guard-based interface. For cores that do not contain such an interface, one must be synthesized. If the functional block is a write-only device (*e.g.* a digital to analog converter), then the guard-based interface is a simple hardware register consisting of an enable port and a data port. However, if the functional block only communicates data back to the processor, then the interface must contain a tri-stateable register. When the block is not being addressed (*i.e.* the processor is not reading its register), then the register outputs must be set to a high impedance value so that the processor or other functional blocks can drive the data bus. If the functional block is a read/write device, then the synthesized interface must contain a bidirectional latch. The read/write lines of the processor are connected to the direction port of the latch to control the flow of data through the latch.

Once the interface has been created for each function block, address decoding logic must be synthesized to control all of the guarding ports. The memory map of the processor must be determined before the address decode logic can be synthesized. A simple way to construct a memory map is to place the physical memory in the lower address space leaving the high address bits to control the address decode logic (see Figure 5).

Another way to view the memory map is shown in Figure 6. In the Chi-

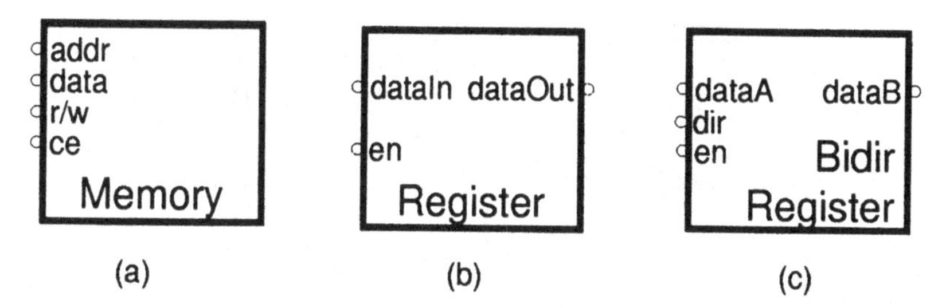

Figure 4. A standard memory interface (a) where the chip enable (CE) port is the guard that disables the remaining ports. Simple interfacing components (b and c) are also shown with guard-based interfaces.

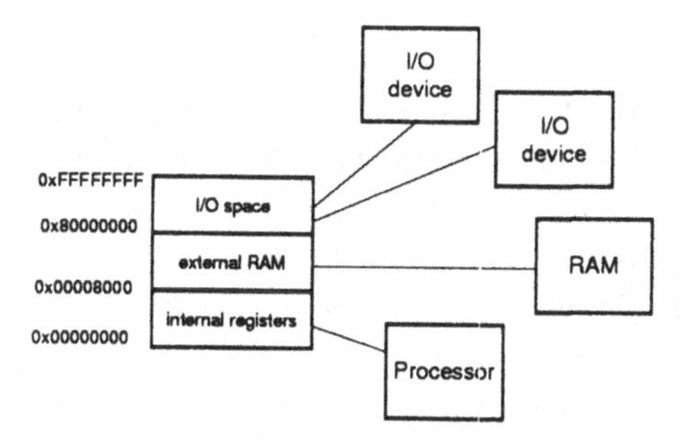

Figure 5. A simple memory map. Addresses with the most significant bit set (0x80000000) are I/O communication instructions.

nook CoSynthesis System's hardware/software interface algorithm [13], the processor's address port is partitioned into three fields: I/O prefix, device-select, and device-control. The fixed width I/O prefix field distinguishes I/O addresses from physical data memory addresses. The device-select field identifies functional blocks in the I/O space that communicate with the processor. The device-control port along with the processor's data port are used to write data to the functional block while only the processor's data port is used to read information from the various devices.

The non-guard ports of the functional block must be assigned to either the device-control ports or the processor's data ports. If the functional block's port can output, then it must be assigned to the processor's bidirectional data port. The remaining non-guard ports are allocated first to the processor's data port in an attempt to maximize the device-select field whose size is bounded by the remaining address bits after all other fields have been assigned.

The device-select field is computed using one of three schemes, from the least expensive to the most expensive in terms of additional glue logic: one-hot, binary, and Huffman encoding. In one-hot encoding, each functional block is selected by one address bit qualified by the I/O prefix. The advan-

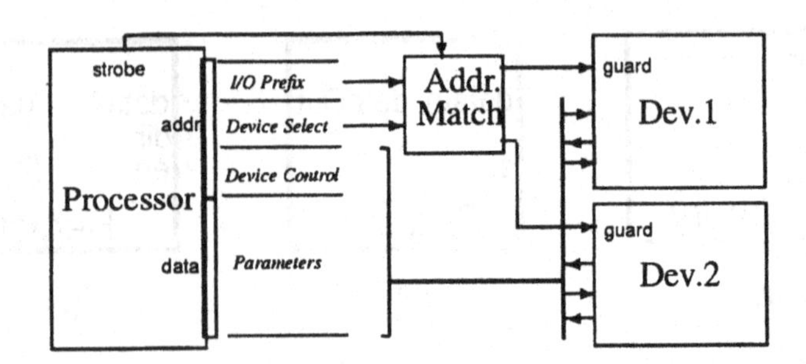

Figure 6. **Another view of a memory map with a focus on the connections of the address bus to external devices. The I/O Prefix field, indicating an I/O instruction is similar to address 0x80000000 in Figure 5.**

tage of one-hot encoding is that it requires very simple address comparator logic (an AND gate). A binary encoding technique of the device-select field encodes n functional blocks with $\lceil log\ n \rceil$ bits. This technique frees $n - \lceil log\ n \rceil$ bits from the one-hot encoding technique by using a $\lceil log\ n \rceil$ input decoder to implement n address comparators.

If both the one-hot and binary techniques are not able to synthesize a solution, then Huffman encoding [17] is attempted on the device-select field. Huffman encoding uniquely identifies each functional block with a variable number of address bits and exploits situations where the blocks require different numbers of bits in the device-control field. A block requiring more device-control bits is addressed by a shorter device-select pattern than a block which requires fewer device-control bits. The widths of the device-control fields are used as cost parameters to a Huffman encoding algorithm. If a block has more available device-select bits, then it is assigned a smaller weight, which yields a longer Huffman encoding. Address comparators of different widths must be synthesized to implement the hardware interface between the processor and the other IP blocks.

After the hardware interface has been synthesized between the processor and the other functional blocks, the processor's device-drivers must be modified to reflect the memory-map and introduced glue logic. The lowest-level device-driver subroutines (those that directly address the functional block's ports) are called SEQs. The SEQs are transformed to reflect the synthesized interface logic by ORing together the address fields (I/O prefix, device-select, device-control) to generate an entire address. Writes to the functional block are transformed into assembly language store instructions. Reads from the functional block and transformed into assembly language load instructions.

Figure 7 shows an example of updating a low-level device-driver SEQ after the address decode logic has been synthesized. The original SEQ is one of a set of SEQs that come with a description for an LCD controller. The

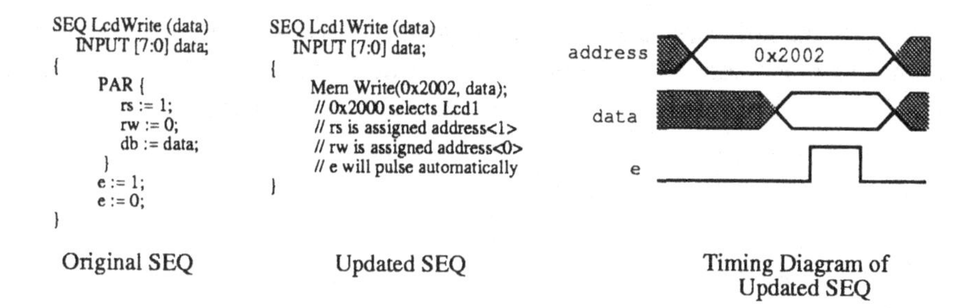

Figure 7. The lowest level device-driver calls (SEQs) must be transformed to reflect the address matching logic for MMIO.

generic SEQ, *LcdWrite* , is transformed to the call *Lcd1Write* indicating that it has been customized for this particular LCD. The LCD is mapped to memory location 0x2000. The control signals *RS* and *RW* are encoded in the device control field (the lower address bits). When the processor executes the routine, the appropriate waveform will be generated that writes the data to the LCD.

3.3. ADDITIONAL HARDWARE TO SUPPORT TIMING CONSTRAINTS

Because a processor is limited to instruction-cycle timing, it may not be possible for the device-driver to satisfy the timing constraints (generate the appropriate waveforms) using the interface synthesis techniques previously mentioned. An I/O sequencer is a hardware finite state machine that interfaces the processor with the functional block. In the simplest case the I/O sequencer is a slave FSM that reads parameters from the processor, generates the appropriate waveforms, and returns any results back to the processor. This requires the processor to initiate all communication with the functional block. A more complex sequencer autonomously interacts with the functional block. This type of sequencer must communicate in parallel with both the processor and the IP block.

An I/O sequencer is synthesized by realizing the appropriate waveforms contained in the library description of the IP block. If this description contains device-drivers, then the SEQs, the lowest-level subroutines that directly address the IP's ports, must be transformed into finite-state machines using behavioral synthesis. These transformed subroutines comprise the sequencer interface to the IP block. Using the hardware-hardware interface techniques described above, a hardware interface between the sequencer and the IP block can be synthesized.

The interface between the processor and the I/O sequencer is synthesized by first creating a simple communication protocol. The protocol involves selecting the appropriate entry point (transformed subroutine), passing the parameters, and synchronization between the processor and the sequencer. Protocol synthesis for the sequencer generates new low-level device-driver subroutines for the processor as well as a hardware interface for the sequencer which can be memory-mapped using the MMIO interfacing techniques described previously.

4. Software Architectures

The increasing computational power of processors coupled with their shrinking feature sizes makes it attractive for system on a chip designers to select multiple processor architectures. The flexibility of core-based technology and the application-specific nature of SOC designs make distributed heterogeneous processor architectures cost effective solutions. This trend is causing more and more of the functionality of an embedded system to be implemented in software. Because timing constraints are an integral part of the correctness of a system design, the software architecture of the system must be aware of the hardware architecture.

Communication can occur between the processor and a functional block either by polling or interrupts. The two methods correspond to a time-driven programming model or an event-driven one [40]. In a time-driven model, the passage of time indicates that certain actions should be performed. Another way to view this is that only timer interrupts are allowed. For instance, an IP block may require that it be given attention every 10ms. A timer interrupt set at 10ms intervals can call a service routine that communicates with the IP block at the given rate. As this example demonstrates, under polling, the functional block must wait for the processor (software) to initiate the communication.

An event-driven model allows for a more reactive style of programming. The generation of events cause certain actions to be performed. With interrupt-based communication, the functional block generates a processor interrupt indicating that it requires attention. A software interrupt service routine is triggered and performs some action such as polling the functional block or setting a flag indicating that the functional requires attention but delaying communication with the block until some time in the future. Using interface synthesis, it is possible to transform an interrupt-driven interface into a polled interface. This entails connecting the interrupt line from the IP block to a synthesized register. The register captures the interrupt signal until the processor polls for this particular event.

Both the time-driven and event-driven models have appropriate domains of applicability. For systems that require high predictability and determinism, the time-driven approach can make stronger timing guarantees about a distributed heterogeneous system architecture. Typically hard real-time and safety-critical systems adopt this methodology. For single processor systems or systems that do not require such a high degree of predictability, the event-driven model allows for adaptability to respond to a highly dynamic environment.

4.1. LOW-LEVEL SOFTWARE ARCHITECTURES

The selection of a time-driven or event-driven model has an impact on the low-level software architecture of the system. The structure of the device-drivers must be consistent with the model chosen. Because of this consistency requirement, the device-drivers accompanying an IP functional block must be flexible enough to support either model. Under a time-driven model, the low-level software architecture consists of subroutines that are called either directly by a software scheduler process running on the processor or by other high-level software processes.

The low-level software architecture for an event-driven model is slightly more complicated than a time-driven one. The device-driver not only consists of subroutines, but also interrupt service routines (ISR). Modern microprocessors have vectored interrupts that determine the priority (order) that any ISR will be called if multiple interrupts occur simultaneously. The frequency of an interrupt along with the time an ISR will take to execute are factors in determining the priority of the interrupts [31]. A typical scenario is for an IP block to generate an interrupt when it requires attention. The processor saves the state of the currently executing code and then calls the appropriate ISR. When the ISR terminates, the processor restores the state of the preempted code and resumes execution. Preemption causes the execution of the system to be less predictable than a time-driven model. The maximum frequency of interrupts generated from the IP blocks must be considered so that timing constraints are not violated.

4.2. RUN-TIME SYSTEMS AND REAL-TIME OPERATING SYSTEMS

The device-drivers provide an interface between the low-level software architecture and the high-level software architecture. This higher-level is called a run-time system. It is the run-time system that provides an abstraction to the user-level software processes and allows them to communicate with each other and the IP blocks surrounding the processor (see Figure 8).

These communication abstractions include shared-memory, shared queues, and message send and receive subroutine calls. Consider the example in Figure 9 of two software processes, running on different processors, that exchange messages. The processors communicate through shared memory. Both user-level processes create a logical channel named *chan* at address *chanAddr*. The sender process creates a message and sends it out on *chan*.

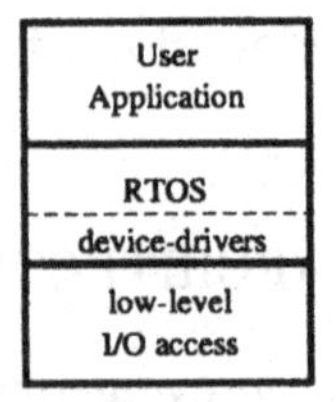

Figure 8. Software architecture layers. The RTOS (run-time system) provides an API that allows the user application to control I/O devices.

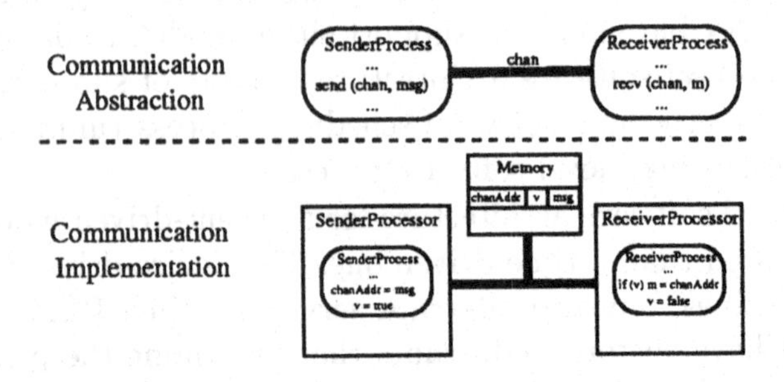

Figure 9. Two processes that communicate through shared memory. The user-level code calls the *send* and *recv*functions which are implemented by the run-time system. Note that additional synchronization primitives are needed for a correct implementation.

The receive process waits for a message to be delivered on channel *chan*. The run-time system delivers the messages sent out on *SenderProcessor*'s channel *chan* and updates the appropriate data structures (the message valid (v) bit in Figure 9) so that *ReceiverProcess* can detect the arrival of the message.

If the run-time system has mechanisms to support timing requirements, then it is called a real-time operating system (RTOS). The two main functions of an RTOS are scheduling the processes and supporting communication. The real-time community has investigated many scheduling techniques to meet various types of timing requirements. Four common real-time scheduling techniques are Rate Monotonic (RM) [45] [2], Earliest Deadline First (EDF) [45], cyclic executive or round-robin, and complex or custom static scheduling [60].

RM scheduling was proposed in 1973 by Liu and Layland [45]. All processes are assumed to be periodic. The processes are sorted by increasing periods with priorities assigned in rank order. The process with the shortest period is given the highest priority, the process with the next shortest period is given the highest unassigned priority, *etc.* until all processes have been assigned a priority. RM has the desirable feature that based on processor utilization calculations it can be determined if all timing requirements will be respected for systems composed of a single processor (and meeting a few other criteria).

Instead of using the static priorities of RM scheduling, EDF dynamically determines the priority for each process executing on the processor. Under EDF, the process with the closest deadline to the current time is allowed to execute on the processor. EDF scheduling allows processor utilizations to approach 100% while still respecting all deadlines.

Cyclic executive scheduling is a very simple technique. The scheduler gives the processor to each process in a predetermined order. The technique works well when the execution time and timing requirements of all the processes are very similar or if the processor is under-utilized and the timing requirements are coarse-grained.

When the above techniques are insufficient to make the needed real-time guarantees, designers must resort to either complicated static scheduling algorithms [60] or create their own custom static schedule. Such techniques require a thorough analysis of the execution behavior of the processes along with their timing requirements. This type of static scheduling is appropriate for hard real-time systems (*i.e.* those systems where all timing constraints must always be respected) whose architectures violate the assumptions made by other the scheduling techniques.

The IP block description must contain the necessary information regarding timing constraints to interact with the given real-time scheduler. Because an IP block can be used with different processor cores, the execution time requirements of the device-drivers on the given processor cannot be in the block description. Instead, estimation tools such as Cinderella [43] must be used to translate code segments into timing estimates. These code segments may also need to be transformed by a scheduling algorithm such as proposed by Chou and Borriello [11] so that low-level timing requirements are not violated. Consider the case of a code fragment executing on a 100 MHz processor and the same code fragment executing on the same processor running at 10 MHz. The IP block may be unable to communicate with the faster processor requiring the code fragment to either be preempted or NOP instructions executed to slow down the communication. From the above example it is clear that both minimum and maximum timing constraints must be specified in the interfacing software accompanying an IP block.

5. High-level Protocols and Communication Synthesis

There are many different types of communication protocols used in system designs and at different stages in the design process. Consider the following

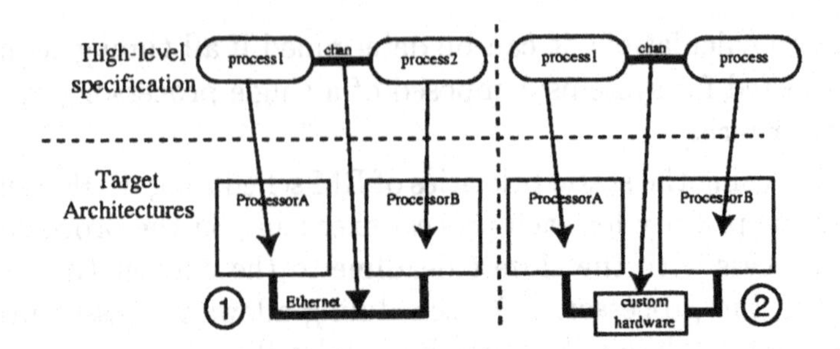

Figure 10. An IP-based design environment where a system architect evaluates mapping the same specification to different architectures.

design scenario shown in Figure 10. A designer has two processes that need to communicate with each other. The high-level or abstract communication protocol chosen is blocking message-passing. This decision is made independent of the system architecture. The designer now wishes to compare two different system architectures. The first architecture consists of two processors connected by an Ethernet bus. The second architecture consists of two processors connected by custom hardware. For each architecture the designer maps (assigns) a process to each processor (see Figure 10). The blocking message-passing semantics must be realized for both architectures. This involves determining the appropriate parameters for the Ethernet protocol used in architecture 1 and a direct hardware implementation of the blocking message passing protocol for architecture 2. Communication synthesis, the automatic generation or customization of the communication protocol details, is necessary to support a design environment that allows system architects to select various IP blocks for a particular implementation.

The communication synthesis research efforts can be divided based on their treatment of communication protocols. Most efforts have focused on custom protocols or a direct implementation of the abstract protocols while other efforts have attempted to use standard bus protocols. Another main difference in the various research efforts is the amount of designer intervention required ranging from fully automatic systems that determine the system architecture to systems that are more interactive.

5.1. DIRECT IMPLEMENTATION AND CUSTOM PROTOCOLS

Direct implementation of a high-level communication protocol can be as simple as having the device-driver implement the appropriate semantics or may involve custom hardware such as the I/O sequencers mentioned earlier. The following research efforts support either direct implementation of the abstract protocols or implement custom protocols to interface IP blocks.

5.1.1. *Polis*

The Polis project [10] has focused on control-dominated applications with system architectures composed of a single processor surrounded by custom or library hardware (*e.g.*,micro-controller peripherals and other IPs). Polis uses Co-design Finite State Machines (CFSM) as the internal representation for a system description, separating communication, function and timing of the system.

1. The functional model of each CFSM is a Finite State Machine extended with a data path operating on internal variables.
2. The communication model is globally asynchronous, locally synchronous, with non-blocking finite buffers between CFSMs.
3. The timing model is derived by *mapping* each CFSM to an architectural entity (a processor, special-purpose hardware or custom hardware), and takes the the form of

   – an estimated delay for an execution of each transition,

   – possibly a scheduling mechanism, for CFSMs mapped to a processor.

   Before a mapping is defined (*i.e.*, at the purely functional level), each transition edge in the state machine has a non-deterministic finite, non-zero reaction time.

By restricting the communication semantics, Polis is able to verify system specifications using programs such as VIS [9] and to synthesize a complete mixed hardware/software implementation without any manual intervention after partitioning.

The CFSMs mapped to software are translated into the C programming language. An application-specific operating system consisting of a scheduler and I/O drivers (both based on polling and interrupts) is generated to run on the processor. The CFSMs mapped to synthesized hardware are translated into a synchronous implementation, modeled using an HDL (VHDL or Verilog), while those mapped to special-purpose (library) hardware are simply instantiated.

All communication within software or between software and hardware occurs through shared memory, I/O ports or memory-mapped I/O. Multiple busses and bus bridges are not supported. The synthesized hardware includes the address decoders, multiplexors, latches and glue logic to interface a hardware CFSM with another hardware CFSM or a software CFSM. Special-purpose hardware must follow a simple, data/strobe based protocol in order to be interfaceable with other CFSMs.

5.1.2. *Interface-Based Design Methodology*

In [52] a new design methodology, Interface-Based Design, was proposed that treated behavior and communication as orthogonal elements. The goal of this research is to make it easier to reuse communication components (hardware and software). Their approach is to continually refine the communication between functional blocks until an implementation is realized. The communication model is based on token passing. A token is a complete communication between two functional blocks. For example, a token can be an entire web page, Ethernet packet, IP packet, or header field of a bus packet. The token is successively refined to finer and finer granularity. At intermediate refinement steps the system can be simulated to ensure the correctness of the refinements and measure the impact on system performance.

5.1.3. *CoWare*

The CoWare co-design environment supports heterogeneous architectures composed of multiple processors and application-specific IP blocks. [5]. The focus has been on digital telecommunication (data-flow) applications. A system description is composed of processes containing multiple threads which communicate through ports connected by point to point links called channels. The communication model for interprocess communication is based on Remote Procedure Call (RPC) [4]. RPC semantics are the same as the abstract communication protocol of blocking message passing. A protocol can be associated with a hierarchical channel to achieve other abstract protocols. For example, adding a FIFO process to a channel gives the abstract protocol of non-blocking message passing. The protocol itself is hierarchical and can be refined down to the physical terminals that require particular timing diagrams to implement the communication. Hierarchical ports can be refined to include protocol information such as a parity bit or cyclic redundancy check. Intraprocess communication occurs via shared memory with the system designer responsible for the appropriate synchronization semantics. All of the processes mapped to the same processor are merged together into a single process. During the merge, all of the threads and data structures of the various processes are grouped into the same process. Merging allows various optimizations such as inlining communication calls and optimizations that span the original processes. Another benefit of merging is that the processor only needs a simple run-time system to execute the process. An algorithm determines if the processor communicates with other blocks through I/O ports, memory-mapped I/O, or interrupt-driven control.

5.1.4. *Solar*

SOLAR is an intermediate language to support system level modeling and synthesis [38]. Recent work has focused on protocol selection and interface generation for a system composed of processes communicating through abstract channels and executed on multiple processors [18]. The communication semantics are based on RPC (as in CoWare). A library of communication units provides different types of communication services and protocols along with their implementations (a mixture of hardware and software). The types of protocols supported include blocking (handshake) and non-blocking (FIFO). Based on the communication requirements of the system description, the appropriate communication units are selected from the library. This step fixes the protocol used by each abstract channel. The abstract channels can be merged (time-multiplexed) onto a communication unit. Interface synthesis techniques are used to permit communication between the processor and the chosen communication unit.

5.1.5. *SpecSyn*

SpecSyn [26] generates busses for communication between two processes using a technique called interface refinement. By analyzing the size of data communicated along with rate of data generation, they determine the bandwidth of the generated bus (number of wires and rate of transfer) and merge the communication channels onto the bus. Next, an abstract protocol is directly implemented using additional control wires. For instance, a blocking protocol will need control lines to implement the desired semantics. Arbiters are synthesized to control access to the bus. The software send and receive subroutines are modified to use the synthesized bus.

5.1.6. *Communication synthesis with arbitrary topology*

Yen and Wolf have looked at communication synthesis in the context of deriving a heterogeneous distributed system architecture with an arbitrary topology [61]. The system behavior is modeled as a set of communicating processes which are organized in partially ordered sets called tasks. Each task has associated with it an invocation rate and a deadline. Because tasks may execute at different rates, inter-task communication does not need to be synchronized. This is the abstract protocol of non-blocking message-passing. A library contains processing elements (PEs) and busses. Each PE has associated with it a component cost in dollars and an execution performance number from which process computation times can be calculated. Bus attributes include cost in dollars and the communication time per unit of data. The co-synthesis algorithm determines the number and type of PEs and busses, the interconnect topology, and the scheduling of tasks on the PEs and messages on the busses.

5.2. STANDARD PROTOCOLS

The research efforts examined so far have focused on a direct implementation of the abstract protocols or have synthesized communication for a custom protocol. Another interesting area of research concerns the problem of synthesizing communication for a fixed protocol. Standards committees spend a tremendous amount of effort to precisely define the semantics for a particular bus protocol. The advantage of a standard bus protocol is that IP designers need only support a limited number of interfaces. Moreover, the software to control these IP blocks can be written using a standard API. However, an API by itself is insufficient to realize system communication. Consider the case of different IP blocks connected to a processor via a standard bus. The arbitration policy of the bus must be considered when assigning protocol attributes so that the timing requirements of the IP blocks are respected. That is, simply having a *send* subroutine is insufficient because the timing information must also be represented. Certain bus protocols such as I^2C [47] arbitrate for the bus based on the processor's priority. Other bus protocols such as Controller Area Network [36] arbitrate for the bus based on the individual message priority. To allow a designer to write a high-level specification independent of the final system architecture, the *send* subroutine must be transformed to incorporate architecture specific details such as which processor is executing the sending and receiving processes and the communication protocol details such as the priorities and bus message formats. Such information is necessary for communication to occur for a particular implementation but is not available until after the specification has been mapped to an architecture as in Figure 10.

5.2.1. *Object-oriented Communication Library (OOCL)*

OOCL is an object-oriented communication library that provides pre-implemented channel-based send/receive communication primitives [57]. This approach is intended for system architectures where at least one of the communicating components has a flexible interface (*e.g.*, a processor or FPGA). The designer chooses the OOCL channel for a particular protocol to realize system communication. This channel encapsulates the protocol's implementation details for a particular architecture. In order to support new architectures OOCL must be rewritten or ported.

5.2.2. *Chinook*

The Chinook project has focused on the problems of mapping and composition for real-time control-dominated applications executed on distributed heterogeneous architectures [49] [12]. A designer specifies a behavioral description consisting of communicating processes. Each process exports a set of modes which are mappings of input events to the subroutines (handlers) that consume them. During a system assembly phase, the designer assigns control composition constraints to the various modes. An example of a constraint is mutual exclusion which mandates that two modes may never be active at the same time. The processes communicate with each other via messages sent through ports connected by channels. The communication is uni-directional one to many. That is, an output port can be connected to multiple input ports but an input port can only have one source of data. For each output port, the designer selects the high-level abstract communication protocol, either blocking or non-blocking message passing, along with a deadline constraint which is the maximum delay to deliver the message to its destination. For each input port, the designer selects the appropriate queuing semantics. The choices include the size of the input FIFO and the action to take when the FIFO is full. Once all of the communication parameters have been chosen, the designer can simulate the behavioral description.

The designer creates a system architecture that may consist of heterogeneous processors connected by standard bus protocols in any topology. The designer then maps the processes to the processors and the messages to the busses. The mode constraints are automatically transformed into communications and distributed among the processors. If there is not a direct connection between two communicating processes, intermediate hop processes are automatically inserted to route a message from one bus onto another bus. A heuristic partitions the deadline among the various hops. Next the protocol attributes for the particular bus protocols are synthesized. Customized software is generated to create a message suitable for transmission over the given bus. This new message incorporates the synthesized protocol attributes along with additional routing information so that the original message can be reconstructed and delivered to its destination. For each processor, a customized real-time operating system is generated that includes a real-time scheduler (selected by the designer), a message routing process for message delivery, the appropriate device-drivers for the processor, and initialization code to customize the device-drivers and instantiate the kernel and user processes.

5.3. OTHER APPROACHES

Communication and interface synthesis is an active area of research. Unfortunately, we are unable to cover all the approaches but refer the reader to the following further readings [23] [20] [27] [24] [37] [46].

6. Co-Simulation and Co-Verification

Verification has traditionally been a very important aspect of interface design. In particular, given a set of partners, each one following its own protocol:

1. If the protocols are different (an interface must still be inserted), one would like to answer questions such as:
   - Can these partners communicate? Are there any intrinsic incompatibilities between the protocols? For example, if the sender is synchronous without flow-control mechanisms and the receiver is asynchronous, there can be a need for (unimplementable) infinite buffering.
   - What is the achievable performance level, with or without resource constraints?
2. If the protocols are the same, or at least directly compatible (e.g., if the receiver has a higher potential bandwidth or supports a richer functionality than the sender):
   - Are the interfaces logically correct, i.e. can they deadlock, do they provide fair access to all the partners, ...?
   - Does the performance (latency and throughput) satisfy all the requirements?

The answers to these questions can come from different sources, with different levels of accuracy, completeness and computational cost. A key factor in obtaining such answers is to use the *abstraction level* that is most appropriate for the task at hand. An orthogonal issue is the mechanism, simulation or exhaustive formal verification, that is used to obtain the answer. In all cases, a sound formal Model Of Computation (MOC) is a basic prerequisite for any successful verification task [19].

It is unfortunately impossible, due to time and space constraints, to give a fair treatment to all the approaches that have been proposed and used, even in the restricted domain of interface verification. We will just review the main concepts, and refer the interested reader to the original papers.

6.1. SIMULATION

Simulation of user-given patterns has traditionally been the method of choice to validate almost any digital system, in particular interfaces. Building testbenches (sets of stimuli and monitors that analyze the system responses) has become one of the most important and time-consuming tasks of embedded system design (up to 70-80% of the total Non-Recurring Engineering Costs in some cases). Even though there are signs that these rising costs will require new and more effective techniques (such as formal verification or hybrid formally directed simulation methods [33]), simulation will conceivably remain the workhorse for some time. Hence, researchers are continuously looking for ways of making simulation faster and faster, in order to keep up with the growing complexity of the circuits and systems that are being designed.

In the case of interfaces, the current view is that the level of abstraction provided by Hardware Description Languages, that at most allows one to declare that an architectural component has interface signals of a structured data type [48], is insufficient to cope with the growth in complexity mentioned above. Interface verification requires a layered view, similar to the ISO/OSI standard in the telecommunications world. While there have been some recent proposals to extend standard HDLs to support this layered view [35], most recent research has focused on custom simulation engines.

Both Rowson *et al.* [52] (described in Section 5.1.2) and Hines *et al.* [32] advocate decomposing the analysis of the performance and correctness of a complex embedded system, and in particular of its interfaces, into abstraction layers, such as token-passing, abstract bus, detailed signal.

6.1.1. *Selective focus in co-simulation*

In [32], the cosimulation environment Pia (Processor Interface Accurate) simulator is described. A technique called *selective focus* is used to increase simulation performance. Consider the simulation of a complex SOC architecture that implements a portable web browser composed of multiple IP blocks communicating with each other through various interfaces. If a designer wants to debug the interface between the processor and the display, the traditional approach has been to simulate every functional block at the highest possible degree of accuracy. However, the designer is not interested in every interface. Hence simulating the entire web browser with cycle-accuracy is a tremendous waste of computation time. Instead, selective focus allows the designer to dynamically select the level of accuracy for each interface. The processor/display interface being debugged can be set to a high level of accuracy while the surrounding interfaces, such as the wireless RF transceiver can be set to a lower degree of accuracy. Pia keeps track of the timing information so that the surrounding components still provide realistic workloads to the interfaces being simulated with high accuracy.

6.1.2. *Commercial co-verification solutions*

In the commercial arena, both Mentor Graphics' Seamless and Synopsys' Eagle co-verification environments support selective focus as a mechanism to reduce simulation time. These approaches are oriented to the co-simulation of embedded hardware and software, by using a clock cycle-accurate Instruction Set Architecture model of the processor (faster, easier to develop, and safer to distribute than a full-blown RTL model). The HDL model of the hardware runs on a commercial VHDL simulator, connected via a special simulation backplane to the ISA model. The speed-up of the HDL simulation is obtained by *filtering* the data that is being passed between the processor and the hardware, in order to reduce simulation time.

In particular, the first phase of the design is to debug the memory interfaces, by allowing instruction and data fetches to be exchanged between the ISA and HDL simulators. After the decoders, tri-state buffers and so on are verified, most memory traffic, except for reads and writes that refer to I/O addresses, can be executed only by the ISA without being passed to the hardware simulator, thus saving a significant amount of CPU time.

6.2. FORMAL VERIFICATION

Formal verification has complementary capabilities with respect to simulation, and hence should be used together, in order to successfully tackle the interface verification problem.

A key aspect of this synergy, as mentioned above, is the use of a sound, verifiable Model Of Computation to describe:

- the specification (the protocol followed by the partners in the communication, and the properties, such as absence of deadlock, that the communication must satisfy),
- the implementation (the interface circuitry and/or software),
- the environmental conditions (*e.g.*, higher-level flow control protocols, assumptions on maximum bus loads due to other sources, and so on).

The most common MOC used in the area of interfaces and protocols is that of Communicating Finite State Machines. Finite State Machines are a convenient model for control-dominated applications, and are suitable to specify complex handshaking, error handling, and so on.

Communication between the FSMs can be either synchronous or asynchronous (regardless of the underlying implementation). In the synchronous case, each FSM makes a transition per clock tick, while in the asynchronous case only transitions labeled with the same event are "shared" by the communicating partners (see below for an example).

Verification algorithms for FSMs are based on *Model Checking* (MC) [21, 16]. MC can be used to verify if a system, modeled as a Finite State Machine, satisfies one or more properties, specified using a temporal logic,

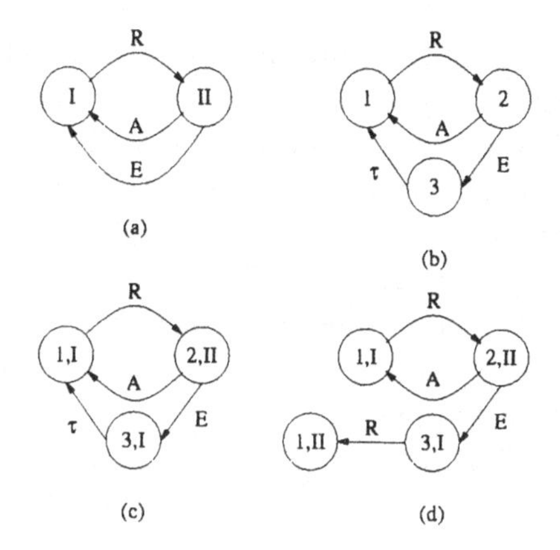

Figure 11. Example of Model Checking

e.g., Computational Tree Logic (CTL)[1].

The basic notion is that of a finite state system, whose states are labeled with "facts" about the system (technically, atomic CTL propositions) such as "in this state the receiver replies to a request" (often a desirable condition...) or "in this state both masters access the bus simultaneously" (often an undesirable condition...). CTL formulae are composed recursively of:

- atomic propositions,
- logical connectors, such as *and* ($\wedge$), *or* ($\vee$), *not* ($\neg$),
- temporal connectors such as *until* (U), *along all possible paths in time from now* ($\forall G$), *along some possible path in time from now* ($\exists G$), *for all paths eventually* ($\forall F$), *for some paths eventually* ($\exists F$).

Let us consider a simple case of an interface with error detection. The protocols followed by the two partners are shown in Figure 11.(a-b).

1. The sender has two states, first sending a *request* on signal R, then waiting for either an *acknowledgment* of correct reception on signal A, or an *error* indication on signal E.
2. The receiver has a similar behavior, but in case of error, it requires one *internal action* (labeled τ) to resynchronize, and hence it has a third state.

Let us consider first an *asynchronous* communication mechanism, which leads *in this case* to satisfying the required properties[2]. The composed FSM, using the CSP-style rendezvous mechanism [34], is shown in Figure 11.(c).

[1]Model checking of other logics, as well as formal verification of finite state systems based on automata and language containment [41] are also possible.

[2]The example is used to illustrate that both function and communication are significant when one considers properties of a formal model of a composite system.

Note how the state space of the composition is the product of the two state spaces, and the two FSMs synchronize on common edge labels.

We can now verify a *liveness* property of this composition. Namely, that in any state eventually a transmission (signal R) will happen again. The state in which R occurs is state $(1, I)$, so we want to check if no matter what happens, we will eventually go through this state. This excludes both deadlock conditions (the composed FSM gets into a state from which it cannot exit) and livelock conditions (the composed FSM can loop forever in states where R never happens).

The CTL formula that represents this property is

$$\forall G \forall F(1, I)$$

it means that it is globally true in all successors of the initial state ($\forall G$) that for all futures eventually ($\forall F$) atomic formula $(1, I)$ (meaning that signal R will happen again) will hold. We can see by inspection that it is satisfied. The MC algorithm would proceed from inside of the CTL formula out, and backward in time over the FSM, as follows:

1. label state $(1, I)$ with the formula $(1, I)$ (trivially true),
2. label $(1, I)$ with $\forall F(1, I)$ (also trivial),
3. label $(3, I)$ with $\forall F(1, I)$ (all paths from $(3, I)$ pass through $(1, I)$),
4. label $(2, II)$ with $\forall F(1, I)$ (same as above),
5. now all states are labeled with $\forall F(1, I)$, and hence all states can also be labeled with the final formula $\forall G \forall F(1, I)$,
6. the initial state is labeled with the outermost formula, and hence the property is satisfied.

A *synchronous* composition of the two interface FSMs would not, however, satisfy the property. The composed FSM is shown in Figure 11.(d). It cannot leave state $(3, I)$ because the sender emits R only in the (global) time instant when the receiver performs its internal action (waits for one clock tick). From then on, the sender is stuck forever, waiting for an acknowledge or an error, while the receiver is stuck waiting for a request.

7. Conclusions

In this paper we have outlined the main problems in interface design, with special attention to issues like System-On-a-Chip and Intellectual Property interfacing. We also discussed several current approaches to interface design, for various abstraction levels and various types of interacting partners (hardware, software, ...) and protocols (custom, standard, ...).

We advocated a formal methodology, based on sound models, synthesis and verification, in order to maximize designer productivity and minimize interfacing problems. Verification of protocol properties at the highest possible abstraction level is a key to achieve first-time-right implementations.

For the future, we believe that designers should be able to specify the system function without caring too much about the decomposition at the implementation level, and be able to perform such decomposition while preserving important functional properties. Functional interfaces will then be mapped (by hand or automatically) on architectural interfaces, while drivers and glue logic will be synthesized automatically.

A design environment supporting this design flow requires system-level tools from specification at the high-level, to hardware and software synthesis (to realize implementations), to formal verification and co-simulation (to guarantee correct systems). Its benefits include: a more thorough analysis of a larger portion of the design space of feasible architectures; a significant reduction in the time-to-market; and a greater confidence that the systems will actually work as expected after being built.

There are many problems that need to be solved before such a design environment can be created. For instance, more work is needed to determine the type of information necessary in an IP library so that it can be fully integrated with the CAD tools. Such information, in addition to the functional and interfacing requirements, might include testing and other high-level semantic data. An interesting topic is providing an IP library whose elements can be implemented in hardware, software, or in both. The actual implementation choice would be made dependent on the performance requirements of a given system. The current generation of hardware synthesis tools need to be made more aware of the software environment that will execute on the generated hardware. Finally, the approaches and tools described in this paper are of course open to integration, improvements, and optimizations.

References

1. P. A. Beerel and T. H-Y. Meng. Automatic gate-level synthesis of speed-independent circuits. In *Proceedings of the International Conference on Computer-Aided Design*, November 1992.
2. A. A. Bertossi and A. Fusiello. Rate-monotonic scheduling for hard-real-time systems. *European Journal of Operational Research*, 96(3):429–43, February 1997.
3. K. Bilinski and E. Dagless. High level synthesis of synchronous parallel controllers. In *Application and Theory of Petri Nets*, June 1996.
4. A. D. Birrell and B. J. Nelson. Implementing remote proceduce calls. *ACM Transactions on Computer Systems*, 2(1):39–59, February 1984.
5. I. Bolsens, H. J. DeMan, B. Lin, K. Van Rompaey, S. Vercauteren, and D. Verkest. Hardware/software co-design of digital telecommunication systems. *Proceedings of the IEEE*, 85(3):391–418, March 1997.
6. Gaetano Borriello. *A New Interface Specification Methodology and its Application to Transducer Synthesis.* PhD thesis, University of California, May 1988. Report No. UCB/CSD 88/430.

7. Gaetano Borriello. Specification and synthesis of interface logic. In R. Camposano and W. Wolf, editors, *High Level VLSI Synthesis*, pages 153–176. Kluwer Academic Publishers, Dordrecht, Netherlands, 1991.
8. Gaetano Borriello. Formalized timing diagrams. In *Proceedings of the European Conference on Design Automation*, pages 372–7, March 1992.
9. R. K. Brayton, G. D. Hachtel, A. Sangiovanni-Vincentelli, F. Somenzi, A. Aziz, S-T. Cheng, S. Edwards, S. Khatri, Y. Kukimoto, A. Pardo, S. Qadeer, R. K. Ranjan, , S. Sarwary, T. R. Shiple, G. Swamy, and T. Villa. VIS: a system for verification and synthesis. In *Computer Aided Verification Proceedings*, pages 428–32, August 1996.
10. M. Chiodo, D. Engels, P. Giusto, H. Hsieh, A. Jurecska, L. Lavagno, K. Suzuki, and A. Sangiovanni-Vincentelli. A case study in computer-aided co-design of embedded controllers. *Design Automation for Embedded Systems*, 1(1-2):51–67, January 1996.
11. Pai Chou and Gaetano Borriello. Interval scheduling: Fine-grained software scheduling for embedded systems. In *Proceedings of the Design Automation Conference*, June 1995.
12. Pai Chou, Ken Hines, Kurt Partridge, and Gaetano Borriello. Control generation for embedded systems based on composition of modal processes. In *to appear in Proceedings of the International Conference on Computer Aided Design*, 1998.
13. Pai Chou, Ross B. Ortega, and Gaetano Borriello. Interface co-synthesis techniques for embedded systems. In *Proceedings of the International Conference on Computer Aided Design*, pages 280–287, November 1995.
14. T.-A. Chu. On the models for designing VLSI asynchronous digital systems. *Integration: the VLSI journal*, 4:99–113, 1986.
15. T.-A. Chu. *Synthesis of Self-timed VLSI Circuits from Graph-theoretic Specifications*. PhD thesis, MIT, June 1987.
16. E. M. Clarke, D. E. Long, and K. L. McMillan. A language for compositional specification and verification of finite state hardware controllers. *Proceedings of the IEEE*, 79(9), September 1991.
17. Thomas H. Cormen, Charles E. Leiserson, and Ronald L. Rivest. *Introduction to Algorithms*. The MIT Press, 1990.
18. Jean-Marc Daveau, Gilberto Fernandes Marchioro, Tarek Ben-Ismail, and Ahmed Amine Jerraya. Protocol selection and interface generation for hw-sw codesign. *IEEE Transactions on VLSI Systems*, 5(1):136–144, March 1997.
19. S. Edwards, L. Lavagno, E.A. Lee, and A. Sangiovanni-Vincentelli. Design of embedded systems: formal models, validation, and synthesis. *Proceedings of the IEEE*, 85(3):366–390, March 1997.
20. Michael Eisenring and Jurgen Teich. Domain-specific interface generation from dataflow specifications. In *Proceedings of the Sixth International Workshop on Hardware/Software Codesign*, pages 43–7, March 1998.
21. E. A. Emerson and E. M. Clarke. Using branching-time temporal logic to synthesize synchronization skeletons. *Science of Computer Programming*, 2(3):241–266, 1982.
22. D. Filo, D. Ku, C. Coelho, and G. De Micheli. Interface optimization for concurrent systems under timing constraints. *IEEE Transactions on Very Large Scale Integration*, 1(3):268–281, September 1993.
23. Franz Fischer, Annette Muth, and Georg Färber. Towards interprocess communication and interface synthesis for a heterogeneous real-time rapid prototyping environment. In *Proceedings of the Sixth International Workshop on Hardware/Software Codesign*, pages 35–9, March 1998.
24. Laurent Freund, Denis Dupont, Michel Israel, and Frederic Rousseau. Interface optimization during hardware-software partitioning. In *Proceedings of the Fifth International Workshop on Hardware/Software Codesign*, pages 75–9, March 1997.
25. Steve B. Furber. *ARM System Architecture*. Addison Wesley Longman, 1996.
26. Daniel D. Gajski, Frank Vahid, Sanjiv Narayan, and Jie Gong. *Specification and Design of Embedded Systems*. P T R Prentice Hall, 1994.

27. Guy Gogniat, Michel Auguin, Luc Bianco, and Alain Pegatoquet. Communication synthesis and hw/sw integration for embedded system design. In *Proceedings of the Sixth International Workshop on Hardware/Software Codesign*, pages 49–53, March 1998.
28. Rajesh K. Gupta and Giovanni DeMicheli. System-level synthesis using reprogrammable components. In *Proceedings of the European Conference on Design Automation*, pages 2–7, March 1992.
29. N. Halbwachs. *Synchronous Programming of Reactive Systems*. Kluwer Academic Publishers, 1993.
30. D. Harel. StateCharts: a visual formalism for complex systems. *Science of Programming*, 8, 1987.
31. Thomas L. Harman. *The Motorola MC68332 Microcontroller:product design, assembly language programming, and interfacing*. Prentice-Hall, 1991.
32. K. Hines and G. Borriello. Dynamic communication models in embedded system co-simulation. In *Proceedings of the Design Automation Conference*, pages 395–400, June 1997.
33. R.C. Ho, C. Han-Yang, M.A. Horowitz, and D.L. Dill. Architecture validation for processors. In *Proceedings of the 22nd Annual International Symposium on Computer Architecture*, June 1995.
34. C. A. R. Hoare. *Communicating Sequential Processes*. Prentice-Hall, 1985.
35. ICL Design Automation. *The VHDL+ language*. website: http://www.icl.com/da/svdocs/vhdlplus.htm.
36. ISO 11898. *Road vehicles - Interchange of Digital Information - Controller Area Network (Can) for High-Speed Communication*, 1st edition, 1993.
37. Dan C. R. Jensen, Jan Madsen, and Steen Pedersen. The importance of interfaces: A hw/sw codesign case study. In *Proceedings of the Fifth International Workshop on Hardware/Software Codesign*, pages 87–91, March 1997.
38. A. A. Jerraya and K. O'Brien. SOLAR: An intermediate format for system-level modeling and synthesis. In J. Rozenblit and K. Buchenrieder, editors, *Computer Aided Software/Hardware Engineering*, pages 147–175. IEEE Press, Piscataway, NJ, 1994.
39. A. Kondratyev, M. Kishinevsky, B. Lin, P. Vanbekbergen, and A. Yakovlev. Basic gate implementation of speed-independent circuits. In *Proceedings of the Design Automation Conference*, 1994.
40. H. Kopetz. Should responsive systems be event-triggered or time-triggered? *IEICE Transactions on Information and Systems*, E76-D(11):1325–32, November 1993.
41. R. P. Kurshan. *Automata-Theoretic Verification of Coordinating Processes*. Princeton University Press, 1994.
42. L. Lavagno and A. Sangiovanni-Vincentelli. *Algorithms for synthesis and testing of asynchronous circuits*. Kluwer Academic Publishers, 1993.
43. Y-T-S. Li and S. Malik. Performance analysis of embedded software using implicit path enumeration. *IEEE Transactions on Computer-Aided Design of Integrated Circuits and Systems*, 16(12):429–43, December 1997.
44. B. Lin and S. Vercauteren. Synthesis of concurrent system interface modules with automatic protocol conversion generation. In *Proceedings of the International Conference on Computer-Aided Design*, pages 101–108, 1994.
45. C. L. Liu and James W. Layland. Scheduling algorithms for multiprogramming in a hard-real-time environment. *Journal of the Association for Computational Machinery*, 20(1), January 1973.
46. J. Madsen and B. Hald. An approach to interface synthesis. In *Proceedings of the Eighth International Symposium on System Synthesis*, pages 16–21, September 1995.
47. R. Mitchell, N. G. Damouny, C. Fenger, and A. P. M. Moelands. An integrated serial bus architecture: principles and applications. *IEEE Transactions on Consumer Electronics*, CE-31(4):687–99, November 1985.

48. Institute of Electrical and Electronics Engineers. *IEEE standard VHDL language reference manual.* IEEE, 1994.
49. Ross B. Ortega and Gaetano Borriello. Communication synthesis for distributed embedded systems. In *to appear in Proceedings of the International Conference on Computer Aided Design*, 1998.
50. R. Passerone, J. Rowson, and A. Sangiovanni-Vincentelli. Automatic synthesis of interfaces between incompatible protocols. In *Proceedings of the Design Automation Conference*, June 1998.
51. L. Y. Rosenblum and A. V. Yakovlev. Signal graphs: from self-timed to timed ones. In *International Workshop on Timed Petri Nets, Torino, Italy*, 1985.
52. James A. Rowson and Alberto Sangiovanni-Vincentelli. Interface-based design. In *Proceedings of the Design Automation Conference*, pages 178–83, June 1997.
53. A. Seawright, U. Holtmann, W. Meyer, B. Pangrle, and J. Buck. A system for compiling and debugging structured data processing controllers. In *Proceedings of the European Design Automation Conference (EURO-DAC)*, pages 86–91, September 1996.
54. A. Seawright and W. Meyer. Partitioning and optimizing controllers synthesized from hierarchical high-level description. In *Proceedings of the Design Automation Conference*, pages 770–775, June 1998.
55. M. B. Srivastava and R. W. Brodersen. Siera: a unified framework for rapid-prototyping of system-level hardware and software. *IEEE Transactions on Computer-Aided Design*, 14(6):676–693, June 1993.
56. Jane S. Sun and Robert W. Brodersen. Design of system interface modules. In *Proceedings of the International Conference on Computer Aided Design*, pages 478–481, November 1992.
57. Frank Vahid and Linus Tauro. An object-oriented communication library for hardware-software codesign. In *Proceedings of the Fifth International Workshop on Hardware/Software Codesign*, pages 81–6, March 1997.
58. V. I. Varshavsky, M. A. Kishinevsky, V. B. Marakhovsky, V. A. Peschansky, L. Y. Rosenblum, A. R. Taubin, and B. S. Tzirlin. *Self-timed Control of Concurrent Processes.* Kluwer Academic Publisher, 1990. (Russian edition: 1986).
59. Virtual Socket Interface Alliance (VSI). *website: http://www.vsi.org.*
60. Jia Xu and David Lorge Parnas. On satisfying timing constraints in hard-real-time systems. *IEEE Transactions on Software Engineering*, 19(1):70–84, January 1993.
61. Ti-Yen Yen and Wayne Wolf. Communication synthesis for distributed embedded systems. In *Proceedings of the International Conference on Computer Aided Design*, pages 288–294, November 1995.

INDEX

ACPI 263, 265, 266, 268-273, 275, 278, 290
Allocation 24-25, 30, 80, 98, 189, 195, 198, 202, 203, 243, 259, 300, 319, 327, 332-333, 335-338, 340, 356
Architecture 1, 3-5, 10-13, 16-19, 21, 23, 25-35, 38-42, 47-50, 97, 99, 104-108, 110-114, 119, 121, 124, 127, 130-132, 134, 136, 149, 157-159, 161, 171, 174, 178, 182, 186, 189, 192, 195, 198-199, 225, 228-235, 237-243, 288, 290-291, 299-303, 306-309, 311, 317-320, 323-324, 326-329, 332-333, 336, 338-339, 341-342, 356-358, 365, 381, 385, 393, 400
Architectural exploration 186
ASIP 35-36, 39, 308-309, 318-320
Asynchronous 12, 15, 61-62, 68, 71, 78-79, 81, 84-86, 93, 99, 103, 108-109, 111, 116, 120, 144, 151, 161-164, 166, 168-170, 180-181, 184, 263, 363, 365-366, 369-371, 375, 401-404
Code optimization 294, 298-300, 303, 307, 319,
Codesign 6, 12, 33, 35, 37-43, 46-47, 50, 78, 82, 99-100, 103-108, 110-114, 121-122, 127-129, 131-132, 134-136, 143, 173-174, 176-177, 183, 189-191, 194-195, 220, 222, 223, 225-226, 229, 233, 242, 265, 321-324, 326, 330-331, 344, 356, 357-358
Cosimulation 103, 105, 107, 121-132, 135, 138, 142, 162, 167-171, 324, 358
Cosmos 43, 104, 114, 131, 136, 160, 323-324, 357-358
Data Flow 5, 10, 16-17, 26-28, 32, 36, 39-42, 59,99, 100-101, 105, 108, 110-112, 115, 133, 167, 193, 194, 199, 202, 310, 324, 358
Dataflow Networks 49
Debugging 25, 125, 219, 222-223, 226, 229, 231, 234, 238-239, 241-245, 395
Design entry 219, 234, 235
Design Reuse 175, 181, 295, 319, 344
Digital signal processor 5-6, 9-10, 16-18, 46, 301, 358, 399
Discrete-Event 50, 61-62, 66-68, 74, 84, 373
Dynamic Power Management 263-268, 273, 275, 282, 285, 289-292
Embedded 4, 39-41, 75, 77, 127-128, 156-157, 162-166, 172, 175-176, 178, 190, 197, 272, 293-295, 298, 301-303, 308, 317-320, 324-325, 335, 358, 360, 362, 366, 373, 376, 395
Embedded system 45-47, 50, 54, 60, 66, 71, 78, 81-83, 97, 99, 100, 122, 132-133, 135-139, 141-143, 156-159, 172-174, 190, 192, 198, 225, 293-294, 303, 317-325, 344, 351, 356-358, 359, 360, 362, 366-368,395, 397, 400
Emulation 219, 221-225, 227-229, 232-238, 240-242, 247, 258-260, 286, 289, 358
Finite state machine 31, 49, 50, 53, 56, 58, 71, 75-76, 79, 81, 82-83, 116, 144, 147, 149, 184-185, 323-324, 328, 345, 348-349, 360, 373, 393, 401, 404
Flexibility 3-5, 11, 21-23, 25, 27-28, 31, 33-35, 37-40, 84-85, 89-90, 172, 179, 233, 288, 359, 362, 365
FPGA 3, 32, 36, 42, 223, 225-230, 232-241, 258, 293
Functional specification 193

GCC (GNU C Compiler) 294, 307, 311
Hardware/software co-synthesis 189-190, 192, 357, 358
Heterogeneous systems 100, 103, 132, 190
High-level synthesis 111, 133, 195, 199, 220, 242, 318, 325-327, 342, 398
High-level estimation 186
HW/SW codesign 37, 100, 131-134, 136, 174, 189, 222, 265, 293-294, 296, 308-309, 317-320
Idleness 263-264, 266-267, 274, 291
Intellectual property (IP) 2-3, 5, 33-36, 38, 47, 180, 237, 295-296, 327-328, 344, 354, 356, 321-322, 325-332, 341-342, 354-356, 397-401, 404, 407-408, 412-414, 418-420, 422
Intellectual property integration 237, 353, 354, 356, 358
Interface synthesis 40, 51, 105, 129, 323, 326, 339, 399, 401-402, 404
Java 110, 117, 121-122, 138, 143-144, 156-157, 173-174, 179, 325, 328, 331, 345, 357, 359-366, 369-382, 384-391, 393-396
Low-power design 291, 292
Microcontroller 2, 4, 6, 12-13, 15-16, 18, 30-31, 33, 228, 233, 236, 240
Models of computation 40, 45-46, 49-54, 58-59, 66-67, 69, 80, 82, 83, 101, 179, 181-182, 324, 360, 362, 376, 391, 395
Multi-formalism specifications 159, 164
Multilanguage specification 103, 105, 121
OnNow 265, 266, 268-269, 290, 292
Partitioning 42, 51, 60, 98, 104-107, 127, 129-131, 135-136, 159-160, 178-179, 182, 190, 197-198, 201, 226, 231-232, 234, 240, 308-309, 318-319, 326-327, 332-338, 356, 358
PEAS system 294, 296, 308-311, 313-314, 317-318, 320
Performance analysis 160, 196
Performance optimization 320, 374
Power manager 264-266, 277-280, 283, 285-286
Protocol verification 115, 397
Rapid prototyping 42, 219
Reactive 11, 46, 99, 100-101, 116, 132, 146, 151, 184, 359, 362-363, 366-371, 391, 395, 396
Real-time scheduling 97, 197
Refinement 45, 52, 98, 103, 107, 110-111, 115, 129, 131, 133, 144, 159, 180, 321, 329, 331-332, 338-339, 344, 359, 362, 366, 372-374, 376, 386
Requirements capture 158
Retargetable compiler 136, 294, 306-307, 317-318, 320, 324, 342
Scheduling 12, 15-16, 30, 39, 41-42, 57-58, 63, 71, 84-88, 94-95, 98-101, 114, 133-134, 154, 168-169, 181, 288, 292
SDL 78, 80, 93, 104-106, 110, 115-116, 118, 120, 129-132, 134, 137-139, 143-148, 159-160, 165-167, 170-171, 174, 184, 191-192, 196-197, 199, 201, 243-244, 299-300, 304-306, 311, 313, 318-320, 326-327, 332, 334-336, 344, 353, 356, 358, 397
Shutdown 263, 266-267, 291-292
Solar 114, 134

Source level emulation 244
SpecC 115, 117, 122, 133, 136, 321-323, 325-326, 331-332, 335, 338-339, 343, 345-347, 349-358
Specialization 3-7, 10-15, 17-18, 21, 24-25, 28, 30-31, 33-34, 36, 38-40, 362
Specification languages 103-105, 107-111, 115-122, 131-135, 137-138, 140, 142-145, 156-157, 159, 171-172, 323, 325, 359, 364, 374, 395
System-level design 103, 134, 136, 175, 177, 264, 266, 290,321-323, 356